Lactose-Derived Prebiotics

Lactose-Derived Prebiotics

A Process Perspective

Andrés Illanes

Cecilia Guerrero

Carlos Vera

Lorena Wilson

Raúl Conejeros

Felipe Scott

ELSEVIER

AMSTERDAM • BOSTON • HEIDELBERG • LONDON
NEW YORK • OXFORD • PARIS • SAN DIEGO
SAN FRANCISCO • SINGAPORE • SYDNEY • TOKYO

Academic Press is an imprint of Elsevier

Academic Press is an imprint of Elsevier
125 London Wall, London EC2Y 5AS, United Kingdom
525 B Street, Suite 1800, San Diego, CA 92101-4495, United States
50 Hampshire Street, 5th Floor, Cambridge, MA 02139, United States
The Boulevard, Langford Lane, Kidlington, Oxford OX5 1GB, United Kingdom

Notices
Knowledge and best practice in this field are constantly changing. As new research and experience broaden our understanding, changes in research methods, professional practices, or medical treatment may become necessary.

Practitioners and researchers must always rely on their own experience and knowledge in evaluating and using any information, methods, compounds, or experiments described herein. In using such information or methods they should be mindful of their own safety and the safety of others, including parties for whom they have a professional responsibility.

To the fullest extent of the law, neither the Publisher nor the authors, contributors, or editors, assume any liability for any injury and/or damage to persons or property as a matter of products liability, negligence or otherwise, or from any use or operation of any methods, products, instructions, or ideas contained in the material herein.

Library of Congress Cataloging-in-Publication Data
A catalog record for this book is available from the Library of Congress

British Library Cataloguing-in-Publication Data
A catalogue record for this book is available from the British Library

ISBN: 978-0-12-802724-0

For information on all Academic Press publications
visit our website at https://www.elsevier.com/

Publisher: Nikki Levy
Acquisition Editor: Megan Ball
Editorial Project Manager: Jaclyn Truesdell
Production Project Manager: Susan Li
Designer: Maria Ines Cruz

Typeset by TNQ Books and Journals

Contents

List of Contributors

A. Illanes Pontificia Universidad Católica de Valparaíso, Chile

L. Wilson Pontificia Universidad Católica de Valparaíso, Chile

C. Guerrero Pontificia Universidad Católica de Valparaíso, Chile

F. Scott Universidad de Los Andes, Chile

C. Vera Pontificia Universidad Católica de Valparaíso, Chile

R. Conejeros Pontificia Universidad Católica de Valparaíso, Chile

Preface

You will write
if you will write without thinking of the result
in terms of a result,
but think of the writing in terms of discovery,
which is to say that creation must take place
between the pen and the paper,
not before in a thought or afterwards in a recasting...
It will come if it is there and if you will let it come.

Gertrude Stein

Hippocrates, a Greek physician of the Age of Pericles, is considered the most outstanding figure in the history of medicine and considered as the "father of medicine." Among its many deep thoughts, his premonitory statement, "Let food be thy medicine and medicine be thy food," has acquired now a social significance that would have certainly surprised Hippocrates himself.

Modern lifestyle has a strong impact on eating habits, with an ever-increasing consumption of highly processed and fast foods, whose adverse effects on health are apparent. Eating in the Western World is tightly related to pleasure, so that modern society is facing the dilemma between gratification and health. Health problems associated with improper eating habits, like diabetes, heart conditions, allergies, obesity, and cancer, are nowadays important causes of morbidity and mortality. In this scenario, functional foods, providing health benefits beyond their nutritional value, have gained increasing importance and captured consumers' attention. This type of food has a long tradition in Asia and Japan has been certainly a leader in this respect. But healthy eating habits are a natural response to confront such problems and there is an increasing awareness all over the world about the importance of diet in health. This is more so as the world population is aging and life expectancy increases. Functional foods and healthy eating habits are now a sustained trend, not only in developed countries but in developing countries as well, where the effects of improper eating habits are exacerbated by economic constraints.

Prebiotics are outstanding ingredients in functional foods that stimulate a healthy intestinal microbiota. Most of them are nondigestible oligosaccharides; those derived from lactose are particularly interesting, being a significant part of health foods related to the dairy industry. Lactose-derived prebiotics are unique in their properties and origin, representing a self-contained system within the dairy industry since both the raw material and most of the targets are dairy products. Lactose is a by-product (and sometimes a waste) from cheese manufacturing that can be upgraded by converting it into valuable prebiotics, representing an appealing option for the cheese industry of bringing together profit with environmental protection. Distinct from other prebiotics, those derived from lactose do not compromise the use of extra arable land, which is sound from the perspective of a life cycle analysis.

This book presents an updated view of lactose-derived prebiotics from a process perspective, emphasizing their technological constraints and opportunities as important ingredients in health-promoting functional foods. The book has eight chapters. "Chapter 1, Lactose: Production and Upgrading" refers to lactose as raw material, proposing a technological platform for its upgrading; "Chapter 2, Functional Foods and Feeds: Probiotics, Prebiotics, and Synbiotics" presents a general background on functional foods and the role of probiotics and prebiotics within that framework; "Chapter 3, Lactose-Derived Nondigestible Oligosaccharides and Other High Added-Value Products" refers to lactose-derived prebiotics from a process perspective analyzing available technologies and market opportunities; "Chapter 4, Enzymatic Production of Galacto-Oligosaccharides" presents an in-depth review of the enzymatic production processes of galacto-oligosaccharides, which are considered as the most relevant lactose-derived prebiotics; "Chapter 5, Enzymatic Production of Lactulose" is an updated review of the technological status of lactulose production by chemical synthesis and the opportunities and challenges for its production by enzymatic biocatalysis; "Chapter 6, Enzymatic Production of Other Lactose-derived Prebiotic Candidates" reviews the technological status and opportunities of other lactose-derived prebiotic candidates; "Chapter 7, Technical and Economic Analysis of Industrial Production of Lactose-Derived Prebiotics With Focus on Galacto-oligosaccharides" presents in the format of a case study the process simulation and economic assessment of an industrial plant for galacto-oligosaccharide production as a stand-alone process and as a production unit within a dairy product facility; and finally, "Chapter 8, Future Trends and Concluding Remarks" presents the future trends for lactose-derived prebiotics within the framework of functional foods.

We have been engaged for some time now in the development of bioprocesses for lactose upgrading, and in the last decade our work has been focused mostly on lactose-derived prebiotics. We expect that our experience in the field, but above all our compromise with it, will be reflected in these pages that we are now presenting for your consideration.

A couple of years ago, when publishing a book on enzyme biocatalysis, I declared that witting a book was "a journey from expectation to consolidation, an act of love from conception to birth." Writing this book has been certainly so. Working with my colleagues and former students has been a remarkably rewarding experience. I have had the privilege of learning from them while building up this collective piece of work. Their knowledge, enthusiasm, and proficiency are the cornerstone of this book. We sincerely hope you will enjoy the product as much as we enjoyed the process; if so, reward will be doubled.

Our special gratitude to the persons at the Pontificia Universidad Católica de Valparaíso who supported and encouraged us during this process: the rector, Professor Claudio Elórtegui, the vice-rector of Research and Graduate Studies, Dr. Joel Saavedra, and the former and present directors of the School of Biochemical Engineering, Dr. Paola Poirrier and Professor María Cristina Schiappacasse. Our deepest appreciation to Jaclyn Truesdell, Karen Miller, and Susan Li from Elsevier for their enthusiastic, warm, and valuable support and advice. And on a personal note, my gratitude to my life partner Dr. Fanny Guzmán for her meaningful support and understanding.

Andrés Illanes
Valparaíso, March 1, 2016

Chapter 1

Lactose: Production and Upgrading

A. Illanes

1.1 WHEY AS RAW MATERIAL FOR LACTOSE PRODUCTION

Milk is not only a basic component of human nutrition but also the raw material for an ever-increasing number of dairy products. It is also a rich source of valuable compounds that are mostly used in foodstuffs but also in nonfood applications. With respect to the latter, applications of casein, whey proteins, lactose, milk fat, and dairy effluents have been thoroughly reviewed by Audic et al. (2003).

Whey (milk whey or cheese whey) is a major by-product from milk processing, being the residual liquid obtained after casein precipitation by the action of acids (Morr and Ha, 1993) or enzymes (Kinsella and Whitehead, 1989). The former, whose pH is 5 or lower, is called acid whey and obtained by direct acidification of milk as in the production of cottage cheese; the latter, whose pH is around 6, is called sweet whey and produced by enzymatic coagulation of milk, as used in the production of most types of cheese. Chymosin is a protease that selectively hydrolyzes the Phe105–Met106 peptide bond of κ-casein triggering its clotting in the presence of calcium ions to yield the curd (Visser et al., 1977). The traditional source of chymosin is calf rennet, obtained as a by-product of veal production. Shortage of rennet as a source of chymosin for cheese making became critical so that now it has been replaced to a considerable extent by recombinant chymosin produced by fermentation with *Aspergillus* (Ward et al., 1990) and *Kluyveromyces* (van den Berg et al., 1990) strains used as hosts of the chymosin gene. Recombinant chymosin performance has been further improved in terms of specificity and pH profile by using protein-engineering strategies (Mantafounis and Pitts, 1990).

On a dry basis, cow milk contains approximately 28% protein, mostly casein, 29% fat, 38% lactose, and 5% minerals, while whey contains approximately 12% protein (less than 2% is casein), about 2% fat, approximately 77% lactose, and 8% minerals (Illanes, 2011). Whey, a by-product of cheese and casein production, despite being a rich source of lactose and valuable proteins—β-lactoglobulin, α-lactalbumin, immunoglobulins, and lactoferrin (Bottomley

Lactose-Derived Prebiotics. http://dx.doi.org/10.1016/B978-0-12-802724-0.00001-9

et al., 1990)—has been traditionally considered a nuisance rather than a valuable asset. In this context, companies have sought ways of proper management rather than valorization (Marwaha and Kennedy, 1988). This situation has been changing progressively through the years driven by environmental regulations (whey BOD and COD are around 40,000 and 70,000 ppm, respectively, almost 200-fold higher than common sewage), but also because of the economic revenues that may come from the commercialization of whey as a whole, its components, and those derived by its valorization (Smithers, 2008).

1.1.1 Whey as End Product

Whey containing more than 90% water and readily fermented components is highly perishable, so its supply under sanitary acceptable conditions is cumbersome. Therefore drying is necessary for its long-term utilization. Drying becomes an expensive operation, both in terms of equipment and energy consumption, when so much water is to be removed and, in practice, only medium-to-large cheese factories can withstand it. Liquid whey can be used as a liquid feed supplement in animal farming and as soil fertilizer, but transportation cost is high so in most cases its use is locally circumscribed (González Siso, 1996).

Spray-drying, rendering a stable nonhygroscopic product of quite uniform particle size, is the most appealing operation for dry whey production (Písecký, 2005). The principal market for dry whole whey is animal feeding where it is used mixed with molasses and soya flour (Schingoethe, 1976). Whole whey is also used in the formulation of institutional foods (Jelen, 1979), but its use is restricted because of its high lactose and mineral content, so demineralization is required (González Siso, 1996). Nutritional value of whey is related to its caloric value and also to its protein content although in this case whey protein concentrates or isolates are preferred to whole whey.

World production of whey was estimated close to 200,000,000 tons/year with an annual increase of 2% (Smithers, 2008). However, the market is unstable and international prices suffer high fluctuations encouraging or discouraging whey producers to dry it and put it into the market. Production is big enough to cover all present uses of whey and those to come in the near future, highlighting the strategic value of establishing a platform for whey utilization, as will be analyzed in the next section.

1.1.2 Whey Fractionation

Whey contains valuable components that will acquire commercial significance if separated. Therefore, whey fractionation is a key operation for fully exploiting its potential (Atra et al., 2005). Lactose and proteins represent almost 90% of whey on a dry weight basis and membrane separation (ie, ultrafiltration and diafiltration) is nowadays the technology of choice for their recovery (Pouliot, 2008). Membrane fractionation has the advantages of reduced cost, high throughput, gentleness, and neat separation of salts from the protein fraction (González Siso, 1996). The retentate, called whey protein concentrate (WPC), contains most of the whey proteins, while the permeate contains most of the lactose and mineral salts. Whey permeate is a main source for lactose production, so that it cannot be considered a waste stream, even though in some instances its recovery may not be profitable. The high mineral content of whole whey (close to 10% on a dry basis) may be objectionable for some of their applications, so desalting is necessary, which can be accomplished by membrane separation (nanofiltration) (Yorgun et al., 2008; Cuartas-Uribe et al., 2009), electrodialysis, and ion-exchange (Greiter et al., 2002; Nagarale et al., 2006). A thorough analysis of whey fractionation was presented by Tsakali et al. (2010).

1.1.3 Whey Proteins and Peptides

Whey proteins represent about 20% of the total protein content in milk, and its biological value has been estimated to be 15% higher than egg proteins (Smithers, 2008). Its protein quality score is much higher than casein's, with a protein efficiency ratio of 3.4 versus 2.8, having a significantly higher proportion of essential amino acids (Evans and Gordon, 1980; Ha and Zemel, 2003). Its lysine content is remarkably high, thus making it a matching complement for lysine-deficient cereals (Delaney, 1976; Ibrahim et al., 2005; Jooyandeh, 2009). Content of sulfur amino acids, methionine and cysteine, is also high (around 20 mg/g), which is important for them being precursors of glutathione (Shoveller et al., 2005); content of branched amino acids (isoleucine, leucine, and valine) is high as well, being those amino acids regulators of cell metabolism and important for body weight control (Zemel, 2004). Protein concentration in WPC produced by whey ultrafiltration (to concentrate proteins) and diafiltration (to remove most lactose and minerals) may vary from 35% to 80% depending on the operation conditions of ultrafiltration, typical values being around 50% (McDonough et al., 1974). Advances in membrane technology have been a powerful driving force for developing more refined and productive processes to recover whey proteins, making it possible to obtain high-purity and high-quality products with more than 90% protein, termed whey protein isolates (WPI). In the field of formulated foods, WPCs and WPIs are valuable ingredients for conferring both nutritional and functional properties (gelation, water-binding, emulsification) and are increasingly being used in protein rich beverages (Foegeding et al., 2002).

Beyond their nutritional value, whey proteins are endowed with important metabolic effects, such as antimicrobial and antiviral effects (Clare et al., 2003), immunomodulation (Low et al., 2003), anticarcinogenic activity (Gill and Cross, 2000; Yalcin, 2006), reduction of cholesterol and triglyceride levels (Madureira et al., 2007), and muscle (Tipton et al., 2007) and bone (Toba et al., 2000) strengthening. These properties are attributed to lactoglobulin and lactalbumin, which are the major whey protein components. A thorough review of the nutritional and physiological effects of whey proteins has been reported by Madureira et al. (2007).

Aside from these well-established applications, whey proteins have been successfully tested as hydrogels and nanoparticles for encapsulation and controlled delivery of bioactive compounds (Gunasekaran et al., 2007), like essential oils (Djordjevic et al., 2004), probiotics (Crittenden et al., 2006), and drugs (Gunasekaran et al., 2006).

Peptides are molecules of paramount importance, especially in the fields of medicine and nutrition, being currently produced by chemical synthesis and, in some cases, by biocatalysis (Guzmán et al., 2007); however, extraction of peptides from natural sources is encouraged mostly by environmental considerations (Newman and Cragg, 2007). Bioactive peptides can be obtained from whey proteins, mostly by controlled enzymatic hydrolysis (Welderufael and Jauregi, 2010), being a rich source of them (Korhonen and Pihlanto, 2003). Worthwhile mentioning are the opioid peptides, α-lactorphin and β-lactorphin, produced by enzymatic proteolysis of whey proteins, the former being endowed with angiotensin-converting enzyme (ACE) inhibitory activity (Pihlanto-Leppälä, 2000; Guo et al., 2009). Comprehensive reviews on bioactive peptides derived from whey proteins have been reported by Korhonen and Pihlanto (2006) and Nagpal et al. (2011).

Modification of whey proteins by heat-induction polymerization and enzymatic hydrolysis has been the subject of considerable research aiming to expand the functional properties of whey proteins, namely gelation, foaming, and emulsification (Foegeding et al., 2002).

1.1.4 Lactose

Lactose (O-β-D-galactopyranosyl-(1–4)-β-D-glucopyranose) is an abundant natural disaccharide that, interestingly, is only found in mammal's milk where it is the principal carbohydrate and energy source. Lactose is synthesized in the mammary epithelial cells from the glucose and galactose absorbed from blood. Synthesis is catalyzed by a protein complex called lactase synthase, in which the catalytic component, UDP-galactosyltransferase, catalyzes the transferring of the galactose moiety from UDP-galactose (formed via the Leloir pathway) to glucose, and the noncatalytic component, α-lactalbumin, acts by reducing the Michaelis constant of the transferase by several hundredfold (Lehninger, 1975), thus giving glucose a definite advantage over other acceptors to be transgalactosylated.

Lactose exists in the form of two anomers, α and β, differing in the configuration of the substituent groups (OH and H) on carbon atom 1 of the glucose moiety. Both anomers differ in their properties, with their solubility in water being the most important: β anomer is about sevenfold more soluble than α anomer at 20°C (500 vs. 70 g/L), but the solubility of the latter is highly temperature dependent, so that at 93.5°C it becomes more soluble than the β anomer (Fox, 2009). Upon dissolution, mutarotation occurs producing a solution at equilibrium with 63% in the β form. The α anomer crystallizes as monohydrate, while β crystallizes in anhydrous form. Therefore when lactose is dried, a mixture of α-lactose, β-lactose, and α-lactose monohydrate is formed, the proportion of them varying according to the conditions of drying (Zadow, 1984). Alpha lactose is highly hygroscopic, while the other forms are not, so caking and lumping in dairy products is due to the former, which needs to be crystallized prior to drying if a nonhygroscopic product is to be obtained (Schuck and Dolivet, 2002).

Lactose is barely sweet (about 15% sucrose), less soluble than its monosaccharide components, and less soluble than most sugars. Lactose presents problems of intolerance to a significant part of the world population. It is estimated that about two-thirds of the population suffer from lactose maldigestion and intolerance, but the condition is strongly ethnic and age dependent (Schaafsma, 2008), so that there is a marked geographical distribution, as seen in Fig.1.1 (Heyman, 2006; Latorre et al., 2014), and several strategies have been proposed to deal with this problem (Shaukat et al., 2010).

These properties of lactose represent severe constraints for its widespread use, so its removal or breakdown is in many cases desirable, as will be analyzed in the following sections. Notwithstanding, in some cases the low solubility of lactose may be an asset when the purpose is using it as a bulking agent or when excessive sweetness is undesirable, as in the case of low-lactose milk (Kailasapathy, 2008); its replacement for sugar in infant foods has been considered adequate to reduce the sensation of appetite and need for sweet foods, preventing overeating. Lactose is considered significantly less cariogenic than other sugars (Schaafsma, 2008).

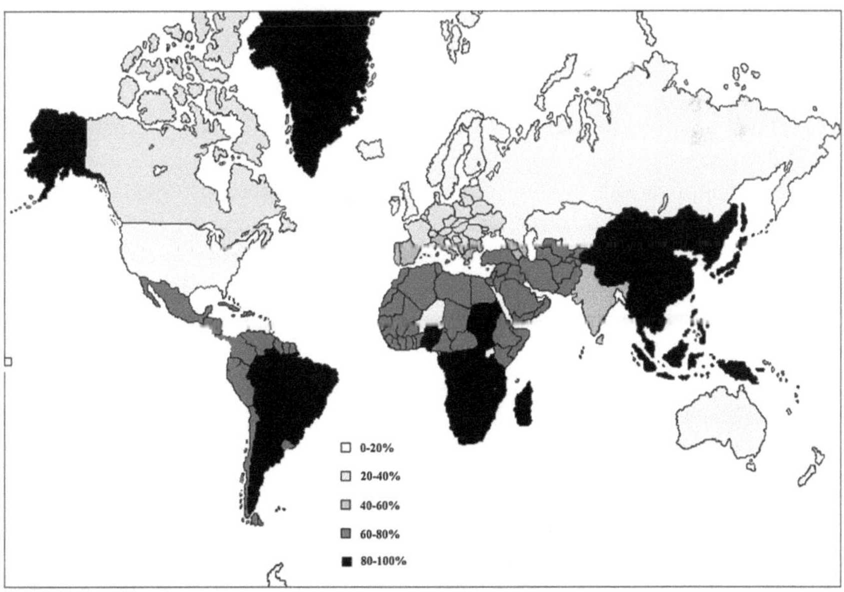

FIGURE 1.1 Worldwide distribution of adult lactose intolerance (% of intolerance of the population).

The only natural source containing significant amounts of lactose is mammal's milk, so the raw material for its production is invariably whey or whey permeate. Whey protein is a valuable asset, so currently lactose is produced using whey permeate as starting material. Whey permeate contains between 5% and 6% solids, typical composition on a dry basis being: lactose 85%, ash 10%, and protein 3%. The production of lactose from whey permeate considers the following unit operations (Yang and Silva, 1995):

- Concentration: usually done by multiple-effect evaporation; in some cases permeate is preconcentrated by reverse osmosis, and preconcentration together with partial demineralization can be done by nanofiltration (Suárez et al., 2009).
- Crystallization: done by cooling from a supersaturated solution; alternatively precipitation by alkaline earth metals (Steffen process) and alcohols have been proposed, but its industrial use is doubtful.
- Recovery: lactose crystals are recovered by decantation or centrifugation; mother liquor containing most of the minerals and about 20% of the lactose is disposed, as lactose recovery from it is not attractive if no prior demineralization has been performed.
- Drying: lactose crystals containing 5–12% moisture are finally subjected to drying in flash or fluidized-bed dryers.
- Purification: depending on its use, further refining may be required. Food-grade lactose does not require polishing operations, but for pharmaceutical-grade lactose, refining is necessary; this involves the redissolution of the lactose crystals, the treatment of the solution with activated carbon to absorb a number of solutes including riboflavin, the removal of traces of residual proteins, and recrystallization.

A schematic representation of lactose production from whey is presented in Fig. 1.2.

Current applications of lactose are many. In the food industry, lactose is used mostly as an energy source and also when low sweetness in the final food product is required. Lactose is used in the confectionery industry to enhance flavor and color in protein-containing goods and in the bakery industry to promote crust browning. Enrichment of cow's milk with lactose is also an important use aiming to mimic the lactose content in human milk (Polberger and Lönnerdal, 1993), and significant lactose demand comes from its use in the standardization of milk products (Rattray and Jelen, 1996). More intense and widespread use of lactose as a food ingredient is precluded by the high incidence of lactose intolerance (Perotti et al., 2012).

As noted previously, lactose for pharmaceutical use requires higher purity than for edible purposes. Typical United States Pharmacopeia (USP) standard lactose must contain less than 5 μg/g of heavy metals, microbial counts lower than 100/g for *Escherichia coli*, yeast, and molds, absence of pathogenic bacteria, less than 0.1% residue upon ignition, less than 0.5% weight loss upon

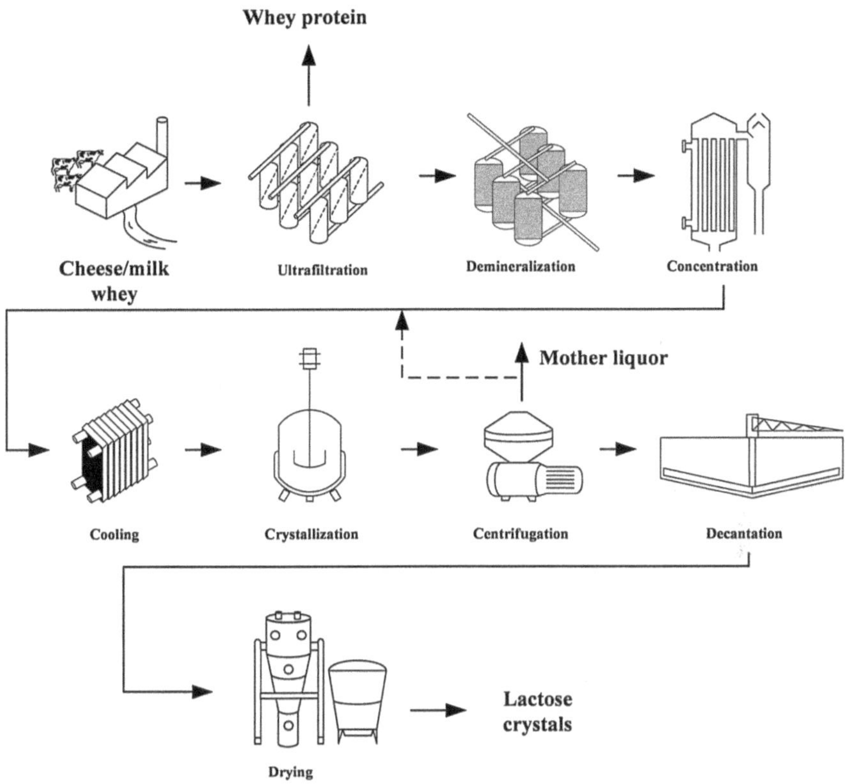

FIGURE 1.2 Schematic diagram of lactose production from whey.

drying, and protein and light-absorbing impurities lower than 0.25 absorbance units at 210–220nm and 0.07 at 270–300nm (Paterson, 2009). Main pharmaceutical use of lactose is as an excipient because of its cost effectiveness, availability, bland taste, low sweetness, low hygroscopicity, compatibility with active ingredients and other excipients, high physical and chemical stability, and acceptable water solubility; these properties make it the most adequate excipient for tablets, where it is estimated to cover about 70% of all such formulations. This is of paramount importance since about two-thirds of the products used in the pharmaceutical industry are in the form of particulate solids (Fages et al., 2004). Lactose is mostly used in the form of α-lactose monohydrate. Anhydrous α-lactose is less prone to Maillard reaction but its lower disintegration rate is a major disadvantage. Anhydrous β-lactose, produced by crystallization of lactose above 93°C and roller drying, acquires the form of extremely fine crystals having high dissolution rate and good compactness, as its low moisture is adequate for moisture-sensitive active pharmaceutical ingredients (Gohel and Jogani, 2005). Tablets prepared by direct compression use anhydrous lactose produced by drying (usually drum drying), which is essentially β-lactose crystals with a smaller amount of α-lactose (Riepma et al., 1993). Finely milled α-lactose monohydrate is used as excipient in dry powder inhalers, which are currently being used in the treatment of asthma and in chronic obstructive pulmonary disease (Sham et al., 2004; Atkins, 2005). Lactose has been also tested as carrier in nanoparticles delivered to the lung, where it dissolves in the aqueous environment of the lung epithelium, delivering the active pharmaceutical ingredient.

Control of processing conditions for lactose production, storage, and particle properties are crucial for controlling its behavior as an ingredient in foods and drugs (Huppertz and Gazi, 2015).

1.2 TECHNOLOGICAL PLATFORM FOR LACTOSE UPGRADING

Undoubtedly, most of the potential use of whey-derived lactose resides in its transformation. Beyond the intrinsic value and increasing opportunities for lactose-derived products, a platform for lactose utilization will serve for the purpose of stabilizing whey market, which has been unstable through the years, with significant changes in price as supply and demand fluctuates. Not to mention that a significant portion of whey is discarded or clearly underutilized, whey fractionation has been envisaged mostly for protein recovery, despite that the lactose amount in whey exceeds protein by a factor of seven. As a consequence, increasing amounts of lactose are being delivered that exceed current uses of lactose per se (Smithers, 2008). Forecasting of lactose demand is uncertain; lactose use as food ingredient is limited by its own properties and by extended intolerance, while its use as cow's milk filler is increasing; pharmaceutical use is steadier because of the leading role of lactose as excipient, but research on lactose substituents for that purpose are ongoing (Vogt et al., 2008). A platform for lactose upgrading is then a strategic opportunity for whey producers.

Fig. 1.3 presents a scheme of technological alternatives for whey utilization and upgrading.

1.2.1 Biogas

Anaerobic digestion of lactose-containing whey and permeate is considered as a low-added-value process for lactose upgrading. Theoretical yield of biogas (with approximately 50% v/v methane) from lactose is 750 L/kg, although actual yields may be significantly lower. It is reported that several plants are producing biogas by anaerobic digestion of lactose in whey and whey permeate (González Siso, 1996; Asplund, 2005). However, a major problem in biogas production by anaerobic fermentation of lactose is the quality of the plant effluent, which still needs to be treated by aerobic polishing before final discharge, making the process hardly profitable. Whey lactose has been sometimes used in codigestion processes with other polluting effluents, like poultry or cattle manure (Gelegenis et al., 2007; Comino et al., 2009; Kavacik and Topaloglu, 2010) and, interestingly, biogas from whey has been considered within the

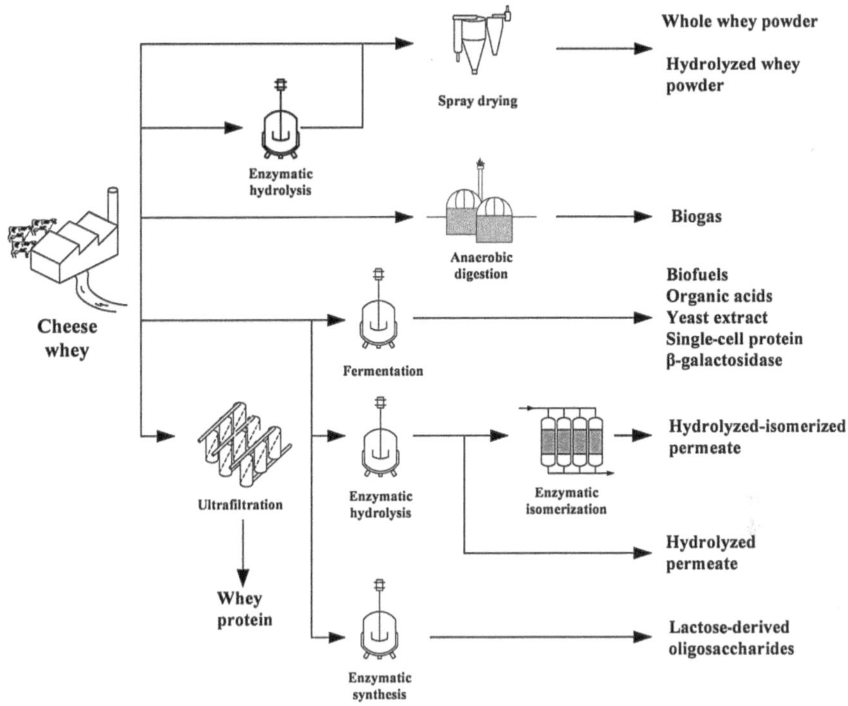

FIGURE 1.3 Technological platform for whey utilization and upgrading.

concept of farm-scale biorefinery (Oleskowicz-Popiel et al., 2012). So, biogas from lactose has to be considered mostly as an option for proper management of surplus whey and whey permeate alleviating waste disposal while producing some energy to aid in fulfilling the requirements of the production facilities.

1.2.2 Bioethanol

Beyond traditional use in beverages and perfumes, ethanol is gaining increasing importance within the energy platform for oil replacement (Tan et al., 2008). About 95% of ethanol is produced from agricultural products, either simple sugars (mostly sucrose from sugarcane or sugar beet) or complex polysaccharides (mostly starch). In the first case, sucrose is directly fermented by yeast to produce ethanol and carbon dioxide; in the second case, starch needs to be hydrolyzed, either chemically or enzymatically to deliver fermentable sugars (mostly glucose and maltose) as substrates for yeast fermentation. Sucrose from sugarcane and sugar beet accounts for about 40% of the total bioethanol produced. At present, about 40% of bioethanol comes from sugar crops and 60% from starch crops (Mussatto et al., 2010). This so-called first-generation bioethanol comes mostly from edible materials produced from arable land, so that 50% or more of the production cost is represented by the raw material (Cardona et al., 2010);

production is severely challenged for compromising food resources for energy production, which is, at least on a global perspective, hardly sustainable (Fritsche et al., 2010). The situation is changing rapidly as second-generation bioethanol, produced from lignocellulosic materials, be it wood (Limayem and Ricke, 2012) or agro-residues (Sarkar et al., 2012), is coming into play and not compromising food resources. Bioethanol production from lignocellulose is challenging because in this case intensive pretreatment is necessary to break down the lignocellulose matrix making cellulose (and hemicellulose) available for enzymatic saccharification. Considerable research has been done in recent years aiming to improve all the unit operations involved: pretreatment, enzymatic hydrolysis, fermentation, and distillation (Margeot et al., 2009). Impressive advances in pretreatment strategies (Kumar et al., 2009; Alvira et al., 2010; Sathisuksanoh et al., 2013), enzyme development (Horn et al., 2012; Zhang et al., 2012; Skovgaard and Jørgensen, 2013), ethanol-producing strains (Matsushika et al., 2009; Chen and Qiu, 2010; Long et al., 2012; Shahsavarani et al., 2012), process optimization strategies (Geddes et al., 2011; Jin et al., 2012; Hoyer et al., 2013; Modenbach and Nokes, 2013; Subhedar and Gogate, 2013; Koppram and Olsson, 2014), and modeling (Cardona and Sánchez, 2007; Piccolo and Bezzo, 2009) have paved the way to large-scale production of fuel ethanol from lignocellulose, which is now a reality and will be more so in the near future.

In this scenario, bioethanol from lactose plays a secondary role. An interesting and simple estimation made by Guimarães et al. (2010) concluded that the world potential of bioethanol production from surplus whey amounts to 2.3 million cubic meters, which roughly represents 3.5% of world ethanol production. This figure, although comparatively small, reveals some potential, especially in connection with waste management of whey. Even though lactose from whey or permeate has been considered for a long time as a carbon and energy source for ethanol production from yeasts, the process will be hardly competitive, at least for biofuel production (Zadow, 1984), with some exceptions, like New Zealand, where a significant part of whey is used for bioethanol production (Mawson, 1994; Antoni et al., 2007). However, in particular cases, bioethanol from surplus whey or permeate, being potable, may find proper markets in the food and beverage, pharmaceutical, and cosmetic sectors (Athanasiadis et al., 2002, 2004; Zafar and Owais, 2006). Since lactose concentration in whey or permeate is low, direct fermentation will produce very low titers of ethanol so that distillation cost will be prohibitive; therefore, concentration by ultrafiltration and even drying is required (Kargi and Ozmıhcı, 2006); otherwise, supplementation with extra lactose or other fermentable sugars will be necessary (Athanasiadis et al., 2002; Silva et al., 2010). Common baker's and brewing yeasts from the genus *Saccharomyces* cannot ferment lactose because they lack the genetic machinery for lactose transport and hydrolysis (Russel, 1986), but closely related *Kluyveromyces* strains can, since they are endowed with such genes (Rubio-Teixeira, 2006), and have been mostly used for ethanol production. Most work on ethanol production from lactose has been conducted with *Kluyveromyces marxianus* (*Kluyveromyces fragilis*) (Zafar and Owais, 2006; Fonseca et al., 2008) and *Kluyveromyces lactis* to a lesser extent (Fukuhara, 2006), despite the problems derived from a strong Crabtree-negative character (González Siso, 1996), low growth rates at high lactose concentrations and low tolerance to ethanol (Bellaver et al., 2004). These drawbacks have to be circumvented, either by environmental (ie, fed-batch, media supplementation, oxygen limitation) and/or genetic manipulations (Kargi and Ozmıhcı, 2006; Ozmıhcı and Kargi, 2007; Fonseca et al., 2008). *Saccharomyces cerevisiae* is the organism of choice for producing bioethanol because of its fermentation capacity, high growth rate under anaerobic conditions, tolerance to ethanol and generally recognized as safe (GRAS) status, making it possible to recover it as a by-product of bioethanol production to be used as feed supplement (it is also one of the most-studied microorganisms both genetically and physiologically); despite not being able to metabolize lactose, it does metabolize galactose more efficiently than *Kluyveromyces*, so that a prehydrolysis of lactose with β-galactosidase is an option for producing bioethanol, even though catabolite repression of galactose metabolism is exerted by glucose, thus affecting productivity. Co-immobilization of *S. cerevisiae* with β-galactosidase also has been proposed as a strategy for ethanol production (Staniszewski et al., 2007, 2009). Genetic manipulation of *S. cerevisiae* has been conducted for decades now, aiming to improve their metabolic performance. In connection with bioethanol production, most strategies are related to confer it the ability to metabolize lactose. Initially work was done constructing hybrid strains of *Saccharomyces* and *Kluyveromyces* by protoplast fusion, the fusant being able to ferment lactose at a higher rate than the parental *Kluyveromyces* strain and produce higher titers of ethanol (Farahnak et al., 1986). Later on, recombinant *S. cerevisiae* strains were constructed expressing LAC genes from *K. lactis* (Guimarães et al., 2008a,b), which is now considered within the ample field of metabolic engineering (Domingues et al., 2010); also recombinant *S. cerevisiae* bearing the β-galactosidase from *E. coli* (Porro et al., 1992) and mixed cultures of *K. marxianus* and *S. cerevisiae* (Guo et al., 2010) have been used for bioethanol production. A complete report on the strategies for constructing such recombinants has been published by Guimarães et al. (2010). Much effort has also been devoted to the development of thermotolerant yeasts allowing bioethanol production at high temperatures with the advantages of energy savings, low contamination, and high productivity (Nolan et al., 1994); screening of natural thermotolerant strains, temperature adaptation, protoplast fusion, mutagenesis, and recombination have been searched as alternatives for strain development for bioethanol production at high temperature (Banat et al., 1998). Several reactor configurations have been proposed for bioethanol production from whey, including continuous processes in membrane reactors (Tin and Mawson, 1993), packed-bed reactors (Ozmihci and Kargi, 2008), UASB-type reactors (Jędrzejewska and Kozak, 2011), suspended flocculating recombinant yeast cells (Domingues et al., 1999), and the use of immobilized yeast cells (Gunasekaran and Kamini, 1991; Roukas et al., 1991; Singh et al., 1998; Kourkoutas et al., 2002). Continuous production of bioethanol has been a major subject of research in recent decades and significant improvements have been obtained; however, despite the benefits of continuous processes, industry has been rather reluctant to adopt this technology mainly due to engineering and microbial metabolism problems (Verbelen et al., 2006).

Bioethanol from whey will hardly contribute significantly to the energy bill. However, being a by-product or a residue, it represents no threat to the productive food chain, as occurs with first-generation bioethanol, so the future is promising at least for certain geographical niches. Improvement of lactose-fermenting organisms to the point of compliance with the high efficiency demanded by industry is still a challenge; advances in strain improvement using cutting-edge technology will certainly contribute to it in the near future.

1.2.3 Other Products of Fermentation

Lactose is a readily fermentable carbohydrate, so its use as a carbon and energy source for the production of metabolites by fermentation is certainly an option. Being metabolized more slowly than glucose or sucrose, lactose may be a good choice to reduce catabolite repression effect. Actually, lactose was the substrate of choice for penicillin production in the early 1950s but was soon replaced by cheaper and more readily available sugars (molasses and starch hydrolyzates) once fed-batch fermentation was established (Yamanè and Shimizu, 1984; Elander, 2003). Only some of the most relevant or promising products obtainable from whey or whey permeate fermentation will be analyzed.

Biobutanol has gained considerable attention in recent years for its potential as a transportation fuel (Niemistö et al., 2013). Production of biobutanol by fermentation with different strains of *Clostridia* (mostly *Clostridium acetobutylicum*) has been thoroughly studied but faced with strong competition from the chemical process because of the low productivity of fermentation and complex product recovery (Napoli et al., 2010). With respect to the raw material for biobutanol production, the situation is not different from that already presented for bioethanol. Whey has been considered as a raw material for biobutanol production with *C. acetobutylicum* and its potential has been assessed (Foda et al., 2010; Stoeberl et al., 2011; Raganati et al., 2013). Solvent toxicity to producing strains is a major drawback so that more solvent-tolerant strains are required. The study of biobutanol production by strains different from *C. acetobutylicum* has been encouraged to overcome problems of solvent toxicity and explore different metabolic pathways; this is the case of *Clostridium tetanomorphum*, which does not produce acetone but an equimolar mixture of butanol and ethanol (Patakova et al., 2013).

Lactic acid is a valuable compound used extensively in the food, chemical, textile, and pharmaceutical industries, which is mostly produced by bacterial fermentation (Schepers et al., 2002; Büyükkileci and Harsa, 2004). Production by fermentation has the advantage over chemical synthesis of producing optically pure lactic acid, whose demand as raw material for the production of the biodegradable polymer polylactic acid has increased considerably (Flieger et al., 2003; Panesar et al., 2007a; Sauer et al., 2008). Whey is one of the raw materials of choice for producing lactic acid, and several strains of *Lactobacillus* have been used for lactic acid production, among which *Lactobacillus helveticus* is outstanding for producing the higher titers (Božanić et al., 2014); however, several agro-industrial by-products and residues are competing sources so that the use of whey for such purposes is globally marginal (John et al., 2007; Zhang et al., 2007).

Citric acid is one of the most used acidulants in the food and pharma sectors; it is a commodity produced mostly by fermentation with selected strains of *Aspergillus*. The raw material of choice has been cane or sugar beet molasses, but a whole set of agro-industrial by-products and residues has been tested both in solid-state and submerged fermentation (Soccol et al., 2006; Dhillon et al., 2011). Whey is seldom considered as raw material for citric acid production, but it has been mentioned in some occasions as a suitable substrate (El-Samragy et al., 1996; El-Holi and Al-Delaimy, 2003).

Lactose has also been tested as substrate for the production of other organic acids, like propionic (Morales et al., 2006), succinic (Song and Lee, 2006; Wan et al., 2008), gluconic (Mukhopadhyay et al., 2005), and acetic acid (Talabardon et al., 2000), and amino acids (Barrett et al., 2004), but its industrial implementation is doubtful.

Other valuable fermentation products, like glycerol (Mostafa, 2001), microbial lipids (single-cell oil) (Vamvakaki et al., 2010), biohydrogen (Davila-Vazquez et al., 2009), xanthan gum (Fernandes Silva et al., 2009), biodegradable polymers (Povolo and Casella, 2003; Koller et al., 2008), plastics (Koller et al., 2012), carotenoids (Aksu and Eren, 2007; Khodaiyan et al., 2008; Nasrabadi and Razavi, 2011; Roukas et al., 2015), biosurfactants (Mukherjee et al., 2006), and secondary metabolites (Ruiz et al., 2010) have been produced on whey, taking advantage of the less repressive effect of lactose when compared to most readily fermented sugars present in other industrial raw materials (Sanchez and Demain, 2002). Whey has also been used for baker's yeast production by feeding dosed β-galactosidase to a molasses-whey mixture (Ferrari et al., 2001).

Carbohydrates are the main components of most industrial fermentations, but within them lactose plays a secondary role (Peters, 2006), so that lactose from whey should demonstrate a clear economic advantage, mostly supported by its condition of a surplus material not requiring pretreatment, which might be the case in certain instances.

1.2.4 Single-Cell Protein

Microbial protein, or single-cell protein (SCP), can be regarded as an important protein reservoir for supplying the nutritional needs of the future

(Nasseri et al., 2011). SCP has been used for a long time as a feed or feed supplement, not to mention unintentional traditional and ethnic uses of microorganisms as part of the human diet (Tusé and Miller, 1984). It reached notoriety in the 1970s mostly related to its production from hydrocarbons (Solomons and Litchfield, 1983). After the oil crisis, attention was driven to the use of agricultural and agro-industrial wastes as sources of protein (Illanes and Schaffeld, 1982; Ravindra, 2000; Villas-Bôas et al., 2002; Bacha et al., 2011). In this context, whey has been considered as a sound raw material for SCP production (Pesta et al., 2007). Several strains have been tested for SCP production from whey, notably yeasts. The most used has been *K. marxianus* (*K. lactis*), but other yeasts, like *Candida pseudotropicalis*, *Candida kefyr*, and *Torulopsis cremoris*, have also been studied. A problem to be solved is the concomitant ethanol production occurring at high-lactose concentrations, but the use of Crabtree-negative mutants has proven effective to avoid it (Schultz et al., 2006) and also the use of mixed cultures (Cristiani-Urbina et al., 2000; Moeini et al., 2004). Yields are typically 0.5 g dry yeast per gram of lactose and batch (Ghaly and Ben-Hassan, 1995), fed-batch (Schultz et al., 2006), and continuous processes (Hosseini et al., 2003; Ghaly et al., 2005) have been developed. Still used mostly for animal feed, the cost of raw material is critical in terms of process economics so that whey will find a place as long as its price remains low. Most strains used for SCP production from whey have GRAS status and are therefore eligible candidates for producing microbial protein for human consumption (Bekatorou et al., 2006). An industrial process for production of SCP from whey has been described by González Siso (1996) considering a mixed culture of *K. lactis* and *Torulopsis bovina*, the later used for ethanol removal. Yeast biomass was recovered by centrifugation, then plasmolyzed by heating at 85°C and finally spray-dried, delivering a product with a protein content of around 50%.

1.2.5 Recombinant Proteins

Expression of foreign genes in suitable hosts is probably the most impressive development in biotechnology (Andersen and Krummen, 2002). For many years *E. coli* was the organism for hosting the recombinant genes (Baneyx, 1999), but yeasts have gained more importance for being nonpathogenic and capable of adequately processing eukaryotic genes and performing posttranslational modifications (Çelik and Çalık, 2012). Recombinant protein genes have to be expressed efficiently, so strong promoters are required; among them, lac promoters have been frequently chosen, using different galactosides, ie, isopropyl-β-D-thiogalactopyranoside, as inducers (Pinsach et al., 2008). Such inducers are the ones mostly used for expression studies, but when it comes to production they are too expensive and eventually toxic, so lactose has been used and proven to be an efficient inducer of the lac promoter genes used in recombinant protein production (Neubauer and Hofmann, 1994; Gombert and Kilikian, 1998;

Menzella et al., 2003; Viitanen et al., 2003; Kim et al., 2007). This application, although important, is not significant in terms of lactose utilization.

1.2.6 β-Galactosidase

β-galactosidase (lactase) is a rather ubiquitous enzyme, but most studies have been conducted with enzymes from *K. lactis*, (*K. marxianus*), *Aspergillus oryzae*, *Aspergillus niger*, and *E. coli*; for safety regulations the latter cannot be used for food or pharmaceutical applications. β-galactosidase synthesis is inducible (Dickson and Markin, 1980; Fekete et al., 2012) and, in some cases, also subjected to catabolite repression (Karaffa et al., 2002), so lactose is the best choice as carbon and energy source for its production (Bailey and Linko, 1990; Manera et al., 2008; Rodrigues Pinho and Lopes Passos, 2011). Whey has also been used as a medium for the production of other enzymes, such as lipases (Lotrakul and Dharmsthiti, 1997), α-amylase (Ferreyra et al., 1998), proteases (Sangeetha et al., 2007), manganese peroxidase (Feijoo et al., 1999), and recombinant chymosin (van den Berg et al., 1990). As lactose is a low-cost and nonrepressing carbon source, whey and whey permeate are appealing media for enzyme production.

1.3 LACTOSE UPGRADING BY ENZYME BIOCATALYSIS

1.3.1 Lactose Hydrolysis

Enzymes have been used as industrial catalysts for over a century. Applications to the food industry are many and significant (Kirk et al., 2002); among them, an important use of β-galactosidases is for the reduction of lactose in milk and dairy products (Panesar et al., 2010). Hydrolysis of lactose to its monosaccharide components brings about several effects that have been largely appreciated by food producers, namely increased solubility, increased sweetness and palatability, and absence of intolerance (Rehman, 2009). The consumption of low-lactose dairy products is now highly appreciated to obtain the nutritional benefits of milk without its contraindications (Perotti et al., 2012). In this way, hydrolysis of lactose expanded its field of use considerably allowing the incorporation of hydrolyzed lactose (from milk or whey) to a great variety of dairy products. Hydrolyzed milk is valuable not only as a product for intolerants (Neuhaus et al., 2006; Lomer et al., 2008) but also to avoid lactose crystallization in concentrated and frozen dairy products (Lindamood et al., 1989; Neelakantan et al., 1999). Reduced-lactose and lactose-free milks are now found in almost any supermarket worldwide. Beyond the benefits already outlined for its use in food, lactose hydrolyzates can be used as raw material for fermentation with a wide variety of microorganisms since only a few can metabolize lactose, but most of them can metabolize the resulting monosaccharides (González Siso, 1996). Reduction of lactose content in milk and whey can be accomplished by membrane fractionation, chromatography, or enzymatic hydrolysis, or combinations of them (Harju et al., 2012). Lactose hydrolysis is the most frequently used operation and with the decreasing prices of commodity enzymes this situation will last.

With the current trend of whey protein recovery, increasing amounts of permeate are being produced containing almost all of the lactose, making it an ideal substrate for hydrolysis with β-galactosidases to produce hydrolyzed lactose syrups for food and feed, but mostly as fermentation feedstock (Yang and Silva, 1995). Lactose can be hydrolyzed chemically with acids (Coté et al., 2004) and ion-exchange resins (Demaimay et al., 1978), or enzymatically with β-galactosidases (Gekas and López-Leiva, 1985), the latter being the technology of choice for processing and environmental considerations. Mesophilic β-galactosidases from GRAS organisms, mostly *Aspergillus* and *Kluyveromyces* strains, are the ones used for lactose hydrolysis (Mlichová and Rosenberg, 2006), but psychrophilic (Coker and Brenchley, 2006; Park et al., 2006; Biaikowska et al., 2009) and thermophilic β-galactosidases (Vasiljevic and Jelen, 2001; Chen et al., 2008; Park and Oh, 2010) have also been tested because of the potential advantages over mesophilic enzymes with respect to asepsis. Efficient use of the catalyst is required to make lactose hydrolysis profitable so that the process is mostly conducted with immobilized enzymes (Knopf et al., 1979; Axelsson and Zacchi, 1990; Illanes et al., 1990; Kosseva, 2013). Many immobilized enzymes and reactor designs have been developed for hydrolyzing lactose in milk and whey (Novalin et al., 2005; Grosová et al., 2008; Illanes, 2011; Akgül et al., 2012; Verma et al., 2012) and process modeling and optimization has been reported (Illanes et al., 2000; Szczodrak, 2000; Fontes et al., 2001; Hatzinikolaou et al., 2005; Şener et al., 2006; Mariotti et al., 2008; Demirhan and Özbek, 2009; Olafadehan et al., 2009). In the case of low-lactose milk, hydrolysis after packaging has been adopted as a convenient procedure, enzyme being dosed to obtain the desired degree of hydrolysis until its inactivation (Gänzle et al., 2008). As an alternative to the use of free or immobilized enzymes, permeabilized yeast cells have also been used for lactose hydrolysis, avoiding enzyme recovery and purification at the expense of lower yield (Panesar et al., 2007b; Kaur et al., 2009). Hydrolysis of lactose from whey permeate with β-galactosidases has been combined with isomerization with glucose isomerase to produce fructose-containing syrups of similar sweetening power as sucrose (Abril and Stull, 1989; Illanes et al., 1999; Lorenzen et al., 2013).

1.3.2 Lactose Esterification

Fatty acid esters of sugars are biodegradable surfactants whose use is well established in the food, pharmaceutical, and cosmetic industries (Devulapalle et al., 2004); they have also been considered as potential antimicrobial agents (Ferrer et al., 2005; Habulin et al., 2008). Enzymatic synthesis with lipases is an attractive technological option to chemical synthesis in terms of selectivity, mild reaction conditions, higher yields, and product quality (Reyes-Duarte et al., 2005; Kennedy et al., 2006). Poor solubility of sugar esters in water and the requirement of low water activity for the transesterification reaction make necessary the use of nonconventional (nonaqueous) reaction media (Shi et al., 2011), to which lipases, as contrary to other enzymes, are well conditioned (Jaeger and Eggert, 2002; Adlercreutz, 2013). Organic solvents (Sakaki et al., 2006; Yoo et al., 2007; Gumel et al., 2011), supercritical fluids (Tai and Brunner, 2009), and ionic liquids (Lee et al., 2008; Adamczak and Bornscheuer, 2009; Yang and Huang, 2012) have been used as reaction media for the synthesis of sugar esters; quite recently deep eutectic solvents have been considered as green media for transesterification reactions with lipases (Durand et al., 2013). Whole cell biocatalysis in nonaqueous media has also been explored for the synthesis of sugar fatty acid esters (Jin et al., 2013). Proteases have also been used for the synthesis of sugar fatty acid esters, but yields and productivities are lower than with lipases because of the rather poor activity and stability of proteases in nonaqueous media (Pedersen et al., 2003).

Although not the most used sugar, lactose has been considered as an option for producing sugar esters (Chang and Shaw, 2009). The synthesis of lactose monolaurate by transesterification of lactose with vinyl laurate was studied using different lipases and organic solvents as reaction media, best results being obtained with *Mucor miehei* lipase in 2-methyl-2-butanol medium (Walsh et al., 2009). However, yields of lactose esters have been in general lower than those obtained with other sugars, like sucrose, fructose, or glucose (Ye and Hayes, 2012). Likewise, reported yields of lactose fatty acid ester synthesis with proteases are low (Wu et al., 2004). Lactose seems not to be the best sugar to produce functional sugar esters, when compared with monosaccharides (fructose, glucose, xylose) and sucrose that are more soluble than lactose in aqueous and nonaqueous media. However, being an underutilized sugar, lactose may turn competitive considering a life cycle analysis, but improvements in biocatalyst and medium engineering are still needed.

1.3.3 Lactose Transgalactosylation

Even though β-galactosidases have been traditionally used for lactose hydrolysis, the enzyme can catalyze transgalactosylation reactions under certain conditions where hydrolysis is depressed. Hydrolysis of lactose proceeds by the formation of an enzyme–galactose complex, glucose being released. However, lactose or other sugars can also act as acceptors leading to the formation of galactose-containing oligosaccharides (Mahoney, 1998). In this way, lactose can act both as donor and acceptor of galactose, leading to the synthesis of galacto-oligosaccharides (GOS) (Rastall, 2006), or else can act only as donor, other sugars, like fructose, acting as acceptors in which case lactulose (Panesar and Kumari, 2011) and fructosyl-galacto-oligosaccharides will be synthesized (Martínez-Villaluenga et al., 2008; Cardelle-Cobas et al., 2011; Guerrero et al., 2013). The fate of the enzyme–galactose complex will be determined by the water activity in the medium: if high, hydrolysis will prevail and oligosaccharide formation will be nil; if low, acceptors other than water will compete so that oligosaccharides will be synthesized in a kinetically controlled fashion and a maximum conversion to oligosaccharide will be reached after which hydrolysis of the product formed will ensue (Gänzle, 2012). In the case of GOS, such maximum yield will occur at around 50% lactose conversion, GOS yield hardly exceeding 30–40% (Vera et al., 2012). Reduction of the hydrolytic potential of the enzyme, as required for oligosaccharide synthesis, is attainable by using very high substrate concentrations (Huerta et al., 2011) or else using nonaqueous media, but the latter is hardly applicable because of the reduced solubility of lactose and the low activity and stability of the enzyme in such media (Bankova et al., 2006; Lang et al., 2006; Srisimarat and Pongsawasdi, 2008).

Synthesis of lactose-derived nondigestible oligosaccharides from whey lactose has been recently reviewed (Nath et al., 2015). Enzymatic synthesis of lactose-derived oligosaccharides is the core subject of this book, therefore it will be analyzed in depth in the following chapters.

REFERENCES

Abril, J.R., Stull, J.W., 1989. Lactose hydrolysis in acid whey with subsequent glucose isomerization. Journal of the Science of Food and Agriculture 48, 511–514.

Adamczak, M., Bornscheuer, U.T., 2009. Improving ascorbyl oleate synthesis catalyzed by *Candida antarctica* lipase B in ionic liquids and water activity control by salt hydrates. Process Biochemistry 44, 257–261.

Adlercreutz, P., 2013. Immobilisation and application of lipase in organic media. Chemical Society Reviews 42, 6406–6436.

Akgül, F.B., Demirhan, E., Özbek, B., 2012. A modelling study of skimmed milk lactose hydrolysis and β-galactosidase stability using three reactor types. International Journal of Dairy Technology 65, 217–231.

Aksu, Z., Eren, T., 2007. Production of carotenoids by the isolated yeast of *Rhodotorula glutinis*. Biochemical Engineering Journal 35, 107–113.

Alvira, P., Tomás-Pejó, E., Ballesteros, M., Negro, M.J., 2010. Pretreatment technologies for an efficient bioethanol production process based on enzymatic hydrolysis: a review. Bioresource Technology 101, 4851–4861.

Andersen, D.C., Krummen, L., 2002. Recombinant protein expression for therapeutic applications. Current Opinion in Biotechnology 13, 117–123.

Antoni, D., Zverlov, V.V., Schwarz, W.H., 2007. Biofuels from microbes. Applied Microbiology and Biotechnology 77, 23–35.

Asplund, S., 2005. The Biogas Production Plant at Umeå Dairy. Evaluation of Design and Start-Up. http://www.diva-portal.org/smash/get/diva2:21340/FULLTEXT01.pdf.

Athanasiadis, I., Boskou, D., Kanellaki, M., Kioesseoglou, V., Koutinas, A.A., 2002. Whey liquid waste of the dairy industry as raw material for potable alcohol production by kefir granules. Journal of Agricultural and Food Chemistry 50, 7231–7234.

Athanasiadis, I., Paraskevopoulou, A., Blekas, G., Kiosseoglou, V., 2004. Development of a novel whey beverage by fermentation with kefir granules. Effect of various treatments. Biotechnology Progress 20, 1091–1095.

Atkins, P.J., 2005. Dry powder inhalers: an overview. Respiratory Care 50, 1304–1312.

Atra, R., Vtai, G., Bekassy-Molnar, E., Balint, A., 2005. Investigation of ultra- and nanofiltration for utilization of whey protein and lactose. Journal of Food Engineering 67, 325–332.

Audic, J.L., Chaufer, B., Dauphin, G., 2003. Non-food applications of milk components and dairy co-products: a review. Lait 83, 417–438.

Axelsson, A., Zacchi, G., 1990. Economic evaluation of the hydrolysis of lactose using immobilized β-galactosidase. Applied Biochemistry and Biotechnology 24-25, 679–693.

Bacha, U., Nasir, M., Khalique, A., Anjum, A.A., Jabbar, M.A., 2011. Comparative assessment of various agro-industrial wastes for *Saccharomyces cerevisiae* biomass production and its quality evaluation as single cell protein. The Journal of Animal and Plant Sciences 21, 844–849.

Bailey, M.J., Linko, M., 1990. Production of β-galactosidase by *Aspergillus oryzae* in submerged bioreactor cultivation. Journal of Biotechnology 16, 57–66.

Banat, I.M., Nigam, P., Singh, D., Marchant, R., McHale, A.P., 1998. Ethanol production at elevated temperatures and alcohol concentrations: part I – yeasts in general. World Journal of Microbiology and Biotechnology 14, 809–821.

Baneyx, F., 1999. Recombinant protein expression in *Escherichia coli*. Current Opinion in Biotechnology 10, 411–421.

Barrett, E., Stanton, C., Zelder, O., Fitzgerald, G., Ross, R.P., 2004. Heterologous expression of lactose- and galactose-utilizing pathways from lactic acid bacteria in *Corynebacterium glutamicum* for production of lysine in whey. Applied and Environmental Microbiology 70, 2861–2866.

Bankova, E., Bakalova, N., Petrova, S., Kolev, D., 2006. Enzymatic synthesis of oligosaccharides and alkylglycosides in water–organic media via transglycosylation of lactose. Biotechnology and Biotechnological Equipment 20, 114–119.

Bekatorou, A., Psarianos, C., Koutinas, A.A., 2006. Production of food grade yeasts. Food Technology and Biotechnology 44, 407–415.

Bellaver, L.H., de Carvalho, N.M.B., Abrahão-Neto, J., Gombert, A.K., 2004. Ethanol formation and enzyme activities around glucose-6-phosphate in *Kluyveromyces marxianus* CBS 6556 exposed to glucose or lactose excess. FEMS Yeast Research 4, 691–698.

Biaikowska, A.M., Ciemlijski, H., Nowakowska, K.M., Kur, J., Turkiewicz, M., 2009. A new β-galactosidase with a low temperature optimum isolated from the Antarctic *Arthrobacter* sp. 20B: gene cloning, purification and characterization. Archives of Microbiology 191, 825–835.

Bottomley, R.C., Evans, M.T.A., Parkinson, C.J., 1990. Whey proteins. In: Harris, P (Ed.), Food Gels. Elsevier Applied Food Science Series, London, pp. 435–466.

Božanić, R., Barukčić, I., Jakopović, K.L., Tratnik, L., 2014. Possibilities of whey utilisation. Austin Journal of Nutrition and Food Sciences 2 (7), 1036.

Büyükkileci, A.O., Harsa, S., 2004. Batch production of L(+) lactic acid from whey by *Lactobacillus casei* (NRRL B-441). Journal of Chemical Technology and Biotechnology 79, 1036–1040.

Cardelle-Cobas, A., Corzo, N., Martínez-Villaluenga, C., Olano, A., Villamiel, M., 2011. Effect of reaction conditions on lactulose-derived trisaccharides obtained by transgalactosylation with β-galactosidase of *Kluyveromyces lactis*. European Food Research and Technology 233, 89–94.

Cardona, C.A., Sánchez, O.J., 2007. Fuel ethanol production: process design trends and integration opportunities. Bioresource Technology 98, 2415–2457.

Cardona, C.A., Quintero, J.A., Paz, I.C., 2010. Production of bioethanol from sugarcane bagasse: status and perspectives. Bioresource Technology 101, 4754–4766.

Çelik, E., Çalık, P., 2012. Production of recombinant proteins by yeast cells. Biotechnology Advances 30, 1108–1118.

Chang, S.W., Shaw, J.F., 2009. Biocatalysis for the production of carbohydrate esters. New Biotechnology 26, 109–116.

Chen, W., Chen, H., Xia, Y., Zhao, J., Tian, F., Zhang, H., 2008. Production, purification, and characterization of a potential thermostable galactosidase for milk lactose hydrolysis from *Bacillus stearothermophilus*. Journal of Dairy Science 91, 1751–1758.

Chen, H., Qiu, W., 2010. Key technologies for bioethanol production from lignocellulose. Biotechnology Advances 28, 556–562.

Clare, D.A., Catignani, G.L., Swaisgood, H.E., 2003. Biodefense properties of milk, the role of antimicrobial proteins and peptides. Current Pharmaceutical Design 9, 1239–1255.

Coker, J.A., Brenchley, J.E., 2006. Protein engineering of a cold-active β-galactosidase from *Arthrobacter* sp. SB to increase lactose hydrolysis reveals new sites affecting low temperature activity. Extremophiles 10, 515–524.

Comino, E., Rosso, M., Riggio, V., 2009. Development of a pilot scale anaerobic digester for biogas production from cow manure and whey mix. Bioresource Technology 100, 5072–5078.

Coté, A., Brown, W.A., Cameron, D., van Walsum, G.P., 2004. Hydrolysis of lactose in whey permeate for subsequent fermentation to ethanol. Journal of Dairy Science 87, 1608–1620.

Cristiani-Urbina, E., Netzahuatl-Muñoz, A.R., Manriquez-Rojas, F.J., Juárez-Ramírez, C., Ruiz-Ordaz, N., Galíndez-Mayer, J., 2000. Batch and fed-batch cultures for the treatment of whey with mixed yeast cultures. Process Biochemistry 35, 649–657.

Crittenden, R., Weerakkody, R., Sanguansri, L., Augustin, M., 2006. Synbiotic microcapsules that enhance microbial viability during nonrefrigerated storage and gastrointestinal transit. Applied and Environmental Microbiology 72, 2280–2282.

Cuartas-Uribe, B., Alcaina-Miranda, M.I., Soriano-Costa, E., Mendoza-Roca, J.A., Iborra-Clar, M.I., Lora-García, J., 2009. A study of the separation of lactose from whey ultrafiltration permeate using nanofiltration. Desalination 241, 244–255.

Davila-Vazquez, G., Cota-Navarro, C.B., Rosales-Colunga, L.M., de León-Rodríguez, A., Razo-Flores, E., 2009. Continuous biohydrogen production using cheese whey: improving the hydrogen production rate. International Journal of Hydrogen Energy 34, 4296–4304.

Delaney, R.A.M., 1976. Composition, properties and uses of whey protein concentrates. International Journal of Dairy Technology 29, 91–101.

Demaimay, M., le Henaff, Y., Printemps, P., 1978. Hydrolysis of lactose on catalytic resin. Process Biochemistry 13, 3–5.

Demirhan, E., Özbek, B., 2009. A modeling study on hydrolysis of lactose recovered from whey and β-galactosidase stability under sonic treatment. Chemical Engineering Communications 196, 767–787.

Devulapalle, K.S., Gómez De Segura, A., Ferrer, B., Alcalde, M., Mooser, G., Plou, F.J., 2004. Effect of carbohydrate fatty acid esters on *Streptococcus sobrinus* and glucosyltransferase activity. Carbohydrate Research 339, 1029–1034.

Dhillon, G.S., Brar, S.K., Verma, M., Tyagi, R.D., 2011. Utilization of different agro-industrial wastes for sustainable bioproduction of citric acid by *Aspergillus niger*. Biochemical Engineering Journal 54, 83–92.

Dickson, R.C., Markin, J.S., 1980. Physiological studies of β-galactosidase induction in *Kluyveromyces lactis*. Journal of Bacteriology 142, 777–785.

Djordjevic, D., McClements, D.J., Decker, E.A., 2004. Oxidative stability of whey protein-stabilized oil-in-water emulsions at pH 3: potential ω-3 fatty acid delivery systems (Part B). Journal of Food Science 69, C356–C362.

Domingues, L., Dantas, M.M., Lima, N., Teixeira, J.A., 1999. Continuous ethanol fermentation of lactose by a recombinant flocculating *Saccharomyces cerevisiae* strain. Biotechnology and Bioengineering 64, 692–697.

Domingues, L., Guimarães, P.M.R., Oliveira, C., 2010. Metabolic engineering of *Saccharomyces cerevisiae* for lactose/whey fermentation. Bioengineered Bugs 1 (3), 164–171.

Durand, E., Lecomte, J., Villeneuve, P., 2013. Deep eutectic solvents: synthesis, application, and focus on lipase-catalyzed reactions. European Journal of Lipid Science and Technology 115, 379–385.

Elander, R.P., 2003. Industrial production of β-lactam antibiotics. Applied Microbiology and Biotechnology 61, 385–392.

El-Holi, M.A., Al-Delaimy, S., 2003. Citric acid production from whey with sugars and additives by *Aspergillus niger*. African Journal of Biotechnology 2, 356–359.

El-Samragy, Y.A., Khorshid, M.A., Foda, M.I., Shehata, A.E., 1996. Effect of fermentation conditions on the production of citric acid from cheese whey by *Aspergillus niger*. International Journal of Food Microbiology 29, 411–416.

Evans, M.T.A., Gordon, J.F., 1980. Whey proteins. In: Grant, R.A. (Ed.), Applied Protein Chemistry. Applied Science Publishers, London. pp. 31–67.

Fages, J., Lochard, H., Letourneau, J.J., Sauceau, M., Rodier, E., 2004. Particle generation for pharmaceutical applications using supercritical fluid technology. Powder Technology 141, 219–226.

Farahnak, F., Seki, T., Ryu, D.D.Y., Ogrydziak, D., 1986. Construction of lactose-assimilating and high-ethanol-producing yeasts by protoplast fusion. Applied and Environmental Microbiology 5, 362–367.

Feijoo, M., Moreira, M.T., Roca, E., Lema, J.M., 1999. Use of cheese whey as a substrate to produce manganese peroxidase by *Bjerkandera* sp. BOS55. Journal of Industrial Microbiology and Biotechnology 23, 86–90.

Fekete, E., Karaffa, L., Seiboth, B., Fekete, E., Kubicek, C.P., Flipphi, M., 2012. Identification of a permease gene involved in lactose utilization in *Aspergillus nidulans*. Fungal Genetics and Biology 49, 415–425.

Fernandes Silva, M., Fornari, R.C.G., Mazutti, M.A., de Oliveira, D., Ferreira Padilha, F., Cichoski, A.J., Cansian, R.L., Di Luccio, M., Treichel, H., 2009. Production and characterization of xanthan gum by *Xanthomonas campestris* using cheese whey as sole carbon source. Journal of Food Engineering 90, 119–123.

Ferrari, M.D., Bianco, R., Froche, C., Loperena, M.L., 2001. Baker's yeast production from molasses/cheese whey mixtures. Biotechnology Letters 23, 1–4.

Ferrer, M., Soliveri, J., Plou, F.J., López-Cortés, N., Reyes-Duarte, D., Christensen, M., Copa-Patiño, J.L., Ballesteros, A., 2005. Synthesis of sugar esters in solvent mixtures by lipases from *Thermomyces lanuginosus* and *Candida antarctica* B, and their antimicrobial properties. Enzyme and Microbial Technology 36, 391–398.

Ferreyra, R., Lorda, G., Balatti, A., 1998. Production of α-amylase in acid cheese whey culture media with automatic pH control. Revista de Microbiología 29, 259–264.

Flieger, M., Kantorová, M., Prell, A., Řezanka, T., Votruba, J., 2003. Biodegradable plastics from renewable sources. Folia Microbiologica 48, 27–44.

Foda, M.I., Dong, H., Li, Y., 2010. Study the suitability of cheese whey for bio-butanol production by *Clostridia*. Journal of American Science 6, 39–46.

Foegeding, E.A., Davis, J.P., Doucet, D., McGuffey, M.K., 2002. Advances in modifying and understanding whey protein functionality. Trends in Food Science and Technology 13, 151–159.

Fonseca, G.G., Heinzle, E., Wittmann, C., Gombert, A.K., 2008. The yeast *Kluyveromyces marxianus* and its biotechnological potential. Applied Microbiology and Biotechnology 79, 339–354.

Fontes, E.A.F., Passos, F.M.L., Passos, F.J.V., 2001. A mechanistic mathematical model to predict lactose hydrolysis by β-galactosidase in a permeabilized cell mass of *Kluyveromyces lactis*: validity and sensitivity analysis. Process Biochemistry 37, 267–274.

Fox, P.F., 2009. Lactose: chemistry and properties. In: Fox, P.F., McSweeney, P.L.H. (Eds.), Advanced Dairy Chemistry. Lactose, Water, Salts and Minor Constituents, vol. 3. Springer, New York. pp. 1–15.

Fritsche, U.R., Sims, R.E.H., Monti, A., 2010. Direct and indirect land-use competition issues for energy crops and their sustainable production – an overview. Biofuels, Bioproducts and Biorefining 4, 692–704.

Fukuhara, H., 2006. *Kluyveromyces lactis* – a retrospective. FEMS Yeast Research 6, 323–324.

Gänzle, M.G., Haase, G., Jelen, P., 2008. Lactose: crystallization, hydrolysis and value-added derivatives. International Dairy Journal 18, 685–694.

Gänzle, M.G., 2012. Enzymatic synthesis of galacto-oligosaccharides and other lactose derivatives (hetero-oligosaccharides) from lactose. International Dairy Journal 22, 116–122.

Geddes, C.C., Nieves, I.U., Ingram, L.O., 2011. Advances in ethanol production. Current Opinion in Biotechnology 22, 312–319.

Gekas, V., López-Leiva, M., 1985. Hydrolysis of lactose; a literature review. Process Biochemistry 20, 2–12.

Gelegenis, J., Georgakakis, D., Angelidaki, I., Mavris, V., 2007. Optimization of biogas production by co-digesting whey with diluted poultry manure. Renewable Energy 32, 2147–2160.

Ghaly, A.E., Ben-Hassan, R.M., 1995. Kinetics of batch production of single-cell protein from cheese whey. Applied Biochemistry and Biotechnology 50, 79–92.

Ghaly, A.E., Kamal, M., Correia, L.R., 2005. Kinetic modelling of continuous submerged fermentation of cheese whey for single cell protein production. Bioresource Technology 96, 1143–1152.

Gill, H.S., Cross, M.L., 2000. Anticancer properties of bovine milk. British Journal of Nutrition 84, 161–166.

Gohel, M.C., Jogani, P.D., 2005. A review of co-processed directly compressible excipients. Journal of Pharmacy and Pharmaceutical Sciences 8, 76–93.

Gombert, A.K., Kilikian, B.V., 1998. Recombinant gene expression in *Escherichia coli* cultivation using lactose as inducer. Journal of Biotechnology 60, 47–54.

González Siso, M.I., 1996. The biotechnological utilization of whey: a review. Bioresource Technology 57, 1–11.

Greiter, M., Novalin, S., Wendland, M., Kulbe, K.D., Fischer, J., 2002. Desalination of whey by electrodialysis and ion exchange resins: analysis of both processes with regard to sustainability by calculating their cumulative energy demand. Journal of Membrane Science 210, 91–102.

Grosová, Z., Rosenberg, M., Rebroš, M., 2008. Perspectives and applications of immobilised β-galactosidase in food industry – a review. Czech Journal of Food Science 26, 1–14.

Guerrero, C., Vera, C., Illanes, A., 2013. Optimization of synthesis of oligosaccharides derived from lactulose (fructosyl-galacto-oligosaccharides) with β-galactosidase of different origin. Food Chemistry 138, 2225–2232.

Guimarães, P.M.R., François, J., Parrou, J.L., Teixeira, J.A., Domingues, L., 2008a. Adaptive evolution of a lactose-consuming *Saccharomyces cerevisiae* recombinant. Applied and Environmental Microbiology 74, 1748–1756.

Guimarães, P.M.R., Teixeira, J.A., Domingues, L., 2008b. Fermentation of high concentrations of lactose to ethanol by engineered flocculent *Saccharomyces cerevisiae*. Biotechnology Letters 30, 1953–1958.

Guimarães, P.M.R., Teixeira, J.A., Domingues, L., 2010. Fermentation of lactose to bio-ethanol by yeasts as part of integrated solutions for the valorisation of cheese whey. Biotechnology Advances 28, 375–384.

Gumel, A.M., Annuara, M.S.M., Heidelberg, T., Chisti, Y., 2011. Lipase mediated synthesis of sugar fatty acid esters. Process Biochemistry 46, 2079–2090.

Gunasekaran, P., Kamini, N.R., 1991. High ethanol productivity from lactose by immobilized cells of *Kluyveromyces fragilis* and *Zymomonas mobilis*. World Journal of Microbiology and Biotechnology 7, 551–556.

Gunasekaran, S., Xiao, L., Ould Eleya, M.M., 2006. Whey protein concentrate hydrogels as bioactive carriers. Journal of Applied Polymer Science 99, 2470–2476.

Gunasekaran, S., Ko, S., Xiao, L., 2007. Use of whey proteins for encapsulation and controlled delivery applications. Journal of Food Engineering 83, 31–40.

Guo, Y., Pan, D., Tanokura, M., 2009. Optimisation of hydrolysis conditions for the production of the angiotensin I-converting enzyme (ACE) inhibitory peptides from whey protein using response surface methodology. Food Chemistry 114, 328–333.

Guo, X., Zhou, J., Xiao, D., 2010. Improved ethanol production by mixed immobilized cells of *Kluyveromyces marxianus* and *Saccharomyces cerevisiae* from cheese whey powder solution fermentation. Applied Biochemistry and Biotechnology 160, 532–538.

Guzmán, F., Barberis, S., Illanes, A., 2007. Peptide synthesis: chemical or enzymatic. Electronic Journal of Biotechnology 10 (2), 279–314.

Ha, E., Zemel, M.B., 2003. Functional properties of whey, whey components, and essential amino acids: mechanisms underlying health benefits for active people. The Journal of Nutritional Biochemistry 14, 251–258.

Habulin, M., Šabeder, S., Knez, Ž., 2008. Enzymatic synthesis of sugar fatty acid esters in organic solvent and in supercritical carbon dioxide and their antimicrobial activity. The Journal of Supercritical Fluids 45, 338–345.

Harju, M., Kallioinen, H., Tossavainen, O., 2012. Lactose hydrolysis and other conversions in dairy products: technological aspects. International Dairy Journal 22, 104–109.

Hatzinikolaou, D.G., Katsifas, E., Mamma, A., Karagouni, D., Christakopoulos, P., Kekos, D., 2005. Modeling of the simultaneous hydrolysis–ultrafiltration of whey permeate by a thermostable β-galactosidase from *Aspergillus niger*. Biochemical Engineering Journal 24, 161–172.

Heyman, M.B., 2006. Lactose intolerance in infants, children, and adolescents. Pediatrics 118, 1279–1286.

Horn, S.J., Vaaje-Kolstad, G., Westereng, B., Eijsink, V.G.H., 2012. Novel enzymes for the degradation of cellulose. Biotechnology for Biofuels 5, 45–56.

Hosseini, M., Shojaosadati, S.A., Towfighi, J., 2003. Application of a bubble-column reactor for the production of a single-cell protein from cheese whey. Industrial and Engineering Chemistry Research 42, 764–766.

Hoyer, K., Galbe, M., Zacchi, G., 2013. The effect of prehydrolysis and improved mixing on high-solids batch simultaneous saccharification and fermentation of spruce to ethanol. Process Biochemistry 48, 289–293.

Huerta, L.M., Vera, C., Guerrero, C., Wilson, L., Illanes, A., 2011. Synthesis of galacto-oligosaccharides at very high lactose concentrations with immobilized β-galactosidases from *Aspergillus oryzae*. Process Biochemistry 46, 245–252.

Huppertz, T., Gazi, I., 2015. Lactose in dairy ingredients: effect on processing and storage stability. Journal of Dairy Science 98, 1–10. Available online 18 September 2015.

Ibrahim, F.S., Babiker, E.E., Yousif, N.E., El Tinay, A.H., 2005. Effect of fermentation on biochemical and sensory characteristics of sorghum flour supplemented with whey protein. Food Chemistry 92, 285–292.

Illanes, A., Schaffeld, G., 1982. Protein enrichment of sugar beet bagasse by fermentation. Revista Argentina de Microbiología 14, 97–104.

Illanes, A., Ruiz, A., Zúñiga, M.E., Aguirre, C., O'Reilly, S., Curotto, E., 1990. Immobilization of lactase for the continuous hydrolysis of whey permeate. Bioprocess Engineering 5, 257–262.

Illanes, A., Wilson, L., Raiman, L., 1999. Design of immobilized enzyme reactors for the continuous production of fructose syrup from whey permeate. Bioprocess Engineering 21, 509–515.

Illanes, A., Wilson, L., Tomasello, G., 2000. Temperature optimization for reactor operation with chitin-immobilized lactase under modulated inactivation. Enzyme and Microbial Technology 27, 270–278.

Illanes, A., 2011. Whey upgrading by enzyme biocatalysis. Electronic Journal of Biotechnology 14 (6). http://dx.doi.org/10.2225/vol14-issue6-fulltext-11.

Jaeger, K.E., Eggert, T., 2002. Lipases for biotechnology. Current Opinion in Biotechnology 13, 390–397.

Jędrzejewska, M., Kozak, K., 2011. Ethanol production from whey permeate in a continuous anaerobic bioreactor by *Kluyveromyces marxianus*. Environmental Technology 32, 37–42.

Jelen, P., 1979. Industrial whey processing technology: an overview. Journal of Agricultural and Food Chemistry 27, 658–661.

Jin, M., Gunawan, C., Uppugundla, N., Balan, V., Dale, B.E., 2012. A novel integrated biological process for cellulosic ethanol production featuring high ethanol productivity, enzyme recycling and yeast cell reuse. Energy and Environmental Science 5, 7168–7175.

Jin, Z., Liang, S., Zhang, X., Han, S., Ren, C., Lin, Y., Zheng, S., 2013. Synthesis of fructose laurate esters catalyzed by a CALB-displaying *Pichia pastoris* whole-cell biocatalyst in a non-aqueous system. Biotechnology and Bioprocess Engineering 18, 365–374.

John, R.P., Nampoothiri, P.J., Pandey, K.M., 2007. Fermentative production of lactic acid from biomass: an overview on process developments and future perspectives. Applied Microbiology and Biotechnology 74, 524–534.

Jooyandeh, H., 2009. Evaluation of physical and sensory properties of Iranian lavash flat bread supplemented with precipitated whey protein (PWP). African Journal of Food Science 3, 28–34.

Kailasapathy, K., 2008. Chemical composition, physical and functional properties of milk and milk ingredients. In: Chandan, R.C., Kilara, A., Shah, N. (Eds.), Dairy Processing and Quality Assurance. Wiley-Blackwell, New York, pp. 75–103.

Karaffa, L., Fekete, E., Sándor, E., Sepsi, A., Seiboth, B., Szentirmai, A., Kubicek, C.P., 2002. Carbon catabolite repression in the regulation of β-galactosidase activity in *Aspergillus nidulans*. Acta Microbiologica et Immunologica Hungarica 49, 261–265.

Kargi, F., Ozmıhcı, S., 2006. Utilization of cheese whey powder (CWP) for ethanol fermentations: effects of operating parameters. Enzyme and Microbial Technology 38, 711–718.

Kaur, G., Panesar, P.S., Bera, M.B., Kumar, H., 2009. Hydrolysis of whey lactose using CTAB-permeabilized yeast cells. Bioprocess and Biosystems Engineering 32, 63–67.

Kavacik, B., Topaloglu, B., 2010. Biogas production from co-digestion of a mixture of cheese whey and dairy manure. Biomass and Bioenergy 34, 1321–1329.

Kennedy, J.F., Kumar, H., Panesar, P.S., Marwaha, S.S., Goyal, R., Parmar, A., Kaur, S., 2006. Enzyme-catalyzed regioselective synthesis of sugar esters and related compounds. Journal of Chemical Technology and Biotechnology 81, 866–876.

Khodaiyan, F., Razavi, S.H., Mousavi, S.M., 2008. Optimization of canthaxanthin production by *Dietzia natronolimnaea* HS-1 from cheese whey using statistical experimental methods. Biochemical Engineering Journal 40, 415–422.

Kim, M., Elvin, C., Brownlee, A., Lyons, R., 2007. High yield expression of recombinant pro-resilin: lactose-induced fermentation in *E. coli* and facile purification. Protein Expression and Purification 52, 230–236.

Kinsella, J.E., Whitehead, D.M., 1989. Proteins in whey: chemical, physical, and functional properties. Advances in Food and Nutrition Research 33, 343–438.

Kirk, O., Borchert, T.V., Fuglsang, C.C., 2002. Industrial enzyme applications. Current Opinion in Biotechnology 13, 345–351.

Knopf, F.C., Okos, M.R., Fouts, D.A., Syverson, A., 1979. Optimization of a lactose hydrolysis process. Journal of Food Science 44, 896–900.

Koller, M., Bona, R., Chiellini, E., Grillo Fernandes, E., Horvat, P., Kutschera, C., Hesse, P., Braunegg, G., 2008. Polyhydroxyalkanoate production from whey by *Pseudomonas hydrogenovora*. Bioresource Technology 99, 4854–4863.

Koller, M., Salerno, A., Muhr, A., Reiterer, A., Chiellini, E., Casella, S., Horvat, P., Braunegg, G., 2012. Whey lactose as a raw material for microbial production of biodegradable polyesters. In: Saleh, H.E.D. (Ed.), Polyester. InTech. http://dx.doi.org/10.5772/48737. Available from: http://www.intechopen.com/books/polyester/whey-lactose-as-a-raw-material-for-microbial-production-of-biodegradable-polyesters.

Koppram, R., Olsson, L., 2014. Combined substrate, enzyme and yeast feed in simultaneous saccharification and fermentation allow bioethanol production from pretreated spruce biomass at high solids loadings. Biotechnology for Biofuels 7, 54–62.

Korhonen, H., Pihlanto, A., 2003. Food-derived bioactive peptides – opportunities for designing future foods. Current Pharmaceutical Design 9, 1297–1308.

Korhonen, H., Pihlanto, A., 2006. Bioactive peptides: production and functionality. International Dairy Journal 16, 945–960.

Kosseva, M.R., 2013. Use of immobilized biocatalyst for valorization of whey lactose. In: Kosseva, M.R., Webb, C. (Eds.), Food Industry Wastes: Assessment and Recuperation of Commodities. Academic Press, New York, pp. 137–156.

Kourkoutas, Y., Dimitropoulou, S., Kanellaki, M., Marchant, R., Nigam, P., Banat, I.M., Koutinas, A.A., 2002. High-temperature alcoholic fermentation of whey using *Kluyveromyces marxianus* IMB3 yeast immobilized on delignified cellulosic material. Bioresource Technology 82, 177–181.

Kumar, P., Barrett, D.M., Delwiche, M.J., Stroeve, P., 2009. Methods for pretreatment of lignocellulosic biomass for efficient hydrolysis and biofuel production. Industrial and Engineering Chemistry Research 48, 3713–3729.

Lang, M., Kamrat, T., Nidetzky, B., 2006. Influence of ionic liquid cosolvent on transgalactosylation reactions catalyzed by thermostable β-glycosylhydrolase CelB from *Pyrococcus furiosus*. Biotechnology and Bioengineering 95, 1093–1100.

Latorre, G., Besa, P., Parodi, C.G., Ferrer, V., Azocar, L., Quirola, M., Villarroel, L., Miquel, J.F., Agosín, E., Chianale, J., 2014. Prevalence of lactose intolerance in Chile: a double-blind placebo study. Digestion 90, 18–26.

Lee, S.H., Ha, S.H., Hiep, N.M., Chang, W.J., Koo, Y.M., 2008. Lipase-catalyzed synthesis of glucose fatty acid ester using ionic liquids mixtures. Journal of Biotechnology 133, 486–489.

Lehninger, A., 1975. Biochemistry, second ed. Worth Publishers Inc., New York.

Limayem, A., Ricke, S.C., 2012. Lignocellulosic biomass for bioethanol production: current perspectives, potential issues and future prospects. Progress in Energy and Combustion Science 38, 449–467.

Lindamood, J.B., Grooms, D.J., Hansen, P.M.T., 1989. Effect of hydrolysis of lactose and sucrose on firmness of ice cream. Food Hydrocolloids 3, 379–388.

Lomer, M.C.E., Parkes, G.C., Sanderson, J.D., 2008. Lactose intolerance in clinical practice – myths and realities. Alimentary Pharmacology and Therapeutics 27, 93–103.

Long, T.M., Su, Y.K., Headman, J., Higbee, A., Willis, L.B., Jeffries, T.W., 2012. Cofermentation of glucose, xylose, and cellobiose by the beetle-associated yeast *Spathaspora passalidarum*. Applied Environmental Microbiology 78, 5492–5500.

Lorenzen, P.C., Breiter, J., Clawin-Rädecker, I., Dau, A., 2013. A novel bi-enzymatic system for lactose conversion. International Journal of Food Science and Technology 48, 1396–1403.

Lotrakul, P., Dharmsthiti, S., 1997. Lipase production by *Aeromonas sobria* LP004 in a medium containing whey and soybean meal. World Journal of Microbiology and Biotechnology 13, 163–166.

Low, P.P.L., Rutherfurd, K.J., Gill, H.S., Cross, M.L., 2003. Effect of dietary whey protein concentrate on primary and secondary antibody responses in immunized BALB/c mice. International Immunopharmacology 3, 393–401.

Madureira, A.R., Pereira, C.I., Gomes, A.M.P., Pintado, M.E., Malcata, F.X., 2007. Bovine whey proteins – overview on their main biological properties. Food Research International 40, 1197–1211.

Mahoney, R.R., 1998. Galactosyl-oligosaccharide formation during lactose hydrolysis: a review. Food Chemistry 63, 147–154.

Manera, A.P., da Costa Ores, J., Amaral Ribeiro, V., Burkert, C.A.V., Juliano Kalil, S., 2008. Optimization of the culture medium for the production of β-galactosidase from *Kluyveromyces marxianus* CCT 7082. Food Technology and Biotechnology 46, 66–72.

Mantafounis, M., Pitts, J., 1990. Protein engineering of chymosin; modification of the optimum pH of enzyme catalysis. Protein Engineering 3, 605–609.

Margeot, A., Hahn-Hagerdal, B., Edlund, M., Slade, R., Monot, F., 2009. New improvements for lignocellulosic ethanol. Current Opinion in Biotechnology 20, 372–380.

Mariotti, M.P., Yamanaka, H., Araujo, A.R., Trevisan, H.C., 2008. Hydrolysis of whey lactose by immobilized β-galactosidase. Brazilian Archives of Biology and Technology 51, 1233–1240.

Martínez-Villaluenga, C., Cardelle-Cobas, A., Olano, A., Corzo, N., Villamiel, M., Jimeno, M.L., 2008. Enzymatic synthesis and identification of two trisaccharides produced from lactulose by transgalactosylation. Journal of Agricultural and Food Chemistry 56, 557–563.

Marwaha, S.S., Kennedy, J.F., 1988. Whey – pollution problem and potential. International Journal of Food Science and Technology 23, 323–336.

Matsushika, A., Inoue, H., Kodak, T., Sawayama, S., 2009. Ethanol production from xylose in engineered *Saccharomyces cerevisiae* strains: current state and perspectives. Applied Microbiology and Biotechnology 84, 37–53.

Mawson, A.J., 1994. Bioconversion for whey utilization and waste abatement. Bioresource Technology 47, 195–203.

McDonough, F.E., Hargrove, R.E., Mattingly, W.A., Posati, L.P., Alford, J.A., 1974. Composition and properties of whey protein concentrates from ultrafiltration. Journal of Dairy Science 57, 1438–1443.

Menzella, H.G., Ceccarelli, E.A., Gramajo, H.C., 2003. Novel *Escherichia coli* strain allows efficient recombinant protein production using lactose as inducer. Biotechnology and Bioengineering 82, 809–817.

Mlichová, Z., Rosenberg, M., 2006. Current trends of β-galactosidase application in food technology. Journal of Food and Nutrition Research 45, 47–54.

Modenbach, A.A., Nokes, S.D.E., 2013. Enzymatic hydrolysis of biomass at high-solids loadings – a review. Biomass and Bioenergy 56, 526–544.

Moeini, H., Nahvi, I., Tavassoli, M., 2004. Improvement of SCP production and BOD removal of whey with mixed yeast culture. Electronic Journal of Biotechnology 7 (3), 6–7.

Morales, J., Choi, J.S., Kim, D.S., 2006. Production rate of propionic acid in fermentation of cheese whey with enzyme inhibitors. Environmental Progress 25, 228–234.

Morr, C.V., Ha, E.Y.W., 1993. Whey protein concentrates and isolates: processing and functional properties. Critical Reviews in Food Science and Nutrition 33, 431–476.

Mostafa, N.A., 2001. Production of acetic acid and glycerol from salted and dried whey in a membrane cell recycle bioreactor. Energy Conversion and Management 42, 1133–1142.

Mukherjee, S., Das, P., Sen, R., 2006. Towards commercial production of microbial surfactants. Trends in Biotechnology 24, 509–515.

Mukhopadhyay, R., Chatterjee, S., Chatterjee, B.P., Banerjee, P.C., Guha, A.K., 2005. Production of gluconic acid from whey by free and immobilized *Aspergillus niger*. International Dairy Journal 15, 299–303.

Mussatto, S.I., Dragone, G., Guimarães, P.M.R., Silva, J.P.A., Carneiro, L.M., Roberto, I.C., Vicente, A., Domingues, L., Teixeira, J.A., 2010. Technological trends, global market, and challenges of bio-ethanol production. Biotechnology Advances 28, 817–830.

Nagarale, R.K., Gohil, G.S., Shahi, V.K., 2006. Recent developments on ion-exchange membranes and electro-membrane processes. Advances in Colloid and Interface Science 119, 97–130.

Nagpal, R., Behare, P., Rana, R., Kumar, A., Kumar, M., Arora, S., Morotta, F., Jain, S., Yadav, H., 2011. Bioactive peptides derived from milk proteins and their health beneficial potentials: an update. Food and Function 2, 18–27.

Napoli, F., Olivieri, G., Russo, M.E., Marzocchella, A., Salatino, P., 2010. Butanol production by *Clostridium acetobutylicum* in a continuous packed bed reactor. Journal of Industrial Microbiology and Biotechnology 37, 603–608.

Nasrabadi, M.R.N., Razavi, S.H., 2011. Optimization of β-carotene production by a mutant of the lactose-positive yeast *Rhodotorula acheniorum* from whey ultrafiltrate. Food Science and Biotechnology 20, 445–454.

Nasseri, A.T., Rasoul-Amini, S., Morowvat, M.H., Ghasemi, Y., 2011. Single cell protein: production and process. American Journal of Food Technology 6, 103–116.

Nath, A., Verasztó, B., Basak, S., Koris, A., Kovács, Z., Vatai, G., 2015. Synthesis of lactose-derived nutraceuticals from dairy waste whey – a review. Food and Bioprocess Technology. http://dx.doi.org/10.1007/s11947-015-1572-2 Published Online: 11 September.

Neelakantan, S., Mohanty, A.K., Kaushik, J.K., 1999. Production and use of microbial enzymes for dairy processing. Current Science 77, 143–149.

Neubauer, P., Hofmann, K., 1994. Efficient use of lactose for the lac promoter controlled overexpression of the main antigenic protein of the foot and mouth disease virus in *Escherichia coli* under fed-batch fermentation conditions. FEMS Microbiology Reviews 14, 99–102.

Neuhaus, W., Novalin, S., Klimacek, M., Splechtna, B., Petzelbauer, I., Szivak, A., Kulbe, K.D., 2006. Optimization of an innovative hollow-fiber process to produce lactose-reduced skim milk. Applied Biochemistry and Biotechnology 134, 1–14.

Newman, D.J., Cragg, G.M., 2007. Natural products as sources of new drugs over the last 25 years. Journal of Natural Products 70, 461–477.

Niemistö, J., Saavalainen, P., Pongrácz, E., Keiski, R.L., 2013. Biobutanol as a potential sustainable biofuel – assessment of lignocellulosic and waste-based feedstocks. Journal of Sustainable Development of Energy, Water and Environment Systems 1, 58–77.

Nolan, A.M., Barron, N., Brady, D., McAree, T., Smith, L., McHale, A.P., 1994. Ethanol production at 45° by an alginate-immobilized, thermotolerant strain of *Kluyveromyces marxianus* following growth on glucose-containing media. Biotechnology Letters 16, 849–852.

Novalin, S., Neuhaus, W., Kulbe, K.D., 2005. A new innovative process to produce lactose-reduced skim milk. Journal of Biotechnology 119, 212–218.

Olafadehan, O.A., Aribike, D.S., Adeyemo, A.M., 2009. Mathematical modeling and simulation of steady state plug flow for lactose–lactase hydrolysis in fixed bed. Theoretical Foundations of Chemical Engineering 43, 58–69.

Oleskowicz-Popiel, P., Kádár, Z., Heiske, S., Klein-Marcuschamer, D., Simmons, B.A., Blanch, H.W., Ejbye Schmidt, J., 2012. Co-production of ethanol, biogas, protein fodder and natural fertilizer inorganic farming – evaluation of a concept for a farm-scale biorefinery. Bioresource Technology 104, 440–446.

Ozmıhcı, S., Kargi, F., 2007. Ethanol fermentation of cheese whey powder solution by repeated fed-batch operation. Enzyme and Microbial Technology 41, 169–174.

Ozmihci, S., Kargi, F., 2008. Ethanol production from cheese whey powder solution in a packed column bioreactor at different hydraulic residence times. Biochemical Engineering Journal 42, 180–185.

Panesar, P.S., Kennedy, J.K., Gandhi, D.N., Bunko, K., 2007a. Bioutilisation of whey for lactic acid production. Food Chemistry 105, 1–14.

Panesar, R., Panesar, P.S., Singh, R.S., Kennedy, J.F., Bera, M.B., 2007b. Production of lactose-hydrolyzed milk using ethanol permeabilized yeast cells. Food Chemistry 101, 786–790.

Panesar, P.S., Kumari, S., Panesar, R., 2010. Potential applications of immobilized β-galactosidase in food processing industries. Enzyme Research. http://dx.doi.org/10.4061/2010/473137.

Panesar, P.S., Kumari, S., 2011. Lactulose: production, purification and potential applications. Biotechnology Advances 29, 940–948.

Park, J.W., Oh, Y.S., Lim, J.Y., Roh, D.H., 2006. Isolation and characterization of cold-adapted strains producing β-galactosidase. The Journal of Microbiology 44, 396–402.

Park, A.R., Oh, D.K., 2010. Effects of galactose and glucose on the hydrolysis reaction of a thermostable β-galactosidase from *Caldicellulosiruptor saccharolyticus*. Applied Microbiology and Biotechnology 85, 1427–1435.

Patakova, P., Linhova, M., Rychtera, M., Paulova, L., Melzoch, K., 2013. Novel and neglected issues of acetone–butanol–ethanol (ABE) fermentation by clostridia: *Clostridium* metabolic diversity, tools for process mapping and continuous fermentation systems. Biotechnology Advances 31, 58–67.

Paterson, A.H.J., 2009. Production and uses of lactose. In: Fox, P.F., McSweeney, P.L.H. (Eds.), Advanced Dairy Chemistry, vol. 3. third ed. Springer, New York, pp. 105–120.

Pedersen, N.R., Wimmer, R., Matthiesen, R., Pedersen, L.H., Gessesse, A., 2003. Synthesis of sucrose laurate using a new alkaline protease. Tetrahedron Asymmetry 14, 667–673.

Perotti, M.C., Wolf, V., Vénica, I., Bergamini, V., 2012. Dairy products modified in their lactose content. Current Nutrition and Food Science 8, 8–18.

Pesta, G., Meyer-Pittroff, R., Russ, W., 2007. Utilization of whey. In: Oreopoulou, V., Russ, W. (Eds.), Utilization of By-products and Treatment of Waste in the Food Industry. Springer, New York, pp. 193–207.

Peters, D., 2006. Carbohydrates for fermentation. Biotechnology Journal 1, 806–814.

Piccolo, C., Bezzo, F., 2009. A techno-economic comparison between two technologies for bioethanol production from lignocellulose. Biomass and Bioenergy 33, 478–491.

Pihlanto-Leppälä, A., 2000. Bioactive peptides derived from bovine whey proteins: opioid and ace-inhibitory peptides. Trends in Food Science and Technology 11, 347–356.

Pinsach, J., de Mas, C., López-Santín, J., 2008. Induction strategies in fed-batch cultures for recombinant protein production in *Escherichia coli*: application to rhamnulose 1-phosphate aldolase. Biochemical Engineering Journal 41, 181–187.

Písecký, J., 2005. Spray drying in the cheese industry 2005. International Dairy Journal 15, 531–536.

Polberger, S., Lönnerdal, B., 1993. Simple and rapid macronutrient analysis of human milk for individualized fortification: basis for improved nutritional management of very-low-birth-weight infants? Journal of Pediatric Gastroenterology and Nutrition 17, 283–290.

Porro, D., Martegani, E., Ranzi, B.M., Alberghina, L., 1992. Lactose/whey utilization and ethanol production by transformed *Saccharomyces cerevisiae* cells. Biotechnology and Bioengineering 39, 799–805.

Pouliot, Y., 2008. Membrane processes in dairy technology—from a simple idea to worldwide panacea. International Dairy Journal 18, 735–740.

Povolo, S., Casella, S., 2003. Bacterial production of PHA from lactose and cheese whey permeate. Macromolecular Symposia 197, 1–9.

Raganati, F., Olivieri, G., Procentese, A., Russo, M.E., Salatino, P., Marzocchella, A., 2013. Butanol production by bioconversion of cheese whey in a continuous packed bed reactor. Bioresource Technology 138, 259–265.

Rattray, W., Jelen, P., 1996. Protein standardization of milk and dairy products. Trends in Food Science and Technology 7, 227–234.

Rastall, R.A., 2006. Galacto-oligosaccharides as prebiotic food ingredients. In: Gibson, G.R., Rastall, R.A. (Eds.), Prebiotics: Development and Applications. John Wiley, New York, pp. 101–110.

Ravindra, A.P., 2000. Value-added food: single cell protein. Biotechnology Advances 18, 459–479.

Rehman, S.U., 2009. Lactose: chemistry and properties. In: Fox, P.F., McSweeney, P.L.H. (Eds.), Advanced Dairy Chemistry, third ed. Springer, New York, pp. 98–104.

Reyes-Duarte, D., López-Cortés, N., Ferrer, M., Plou, F.J., Ballesteros, A., 2005. Parameters affecting productivity in the lipase-catalysed synthesis of sucrose palmitate. Biocatalysis and Biotransformations 23, 19–27.

Riepma, K.A., Vromansb, H., Zuurman, K., Lerk, C.F., 1993. The effect of dry granulation on the consolidation and compaction of crystalline lactose. International Journal of Pharmaceutics 97, 29–38.

Rodrigues Pinho, J.M., Lopes Passos, F.M., 2011. Solvent extraction of β-galactosidase from *Kluyveromyces lactis* yields a stable and highly active enzyme preparation. Journal of Food Biochemistry 35, 323–336.

Roukas, T., Lazarides, H., Kotzekidou, P., 1991. Ethanol production from deproteinized whey by *Saccharomyces cerevisiae* cells entrapped in different immobilization matrices. Milchwissenschaft 46, 438–441.

Roukas, T., Varzakakou, M., Kotzekidou, P., 2015. From cheese whey to carotenes by *Blakeslea trispora* in a bubble column reactor. Applied Biochemistry and Biotechnology 175, 182–193.

Rubio-Teixeira, M., 2006. Endless versatility in the biotechnological applications of *Kluyveromyces* LAC genes. Biotechnology Advances 24, 212–225.

Ruiz, B., Chávez, A., Forero, A., García-Huante, Y., Romero, A., Sánchez, M., Rocha, D., Sánchez, B., Rodríguez-Sanoja, R., Sánchez, S., Langley, E., 2010. Production of microbial secondary metabolites: regulation by the carbon source. Critical Reviews in Microbiology 36, 146–167.

Russel, I., 1986. Will a recombinant DNA yeast be able to solve whey disposal problems? Trends in Biotechnology 4, 107–108.

Sakaki, K., Aoyama, A., Nakane, T., Ikegami, T., Negishi, H., Watanabe, K., Yanagishita, H., 2006. Enzymatic synthesis of sugar esters in organic solvent coupled with pervaporation. Desalination 193, 260–266.

Sanchez, S., Demain, A., 2002. Metabolic regulation of fermentation processes. Enzyme and Microbial Technology 31, 895–906.

Sangeetha, R., Geetha, A., Arulpandi, I., 2007. Optimization of protease and lipase production by *Bacillus pumilus* SG 2 isolated from an industrial effluent. The Internet Journal of Microbiology 5 (2).

Sarkar, N., Ghosh, S.K., Bannerjee, S., Aikat, K., 2012. Bioethanol production from agricultural wastes: an overview. Renewable Energy 37, 19–27.

Sauer, M., Porro, D., Mattanovich, D., Branduardi, P., 2008. Microbial production of organic acids: expanding the markets. Trends in Biotechnology 26, 100–108.

Sathisuksanoh, N., George, A., Zhang, Y.H.P., 2013. New lignocellulosic pretreatments using cellulose solvents: a review. Journal of Chemical Technology and Biotechnology 88, 169–180.

Schaafsma, G., 2008. Lactose and lactose derivatives as bioactive ingredients in human nutrition. International Dairy Journal 18, 458–465.

Schepers, A.W., Thibault, J., Lacroix, C., 2002. *Lactobacillus helveticus* growth and lactic acid production during pH-controlled batch cultures in whey permeate/yeast extract medium. Part I. Multiple factor kinetic analysis. Enzyme and Microbial Technology 30, 176–186.

Schingoethe, D.J., 1976. Whey utilization in animal feeding: a summary and evaluation. Journal of Dairy Science 59, 556–570.

Schuck, P., Dolivet, A., 2002. Lactose crystallization: determination of α-lactose monohydrate in spray-dried dairy products. Lait 82, 413–421.

Schultz, N., Chang, L., Hauck, A., Reuss, M., Syldatk, C., 2006. Microbial production of single-cell protein from deproteinized whey concentrates. Applied Microbiology and Biotechnology 69, 515–520.

Şener, N., Apar, D.K., Özbek, B., 2006. A modelling study on milk lactose hydrolysis and β-galactosidase stability under sonication. Process Biochemistry 41, 1493–1500.

Shahsavarani, H., Sugiyama, M., Kaneko, Y., Chuenchit, B., Harashima, S., 2012. Superior thermotolerance of *Saccharomyces cerevisiae* for efficient bioethanol fermentation can be achieved by overexpression of *RSP5* ubiquitin ligase. Biotechnology Advances 30, 1289–1300.

Sham, J.O.H., Zhang, Y., Finlay, W.H., Roa, W.H., Löbenberg, R., 2004. Formulation and characterization of spray-dried powders containing nanoparticles for aerosol delivery to the lung. International Journal of Pharmaceutics 269, 457–467.

Shaukat, A., Levitt, M.D., Taylor, B.C., MacDonald, R., Shamliyan, T.A., Kane, R.L., Wilt, T.J., 2010. Systematic review: effective management strategies for lactose intolerance. Annals of Internal Medicine 152, 797–803.

Shi, Y.G., Li, J.R., Chu, Y.H., 2011. Enzyme-catalyzed regioselective synthesis of sucrose-based esters. Journal of Chemical Technology and Biotechnology 86, 1457–1468.

Shoveller, A.K., Stoll, B., Ball, R.O., Burrin, D.G., 2005. Nutritional and functional importance of intestinal sulphur amino acid metabolism. Journal of Nutrition 135, 1609–1612.

Silva, A.C., Guimaraes, P.M.R., Teixeira, J.A., Domingues, L., 2010. Fermentation of deproteinized cheese whey powder solutions to ethanol by engineered *Saccharomyces cerevisiae*: effect of supplementation with corn steep liquor and repeated-batch operation with biomass recycling by flocculation. Journal of Industrial Microbiology and Biotechnology 37, 973–982.

Singh, D., Nigam, P., Banat, I.M., Marchant, R., McHale, A.P., 1998. Ethanol production at elevated temperatures and alcohol concentrations: part I – use of *Kluyveromyces marxianus* IMB3. World Journal of Microbiology and Biotechnology 14, 823–834.

Skovgaard, P.A., Jørgensen, H., 2013. Influence of high temperature and ethanol on thermostable lignocellulolytic enzymes. Journal of Industrial Microbiology and Biotechnology 40, 447–456.

Smithers, G.W., 2008. Whey and whey proteins – from "gutter" to gold. International Dairy Journal 18, 695–704.

Soccol, C.R., Vandenberghe, L.P.S., Rodrigues, C., Pandey, A., 2006. New perspectives for citric acid production and application. Food Technology and Biotechnology 44, 141–149.

Solomons, G.L., Litchfield, J.H., 1983. Single cell protein. Critical Reviews in Biotechnology 1, 21–58.

Song, H., Lee, S.Y., 2006. Production of succinic acid by bacterial fermentation. Enzyme and Microbial Technology 39, 352–361.

Srisimarat, W., Pongsawasdi, P., 2008. Enhancement of the oligosaccharide synthetic activity of β-galactosidase in organic solvents by cyclodextrin. Enzyme and Microbial Technology 43, 436–441.

Staniszewski, M., Kujawski, W., Lewandowska, M., 2007. Ethanol production from whey in bioreactor with co-immobilized enzyme and yeast cells followed by pervaporative recovery of product – kinetic model predictions. Journal of Food Engineering 82, 618–625.

Staniszewski, M., Kujawski, W., Lewandowska, M., 2009. Semi-continuous ethanol production in bioreactor from whey with co-immobilized enzyme and yeast cells followed by pervaporative recovery of product – kinetic model predictions considering glucose repression. Journal of Food Engineering 91, 240–249.

Stoeberl, M., Werkmeister, R., Faulstich, M., Russ, W., 2011. Biobutanol from food wastes – fermentative production, use as biofuel and the influence on the emissions. Procedia Food Science 1, 1867–1874.

Suárez, E., Lobo, A., Álvarez, S., Riera, F.A., Álvarez, R., 2009. Demineralization of whey and milk ultrafiltration permeate by means of nanofiltration. Desalination 241, 272–280.

Subhedar, P.B., Gogate, P.R., 2013. Intensification of enzymatic hydrolysis of lignocellulose using ultrasound for efficient bioethanol production: a review. Industrial and Engineering Chemistry Research 52, 11816–11828.

Szczodrak, J., 2000. Hydrolysis of lactose in whey permeate by immobilized β-galactosidase from *Kluyveromyces fragilis*. Journal of Molecular Catalysis B: Enzymatic 10, 631–637.

Tai, H.P., Brunner, G., 2009. Sugar fatty acid ester synthesis in high-pressure acetone-CO_2 system. The Journal of Supercritical Fluids 48, 36–40.

Talabardon, M., Schwitzguébel, J.P., Péringer, P., 2000. Acetic acid production from lactose by an anaerobic thermophilic coculture immobilized in a fibrous-bed bioreactor. Biotechnology Progress 16, 1008–1017.

Tan, K.T., Lee, K.T., Mohamed, A.R., 2008. Role of energy policy in renewable energy accomplishment: the case of second-generation bioethanol. Energy Policy 36, 3360–3365.

Tin, C.S.F., Mawson, A.J., 1993. Ethanol production from whey in a membrane recycle bioreactor. Process Biochemistry 28, 217–221.

Tipton, K.D., Elliott, T.A., Cree, M.G., Aarsland, A.A., Sanford, A.P., Wolfe, R.R., 2007. Stimulation of net muscle protein synthesis by whey protein ingestion before and after exercise. American Journal of Physiology – Endocrinology and Metabolism 292, E71–E76.

Toba, Y., Takada, Y., Yamamura, J., Tanaka, M., Matsuoka, Y., Kawakami, H., Itabashi, A., Aoe, S., Kumegawa, M., 2000. Milk basic protein: a novel protective function of milk against osteoporosis. Bone 27, 403–408.

Tsakali, E., Petrotos, K., D'Allessandro, A.G., Goulas, P., 2010. A review on whey composition and the methods used for its utilization for food and pharmaceutical products. In: Geril, P. (Ed.), Proceeding of the 6th International Conference on Simulation and Modelling in the Food and Bio-Industry. EUROSIS-ETI Publication, Ghent, pp. 195–201.

Tusé, D., Miller, M.W., 1984. Single-cell protein: current status and future prospects. CRC Critical Reviews in Food Science and Nutrition 19, 273–325.

van den Berg, J.A., van der Laken, K.J., van Ooyen, A.J.J., Renniers, T.C.H.M., Rietveld, K., Schaap, A., Brake, A.J., Bishop, R.J., Schultz, K., Moyer, D., Richman, M., Shuster, J.R., 1990. *Kluyveromyces* as a host for heterologous gene expression: expression and secretion of prochymosin. Bio/Technology 8, 135–139.

Vamvakaki, A.N., Kandarakis, I., Kaminarides, S., Komaitis, M., Papanikolaou, S., 2010. Cheese whey as a renewable substrate for microbial lipid and biomass production by *Zygomycetes*. Engineering in Life Sciences 4, 348–360.

Vasiljevic, T., Jelen, P., 2001. Production of β-galactosidase for lactose hydrolysis in milk and dairy products using thermophilic lactic acid bacteria. Innovative Food Science and Emerging Technologies 2, 75–85.

Vera, C., Guerrero, C., Conejeros, R., Illanes, A., 2012. Synthesis of galacto-oligosaccharides by β-galactosidase from *Aspergillus oryzae* using partially dissolved and supersaturated solution of lactose. Enzyme and Microbial Technology 50, 188–194.

Verbelen, P.J., De Schutter, D.P., Delvaux, F., Verstrepen, K.J., Delvaux, F.R., 2006. Immobilized yeast cell systems for continuous fermentation applications. Biotechnology Letters 28, 1515–1525.

Verma, M.L., Barrow, C.J., Kennedy, J.F., Puri, M., 2012. Immobilization of β-galactosidase from *Kluyveromyces lactis* on functionalized silicon dioxide nanoparticles: characterization and lactose hydrolysis. International Journal of Biological Macromolecules 50, 432–437.

Viitanen, M.I., Vasala, A., Neubauer, P., Alatossava, T., 2003. Cheese whey-induced high-cell-density production of recombinant proteins in *Escherichia coli*. Microbial Cell Factories 2, 2. http://dx.doi.org/10.1186/1475-2859-2-2.

Villas-Bôas, S.G., Esposito, E., Mitchell, D.A., 2002. Microbial conversion of lignocellulosic residues for production of animal feeds. Animal Feed Science and Technology 98, 1–12.

Visser, S., van Rooijen, P.J., Schattenkerk, C., Kerling, K.E., 1977. Peptide substrates for chymosin (rennin). Kinetic studies with bovine kappa-casein-(103-108)-hexapeptide analogues. Biochimica et Biophysica Acta 481, 171–176.

Vogt, M., Kunath, K., Dressman, J.B., 2008. Dissolution improvement of four poorly water soluble drugs by cogrinding with commonly used excipients. European Journal of Pharmaceutics and Biopharmaceutics 68, 330–337.

Walsh, M.K., Bombyk, R.A., Wagh, A., Bingham, A., Berreau, L.M., 2009. Synthesis of lactose monolaurate as influenced by various lipases and solvents. Journal of Molecular Catalysis B: Enzymatic 60, 171–177.

Wan, C., Li, Y., Shahbazi, A., Xiu, S., 2008. Succinic acid production from cheese whey using *Actinobacillus succinogenes* 130 Z. Applied Biochemistry and Biotechnology 145, 111–119.

Ward, M., Wilson, L.J., Kodama, K.H., Rey, M.W., Berka, R.M., 1990. Production of chymosin in *Aspergillus* by expression as a glucoamylase-chymosin fusion. Bio/Technology 8, 435–440.

Welderufael, F., Jauregi, P., 2010. Development of an integrative process for the production of bioactive peptides from whey by proteolytic commercial mixtures. Separation Science and Technology 45, 2226–2234.

Wu, Q., Wang, N.A., Xiao, Y.M., Lu, D.S., Lin, X.F., 2004. Regiospecific alkaline protease-catalyzed divinyl acyl transesterifications of primary hydroxyl groups of mono- and di-saccharides in pyridine. Carbohydrate Research 339, 2059–2067.

Yalcin, A.S., 2006. Emerging therapeutic potential of whey proteins and peptides. Current Pharmaceutical Design 12, 1637–1643.

Yamanè, T., Shimizu, S., 1984. Fed-batch techniques in microbial processes. Advances in Biochemical Engineering/Biotechnology 30, 147–194.

Yang, S.T., Silva, E.M., 1995. Novel products and new technologies for use of a familiar carbohydrate, milk lactose. Journal of Dairy Science 78, 2541–2562.

Yang, Z., Huang, Z.L., 2012. Enzymatic synthesis of sugar fatty acid esters in ionic liquids. Catalysis, Science and Technology 2, 1767–1775.

Ye, R., Hayes, D.G., 2012. Lipase-catalyzed synthesis of saccharide-fatty acid esters utilizing solvent-free suspensions: effect of acyl donors and acceptors, and enzyme activity retention. Journal of the American Oil Chemists Society 89, 455–463.

Yoo, I.S., Park, S.J., Yoon, H.H., 2007. Enzymatic synthesis of sugar fatty acid esters. Journal of Industrial and Engineering Chemistry 13, 1–6.

Yorgun, M.S., Akmehmet Balcioglu, I., Saygin, O., 2008. Performance comparison of ultrafiltration, nanofiltration and reverse osmosis on whey treatment. Desalination 229, 204–216.

Zadow, J.G., 1984. Lactose: properties and uses. Journal of Dairy Science 67, 2654–2679.

Zafar, S., Owais, M., 2006. Ethanol production from crude whey by *Kluyveromyces marxianus*. Biochemical Engineering Journal 27, 295–298.

Zemel, M.B., 2004. Role of calcium and dairy products in energy partitioning and weight management. American Journal of Clinical Nutrition 79, 907S–912S.

Zhang, Z.Y., Jin, B., Kelly, J.M., 2007. Production of lactic acid from renewable materials by *Rhizopus* fungi. Biochemical Engineering Journal 35, 251–263.

Zhang, Z., Donaldson, A.A., Ma, X., 2012. Advancements and future directions in enzyme technology for biomass conversion. Biotechnology Advances 30, 913–919.

Chapter 2

Functional Foods and Feeds: Probiotics, Prebiotics, and Synbiotics

A. Illanes and C. Guerrero

2.1 NEW TRENDS IN HUMAN AND ANIMAL FEEDING

Modern lifestyle has a strong impact on eating habits. The increased consumption of processed foods and so-called fast foods is causing perceptible effects on health. Eating is a fundamental aspect of life that goes well beyond nutrition, having strong social and psychological implications. Eating is associated with pleasure, so the compromise between health and indulgence is a profound dilemma in modern society (Niva, 2007). The discrepancy between modern eating habits and body metabolism is most certainly associated with diseases like diabetes, cancer, heart failure, allergies, osteoporosis, and dental problems; obesity, which is clearly related to eating habits, is a major concern not only in adults but in children as well (Ungar and Taeford, 2002). Then, pleasure of eating may promote well-being but at the same time compromise good health. In this context, the world population is increasingly concerned with the impact of eating habits on health, which is the basis of the onset and development of functional foods, which will be analyzed in the next section. Although primarily a problem of developed societies, health problems associated with modern eating habits are having a profound impact on developing countries as well, where deleterious effects are likely to be more evident because of economic constraints (Uauy and Monteiro, 2004; Ezzati et al., 2005).

Animal feeding has also experienced a rapid evolution. In the case of farm animals, intensive farming took over extensive grazing of livestock, even though in a comparison from a life cycle analysis perspective, extensive organic farming may be advantageous (Haas et al., 2001) with the consequent need of a balanced nutrition. Traditional feeding (pasture, forages, crop residues, and agro-industrial byproducts) is moving to a more innovative feeding in which supplements and additives, probiotics, and prebiotics play an important role (Bocquier and González-García, 2010). Pet feeding has also evolved to specialized and

Lactose-Derived Prebiotics. http://dx.doi.org/10.1016/B978-0-12-802724-0.00002-0

nutritionally balanced products. Associated with modern lifestyle, pets are no longer fed on home-made food, and owners very much rely on formulated dry products. However, pets are considered mostly as part of the family, so owners are concerned with their health, looking for high-quality products. Besides looking for adequate nutrition, appealing taste and texture, health is an issue, so the pet food industry has responded by offering a palette of products in which health-promoting components are incorporated. The market for pet food has increased significantly in recent decades with an average annual growth rate of 2–3%/year.

Among health-promoting components, probiotics and prebiotics are now considered key elements in feed formulation both for farm animals and pets, the main effects being the improved resistance to pathogen colonization and enhanced host mucosa immunity, resulting in reduced pathogen load and improved health condition. This is now particularly important since the use of antibiotics as animal growth promoters has been banned in many countries because of the promotion and transference of antibiotic resistance genes from animal to human intestinal microbiota (Gaggia et al., 2010). This aspect will be analyzed in the following sections.

2.2 HEALTH-PROMOTING FOODS: FUNCTIONAL FOODS AND NUTRACEUTICALS

Foods beyond their nutritional and hedonistic value are increasingly perceived in connection with disease prevention and physical and mental well-being (Menrad, 2003). It seems clear now that diet has a direct impact on gut health, which in turn is a major determinant in overall health (Xu et al., 2015). Health-promoting foods are increasingly being demanded, which can be explained by the increasing cost of health care, the increased life expectancy of the population, and the pursuit of good conditions of living by the elderly (Roberfroid, 2000a). Functional foods cannot be considered only as sophisticated developed world assets, as in fact health-enhancing foods are gaining progressive significance in developing countries (Kotilainen et al., 2006).

Health-promoting foods have a long history in Asian countries, where many types of traditional foods have been recognized as health promoters (Westrate et al., 2002). The term *functional food* was coined in Japan in the early 1980s to refer to foods fortified with substances that confer favorable physiological effects (Hardy, 2000). Later on, in 1991, the Japanese Ministry of Health, Labor, and Welfare defined and set rules for approval of a distinctive health-promoting food category termed "food for specified health uses" (FOSHU); this was done mostly to avoid exaggerated and misleading claims of health-promoting foods (Ohama et al., 2006). It refers to foods containing ingredients with functions for health that are officially approved to claim its physiological effects on humans (Shimizu, 2012).

Many definitions have been proposed for functional foods (Kaur and Das, 2011). Some are concise: "foods that may provide health benefits beyond basic nutrition." Others are more explicit: "foods similar in appearance as conventional foods that are intended to be consumed as part of a normal diet, but have been modified to provide a physiological benefit beyond the provision of simple nutrient requirements" (Siró et al., 2008). There is no universally accepted definition and the borderline between conventional and functional food is not always easy to establish; furthermore, in most countries there are no legislative definitions supporting functional foods. Trying to extract the essence of the concept, foods are to be considered functional if, beyond basic nutritional impact, they benefit one or more physiological functions in the human body, be it by improving the general and physical conditions and/or by decreasing the risk of disease, but a key aspect of functional foods is that the amount and form of intake should be as normally expected for dietary purposes. In this respect, the concept moves somehow away from the FOSHU concept, where capsules and tablets, though not being the more common form of food intake, are also included.

Functional food is mostly a concept that pertains to nutrition, not to pharmacology (Doyon and Labrecque, 2008); as clearly pointed out by Roberfroid (2002) they are foods, not drugs, and if a role is played with respect to disease, it will be in most cases a risk-reducing rather than a preventive role. This view is shared by the European Union; in the United States the focus with respect to functional foods has been mostly on controlling health claims on products and marketing (Lupton, 2009). But functional foods can actually aid in the prevention or treatment of diseases, and if so, they are called *nutraceuticals* (Alissa and Ferns, 2012), a term coined by combining the words *nutrition* and *pharmaceutical*. Being functional foods and nutraceuticals both health-promoting foods, a clear distinction between them is that nutraceuticals are formulated and taken in dosage form (El Sohaimy, 2012). The concepts are intertwined, so that a functional food for one consumer may act as a nutraceutical for another.

Functional foods can be either natural foods—foods from which a component has been added, removed, or modified—or foods in which the bioavailability of a component has been modified (Roberfroid, 2002). A pertinent example of a functional food is low-lactose milk, in which one component (lactose) has been removed by converting it into its monosaccharide moieties (glucose and galactose).

Consumers are increasingly aware of the significant role of the diet in disease risk management and well-being, which is exacerbated by the increase in global population aging (Lutz et al., 2008). This represents a tremendous opportunity and a challenge to the food industry, which is eager to supply those increasingly demanded healthy foods. The food industry is viewed as a low-research-intensity sector with a comparatively small investment in research and development. In this scenario, functional foods appear as a leading edge; in fact, innovations in the food industry are closely tied to development and replacement of products based on nutritional directions or compliance of regulations (Annunziata and Vecchio, 2011). Functional food development is a big challenge for the food industry because it demands both technological and cultural innovation. It is clear though that trends in modern society and demographics are in favor of functional foods, these products representing a sustainable trend worldwide, as illustrated by the increasing number of them being launched each year (Bigliardi and Galati, 2013). The biological roles of functional food components are now a major subject of research and a complex task since their effect may depend on the microbial ecology of the site of action (gut microbiota in most cases) and may vary from one person to another (Laparra and Sanz, 2010).

Early developments in functional foods were mostly related to fortification with minerals and vitamins, though controversy arose as to the allowed levels of fortification (Kloosterman et al., 2007); later on, fortification with micronutrients, like soluble fiber, phytosterols, and omega-3 fatty acids took over, and at present food companies are developing food products with compounded health benefits (Siró et al., 2008).

A major issue in functional foods is regulation and claims, particularly health claims. The situation varies from region to region; in the European Union focus is on safety regulations rather than in claims, while in the United States emphasis is mostly on claims. The problem is complex since, even though foods and food components may certainly prevent or ameliorate certain diseases, not enough scientific evidence exists to identify the effective ingredient and substantiate its efficacy and safety. Proper regulatory oversight is mostly lacking, so it is hard to determine the truth behind the claim and avoid quackery (El Sohaimy, 2012). Health claims for functional foods should unambiguously link the product with a specific health-related function; misleading information in this respect is considered illegal in some countries (Lähteenmäki et al., 2010). Soft claims implying health effects without referring to a specific disease have been quite frequent within food manufacturers since little evidence is required. In response, new regulations are being enforced in the European Union in an attempt to protect the consumer from misleading or false claims, prohibit vague claims, and give incentives for hard claims based on solid data and sound scientific evidence (Nocella and Kennedy, 2012); on the other hand, consumers tend to prefer short claims written in plain terms and are more prone to accept them when they refer to products with a previous healthy image (Dean et al., 2012). With the continuous proposals for novel health-promoting compounds the concept of novel foods has arisen differently from functional foods. Differences may have strong implications since novel foods will have to be rigorously evaluated based on their safety, while functional foods will be evaluated with respect to the desired health claims. Unfortunately, boundaries between both are not always clear, and the concept of novel may vary from one country to another (Kumar et al., 2015).

Consumer acceptance is a major issue in functional foods and it may vary significantly. In general, functional foods are viewed as pertaining to a particular food category rather than as a specific type of product, and this may certainly favor perception. However, consumer acceptance is not unconditional, and appearance, taste, and clarity of health claims are main conditions of acceptance. Educational level, geographic origin, and even gender have been considered to influence consumer perception. Dieticians' attitudes are also important; they may perceive risks associated from functional food consumption, but in general this will not preclude dieticians from prescribing them (Cha et al., 2010). Educational campaigns and communication by authorities are major contributors to improving public perception of functional foods (Annunziata and Vecchio, 2011).

Functional foods have been mostly launched by the dairy industry but are also important in the bakery, baby food, confectionery, and soft drink sectors (Menrad, 2003).

A thorough analysis of functional food and its historic development can be found in the book by Heasman and Mellentin (2001); the current impact of functional foods within the food industry has been thoroughly revised by El Sohaimy (2012). A comprehensive book on nutraceutical and functional foods, with special emphasis on technological challenges, was published by Boye (2015).

The market size of functional foods is difficult to establish with precision because the boundaries are rather blurred. The global market for functional foods was valued at US$74 billion in 2005 with a forecast close to US$100 million in 2015. This market is mostly covered by the United States, the European Union, and Japan as seen in Fig. 2.1. However, China, India, and Latin America are emerging functional food markets that have grown considerably in recent years as the concept of health eating habits becomes more widespread (Patel and Goyal, 2012).

Probiotics and prebiotics are outstanding functional food ingredients with a well-established role as health-promoting agents (Choudhari et al., 2008); the following sections will refer to probiotics and prebiotics within the framework of functional foods.

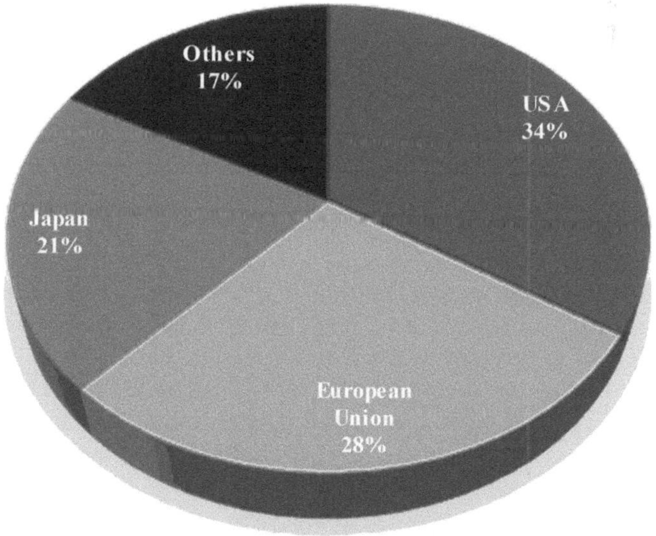

FIGURE 2.1 Market share of functional foods.

2.3 PROBIOTICS: CONCEPT, PRESENT STATUS, TECHNOLOGICAL GOALS, AND CHALLENGES

The word *probiotic* was originally coined as opposed to *antibiotic* (Lilly and Stillwell, 1965). In simple words, probiotics are beneficial microorganisms that are directly added to supplement the intestinal microbiota with live beneficial organisms (Teitelbaum and Walker, 2002; Yan and Polk, 2006). Its definition has evolved through time as more scientific knowledge has been gathered. Probiotics were early-on defined as "growth-promoting factors produced by microorganisms" and some years later as "organisms and compounds with beneficial effects for animals by influencing the intestinal microflora"; the latter adds in precision by relating the effect to the intestinal microbiota, but both of definitions considered indistinctly organisms and compounds as the effectors, so that prebiotics may well be included in such definitions. Then probiotics were unambiguously defined as live microorganisms (Fuller, 1989) and their effect more precisely related to beneficial effects exerted on the intestinal tract. Even more precision was added by specifying the requirement of exerting a beneficial physiological effect beyond nutrition, that is, promoting an improved health condition and/or reducing diseases risks; some definitions are explicit as to the kind of effects produced, such as the modulation of mucosal and systemic immunity and the improvement of nutritional and microbial balance in the intestinal tract. A more concise but robust definition of probiotics has been adopted by the United Nations World Health Organization (WHO) and Food and Agricultural Organization (FAO), considering them as "live microorganisms which, when administered in adequate amounts, confer a health benefit on the host" (Sanders, 2003); the condition of being part of a food has also been proposed (Pineiro and Stanton, 2007). The condition of live organism is now assumed as a necessary condition of a probiotic even though it has been a matter of some debate, some authors considering nonviable microorganisms as potential probiotics as well (Salminen et al., 1999). It is, however, agreed that the ability to remain viable and even proliferate at the site of action is a required condition that has to be verified for each probiotic strain (Quigley, 2010). In this context, microencapsulation of probiotic strains is used as a strategy to account for this (Prakash et al., 2011).

A history of probiotics can be traced back to the use of fermentation for the preservation of milk. Soon it was realized that, beyond preservation, fermented milk also added good sensory properties. Recognizing its effect on the bacterial composition in feces of breast-fed infants and also in adults consuming fermented milk as part of their regular diet—both habits seemed to be related to good health—led to the identification of different strains of lactic acid bacteria. Early in the 20th century, Nobel laureate Metchnikoff suggested that by consuming fermented milk such bacteria will exert a beneficial effect by establishing themselves in the intestinal tract displacing the pathogenic bacterial population. The mechanism proposed was essentially wrong, since it was proven that most of the milk-fermenting microorganisms were unable to establish in the gut (Caplice and Fitzgerald, 1999), but this outstanding early proposal remains the basis of the concept of probiosis, since later on several lactic acid bacteria were isolated that were capable of colonizing the human intestinal tract (Vasiljevic and Shah, 2008).

According to its accepted definition, a wide variety of microorganisms can be considered as potential probiotics, but certainly that condition is mostly associated with those belonging to the genera *Lactobacillus* and *Bifidobacterium*, though notably some strains of yeast, like *Saccharomyces cerevisiae* Boulardii (Czerucka et al., 2007), and bacteria, like *Bacillus pumilus*, *Bacillus clausii*, *Bacillus cereus*, *Bacillus subtilis* (Duc et al., 2004; Cutting. 2011), and *Escherichia coli* Nissle (Henker et al., 2007) also have been considered as potential probiotics. However, in the latter cases safety is a critical issue (Muñoz et al., 2005; Sorokulova et al., 2008) so that *Lactobacillus* and *Bifidobacterium*, having a long history of use whose safety has remained unquestioned, are by far the most prominent probiotics. Despite this, some concern has been raised about their use as probiotics, since many species of *Lactobacillus* and *Bifidobacterium* have been frequently isolated from various types of infective lesions (Ishibashi and Yamazaki, 2001).

Basic requirements for a microorganism to be considered probiotic are its survival in the intestinal ecosystem and the scientifically supported health benefit conferred to the host by harboring it. It must be conclusively proven as nonpathogenic, be able to cope with processing and storage conditions of the product containing it, to remain physiologically active and genetically stable throughout its span of use, and be amenable for large-scale production. It has been suggested also that a good probiotic for human use must be of human origin (O'Sullivan, 2001), but in fact some microbial strains not isolated from humans are also good probiotics for human consumption (Roselli et al., 2006). Requirement of conferring health benefit is vaguely expressed and since the targets for in vivo function of probiotics are broad, it is a formidable task to be more precise about it (Sanders, 2008). Several beneficial effects have been claimed for probiotics, like alleviation of lactose intolerance, prevention and reduction of diarrhea, reduction of risk of carcinogenicity, reduction of blood cholesterol, inhibition of gastric and intestinal pathogens, prevention of inflammatory bowel disease, and immunomodulation (Vasiljevic and Shah, 2008), but for the most part results are preliminary and sound conclusions are lacking. Probiotics may act by allowing a return to a normal condition of the intestinal microbiota after a perturbation (ie, antibiotic therapy or illness) or diminish the change produced by it. It has been claimed that probiotics improve the balance of the intestinal microbiota, but the mere fact that they certainly alter the microbial population does not necessarily imply that this is an improved balance. Other health benefits claimed for probiotics are the strengthening of the immune system to reduce allergies, stress, exposure to toxins, and even reducing alcohol intake (Ogueke et al., 2010; Gourbeyre et al., 2011).

Some mechanisms for human health promotion by probiotics have been proposed: provision of end products of carbohydrate fermentation (mostly short-chain fatty acids) that will lower the pH to a level detrimental for pathogens, competition with pathogens with eventual displacement from the intestinal mucosa, neutralization of enterotoxins produced by pathogens, and stimulation of host immune response associated with the production of specific polysaccharides (Choudhari et al., 2008; Figueroa-González et al., 2011). Many of these mechanisms have not been conclusively proven and for the most part are based on in vitro studies whose predictability of in vivo functioning is arguable (Ouwehand and Salminen, 2003). An overview on current knowledge about probiotic mechanisms has been published recently (Saad et al., 2013).

As said before, *Lactobacillus* and *Bifidobacterium* are the most reputed probiotic microorganisms. However, health benefits are strain specific and no single strain will provide all proposed benefits. Table 2.1 presents a nonexhaustive list of probiotic strains; those marked with an asterisk are the ones already used in commercial applications (Sanders, 2003; Vasiljevic and Shah, 2008; Hoppe and Nexmann Larsen, 2009; Kaur et al., 2010; Figueroa-González et al., 2011; Golowczyc et al., 2011; Foligné et al., 2013).

Selection of probiotic strains for commercial applications should consider several aspects, namely, safety (origin, pathogenicity, and infectivity of the microbial strain), functionality (tolerance to gastric acids and bile, adhesion to intestinal mucosa, validated health effects), physiology (immunomodulation, antagonism toward gastrointestinal pathogens, lactose, and cholesterol metabolism), and technological considerations (genetic and process stability of microbial strains, sensory properties, and large-scale production) (Morelli, 2007).

There are several technological challenges in probiotic production. Delivery of the probiotic in active and viable form and in the dose required to its site of action is a major issue. Levels between 10^6 to 10^8 colony-forming units (cfu) per mL (or g) of product have been suggested, but this varies significantly from one probiotic strain to another, and doses as high as 10^{12} cfu/mL have been suggested in certain cases. Dosage should be based on efficacy as determined in studies with humans, but dosage expressed as above is not the only parameter determining efficacy, since other aspects like probiotic growth during storage and delivery, metabolic state of the cells, and the food matrix of delivery may strongly influence the efficacy of the probiotic (Sanders, 2008). This uncertainty is a major hurdle in the use of probiotics. Probiotic-containing foods have been designed for specific age groups, such as infants (Boehm et al., 2010) and the elderly (Satokari et al., 2010), whose intestinal microbiota is markedly different. Intestine is almost sterile at birth and soon inoculated during the first days of life, being the conformation of the intestinal microbiota determined by milk-feeding, so probiotic strains used in infant formula are mainly *Lactobacilli* and *Bifidobacteria*. Aging strongly affects intestinal microbiota after a long period of solid feeding and eventual antibiotic treatment; this leads to a more diverse population with increase in detrimental bacteria at the expense of more

TABLE 2.1 Most Reported Probiotic Microbial Strains for Human Consumption

Genera	Species
Lactobaciilus	*L. acidophilus*
	L. amylovorus
	*L. casei**
	L. caucasicus
	L. crispatus
	*L. delbrueckii**
	L. fermentum
	L. gallinarum
	L. gasseri
	L. helveticus
	*L. johnsonii**
	L. kefir
	*L. paracasei**
	*L. plantarum**
	*L. reuteri**
	*L. rhamnosus**
	L. sakei
	L. salivarius
Bifidobacterium	*B. adolescentis*
	*B. animalis**
	B. bifidum
	*B. breve**
	B. coagulans
	B. infantis
	*B. lactis**
	*B. longum**

beneficial groups, making the host more prone to immunodeficiency and colon cancer development, aspects that have to be taken into account when designing the product (Saulnier et al., 2009).

Probiotics are not restricted to human consumption and, in fact, the use of probiotics for farm and companion animals has increased significantly in the last two decades. Major changes in the diets of both farm animals and pets have been experienced in recent years, moving from a forage-based to a readily fermentable diet in the former and from cooked home meal and household food scraps to pelleted compound feed in the latter. These changes have altered animal intestinal microbiota so that the use of feed supplements to achieve better animal health by acting upon it has gained importance; particularly relevant is the use of probiotic supplements (Chaucheyras-Durand and Durand, 2010), even more so because of the increased pressure for banning of antibiotics as feed supplements (Simon, 2005; Marshall and Levy, 2011). Probiotics are now being used also in aquaculture (Pieters et al., 2008).

A noncomprehensive list of probiotic strains used as feed additives is presented in Table 2.2 (Lim and Tan, 2009; Gaggia et al., 2010; Vester and Fahey, 2010). As seen, many more bacterial strains are considered than for human consumption, yeasts and molds being also included. Phytase-producing bacteria have been recently proposed as probiotic candidates for animal feeding (Sharma and Trivedi, 2015).

With respect to the safety of probiotic strains, *Lactobacilli* and *Bifidobacteria* are normal inhabitants of the human gut and have been consumed as part of fermented foods for centuries and used in dietary supplements for decades so that there is essentially no risk associated with their oral consumption by healthy individuals (Salminen et al., 1998); in fact, very few cases have been reported of adverse effects caused by their consumption. This is not so clear for other probiotic microbial strains, such as *Enterococcus*, *Bacillus*, and even *Saccharomyces*; in the case of *Enterococcus* strains, it has been advised not to use them as probiotics for human consumption (Itsaranuwat et al., 2003). Above all, safety must be established on a strain basis so that documentation of safety for each candidate strain, properly classified taxonomically, should exist including history of safe use, health status of the consumer, production of potentially deleterious metabolites, association with transferable antibiotic resistance, sensitivity to therapeutic antibiotics, and connection with harmful species (Sanders, 2003). Safety is indeed a major issue for probiotics and concern regarding their use is mostly associated with potential disease causation (bacteremia or endocarditis), toxic metabolic effects on the gastrointestinal tract, and transfer of antibiotic resistance to gut microbiota (Snydman, 2008).

Governmental regulations on the use of probiotics differ among countries, and the status of probiotics as food components has yet to be established on an international basis. However, regulations are in most cases based on FAO/WHO-delivered recommendations for the registration process of probiotics

TABLE 2.2 Most Reported Probiotic Microbial Strains Used as Feed Additives

Genera	Species
Bacillus	*B. brevis* *B. cereus* *B. clausii* *B. coagulans* *B. lentus* *B. licheniformis* *B. pumilus* *B. subtilis* *B. toyoi*
Lactobacillus	*L. acidophilus* *L. amylovorus* *L. brevis* *L. buchneri* *L. bulgaricus* *L. casei* *L. cellobiosus* *L. collinoides* *L. curvatus* *L. farciminis* *L. fermentum* *L. helveticus* *L. johnsonii* *L. lactis* *L. mucosae* *L. murinus* *L. paracasei* *L. pentosus* *L. plantarum* *L. reuteri* *L. rhamnosus* *L. sakei* *L. salivarius*
Bacteroides	*B. amylophilus* *B. capillosus* *B. ruminicola* *B. suis*
Bifidobacterium	*B. adolescentis* *B. animalis* *B. bifidum* *B. breve* *B. infantis* *B. lactis* *B. longum* *B. pseudolongum* *B. thermophilum*

Continued

TABLE 2.2 Most Reported Probiotic Microbial Strains Used as Feed Additives—cont'd

Genera	Species
Clostridium	*C. butyricum* *C. sporogenes* *C. tyrobutyricum*
Enterococcus	*E. cremoris* *E. diacetylactis* *E. faecalis* *E. faecium* *E. mundtii*
Lactococcus	*L. lactis*
Leuconostoc	*L. citreum* *L. lactis* *L. mesenteroides*
Pediococcus	*P. acidilactici* *P. pentosaceus*
Propionibacterium	*P. acidipropionici* *P. freudenreichii* *P. globosum* *P. shermanii*
Streptococcus	*S. cremoris* *S. diacetylactis* *S. faecium* *S. intermedius* *S. lactis* *S. thermophilus*
Candida	*C. glabrata* *C. pinolepessi* *C. utilis*
Kluyveromyces	*K. marxianus* *K. marxianus-fragilis*
Saccharomyces	*S. boulardii* *S. pastorianus*
Aspergillus	*A. niger* *A. oryzae*

considering the phenotypic and genotypic strain identification, functional characterization and safety assessment by in vitro and animal tests, Phase I clinical tests and Phase II human clinical trials to assess probiotic efficacy, Phase III effectiveness trials, and product labeling (FAO/WHO, 2002). A detailed description of legal status and regulatory issues for probiotics in the

United States (Saavedra and Degnan, 2009), European Union (Ouwehand and Lahtinen, 2009), and Asian countries (Tien and Chang, 2009) has been reviewed, the above references being suggested for the interested reader.

Probiotics from genetically modified organisms (GMOs) may enhance the potential of a probiotic strain, endow an organism with probiotic properties, or improve their target to specific pathogens or toxins; however, very few of these GMOs have been used as probiotics mostly because of consumer reluctance to use them (Gupta et al., 2014). GMO probiotics will hardly be accepted as food ingredients, but they may be more acceptable in the case of clinical applications for severe disease treatment (Sreeja and Prajapati, 2013). This consumer perception may vary with time but it is clear that stringent procedures for safety assessment of GMO prebiotics will be enforced for medical reasons as well as for environmental implications (Celec et al., 2005; Reid et al., 2006).

A decade ago there were about 70 probiotic-containing products in the market, and the trend is a sustained increase (Vasiljevic and Shah, 2008) so that the number now may be over a hundred. Probiotic-containing foods were early developed and marketed in Japan and Europe, but soon the United States caught up so that it is estimated that more than 80% of the yogurt sold there contains prebiotic *Lactobacillus acidophilus* and several *Bifidobacterium* strains. The most relevant carrier for probiotics is yogurt (Lourens-Hattingh and Viljoen, 2001), but several other food products have been considered, such as cheese, mayonnaise, spreads, cereals, and meat products (Stanton et al., 2003). However, most probiotics for human consumption are delivered in dairy products. Although the condition of being part of a food has been considered as inherent to the concept of probiotic, probiotics are also marketed for direct consumption in the form of tablets, capsules, powders, and sachets containing the freeze-dried microorganisms (Oguecke et al., 2010; Golowczyc et al., 2011).

The quality of a probiotic is tightly related to its viability, being of major importance from a technological perspective. Viability during the manufacturing process, storage, and distribution is certainly important but, above all, viability during gastrointestinal passage is critical. Viability in the food matrix where the probiotic is delivered is also important and freeze-drying of probiotic strains within food matrices has been proposed as a strategy to improve cell viability (Saarela et al., 2006). Microencapsulation is increasingly being used to ensure viability during gastrointestinal tract passage and proper delivery at the site of action (Chan and Zhang, 2002). Different polymers, waxes, and lipids have been used as coating materials for encapsulation, and interestingly some prebiotics, like fructo-oligosaccharides, also have been used for such purposes (Chen et al., 2007). Many techniques of microencapsulation of probiotics have been proposed (Anal and Singh, 2007), the most popular being spray-drying that, even though reducing viability, confers good protection during transit and controlled release under gut conditions. Survival of probiotic strains after spray-drying can be enhanced by incorporating thermoprotectants such as glycerin, trehalose, mannitol, sorbitol, inulin, dextrin, starch, and nonfat milk solids (Savini et al., 2010). Prebiotic galacto-oligosaccharides (GOSs) have been also shown to protect *Lactobacillus* strains during freeze-drying, setting up the basis of a synbiotic product (Tymczyszyn et al., 2011; Golowczyc et al., 2013).

Effective dosage is one of the most blurring aspects of probiotics. Dosage should rely upon efficacy in human studies (Sanders, 2008), but this depends on many factors whose effect is difficult to determine or predict. Dosage will be different with each strain; even for those tightly linked taxonomically, strain behavior in the food matrix, survival, and proliferation at the site of action are hard to determine, healthy effects are not always clearly identified, and more knowledge is needed on mechanisms of probiotic action. Frequency of administration may also have an effect, and daily doses suggested in the range from 10^7 to 10^{10} cfu (Douglas and Sanders, 2008; Lee, 2009) are to be considered merely as a rule of thumb. Values of viable probiotic bacteria between 10^7 and 10^9 cfu/g were determined at the end of the shelf life of several dairy and nondairy foods containing a probiotic mixture of *Lactobacillus rhamnosus* and *Lactobacillus paracasei* (Coman et al., 2012).

Even though problems of dosage, strain viability, insufficient knowledge on mechanism of action, lack of documentation on health effect based on controlled studies in humans, and also great variability with respect to safety and regulatory issues are still being addressed, the field of probiotics is growing fast in synchrony with the worldwide trend toward healthy eating habits. Probiotics are having an impact not only in the food sector but also in farm and companion animal feed, which is a rapidly expanding market.

Current regulations restrict the microbial species allowed as food ingredients to those having qualified presumption of safety (QPS) or generally recognized as safe (GRAS) status. More than 75 microbial species comply with such regulations, and 25 of them are currently used as probiotics. The probiotics industry is now facing limitations arising from the fact that live bacteria may not induce significant changes in healthy individuals hampering claim recognition, and from a rather limited number of allowable microbial species (Gosálbez and Ramón, 2015). In this regard, nonviable microbial cells have been proposed, since many probiotics have been shown to retain their functionality after being inactivated. A changing pattern is also observed in the method of delivering such probiotics, moving from food ingredients to supplement capsules more closely related to a drug manner of delivery. Maybe in the near future the concept of probiotic will have to be redefined. The concept of "postbiotics" has been recently coined to refer to nonviable bacterial products or metabolites, which makes the boundary between probiotics and prebiotics thinner.

For more detailed information on probiotics, the books by Tamine (2005), Lee and Salminen (2009), and Cho and Finocchiaro (2010) are recommended to the interested reader.

2.4 PREBIOTICS: CONCEPT, PRESENT STATUS, TECHNOLOGICAL GOALS, AND CHALLENGES

Although dietary fiber is not precisely defined, its benefits have been appreciated for a long time (Lunn and Buttriss, 2007). Dietary fiber refers mostly to nondigestible polysaccharides and nondigestible oligosaccharides (NDOs) (Blaut, 2002), the latter being outstanding due to their health-promoting effects. Some NDOs are well-established prebiotics (Bellei and Haslberger, 2012), and many more share some of the requirements (see following) for being considered as such (Fu and Wang, 2013). The concept of prebiotics is new, but natural foods with high content of prebiotic substances, such as asparagus, chicory, Jerusalem artichokes, garlic, onions, wheat, oats, and soybeans, have been consumed since ancient times (Van Loo et al., 1995).

Prebiotics have been defined as nondigestible food ingredients that selectively stimulate the proliferation and/or activity of beneficial microbial populations in the colon (Gibson and Roberfroid, 1995; Crittenden and Playne, 2009). This definition has been expressed more precisely as "a selectively fermented ingredient that allows specific changes, both in composition and/or activity in the gastrointestinal microbiota that confers benefits upon host wellbeing and health" (Gibson et al., 2004; Roberfroid, 2007). A recent proposal of the ILSI Europe Prebiotic Expert Group defines the prebiotic concept as "the selective stimulation of growth and/or activity of one or a limited number of microbial genera/species in the gut microbiota conferring health benefits to the host" (Roberfroid et al., 2010). The effect of a prebiotic is indirect since it is not the prebiotic itself but the stimulatory effects it provokes in the colon microbiota that confer health benefits (Wang, 2009). The premise is that prebiotics stimulate indigenous bacteria inhabiting the gut, while probiotics imply the addition of exogenous microbial species (Scantlebury Manning and Gibson, 2004). In this sense, it is expected that prebiotics would be more effective and that this action is more predictable than probiotics whose fate during passage through the gastrointestinal tract and implantation in the colon adds uncertainty as to their effects.

Most of the intestinal microbiota reside in the colon, so prebiotics are usually targeted to stimulate beneficial bacteria, mostly *Bifidobacteria* and *Lactobacilli*, in that niche (Bouhnik et al., 2004). These bacteria are characterized by producing healthy effects associated with short-chain fatty acid (SCFA) formation, no toxin formation, formation of defensins and vitamin synthesis, but also with inhibition of pathogen persistency by competing with them for nutrients and receptor sites on the intestine walls (Roberfroid et al., 2010). However, as more knowledge is gathered on the complex gut microbiology, other potential beneficial bacterial inhabitants have been suggested; among them, nonpathogenic butyrate-producing bacteria, like *Roseburia* and *Faecalibacteria* and members of some clostridial clusters (Sarbini and Rastall, 2011), are particularly desirable to stimulate, since butyrate is considered to play a key role in maintaining human gut health (Louis and Flint, 2009). Members of other genera, like *Streptococci*, *Enterococci*, *Eubacteria*, and *Bacteroides* may be beneficial or harmful, according to the species, their condition being in some cases still a matter of debate (Roberfroid et al., 2010). Even though prebiotics have been targeted to the colon, it is clear now that beneficial effects beyond the colon may be exerted not only at the intestinal but also at the systemic level (Lenoir-Wijnkoop et al., 2007; Kellow et al., 2014). Positive effects of NDOs on protection against tumor development outside the colon (Taper and Roberfroid, 2002), positive modulation of hepatic lipid metabolism (Delzenne and Williams, 2002), and reduction of total and LDL-cholesterol (Ooi and Liong, 2010) have been observed in test animals and a few studies with humans (Delzenne, 2003). Effects of prebiotics on other microbial ecosystems, like the oral cavity, respiratory tract, and vagina, are under research but no conclusive proof of beneficial effects in such niches exists for the moment (Blatchford et al., 2013). Health benefits associated with NDOs have been reviewed by Qiang et al. (2009).

Several physiological and health-related effects have been claimed for prebiotics as can be seen in Table 2.3.

An overview of current knowledge about prebiotic health-promoting effects has been recently published (Saad et al., 2013). The specific effect of each prebiotic or candidate prebiotic will differ to a great extent, and none will exert all of the effects listed in Table 2.3. A thorough analysis of this aspect can be found in a compilation of contributions from experts both from academia and industry (Roberfroid et al., 2010). A complete revision of health responses associated to prebiotic consumption has been published by Macfarlane et al. (2008). Many substances are known that stimulate the growth and/or activity of the endogenous intestinal *Bifidobacteria* population (the so-called bifidus or bifidogenic factor); it is clearly insufficient to consider them as potential prebiotics if their physiological health effects are not clearly demonstrated (Pineiro et al., 2008).

The criteria accepted for an ingredient to be considered as a prebiotic are:

- Resistance to degradation by acids and enzymes through the gastrointestinal passage, and resistance to gastrointestinal absorption;
- Fermentation by gut microbiota;
- Selective stimulation of the growth and/or activity of colonic bacteria associated with health and well-being.

Most of these effects are associated with colon functioning and metabolism, and changes in the colonic microbiota, their metabolic products, and their effect on host biochemistry are the basis of the rationale of the association between prebiotics and health benefits (Douglas and Sanders, 2008).

In addition, prebiotics must withstand the harsh processing conditions of the food that contains it, like high temperature, low pH, and Maillard reaction conditions (Charalampopoulos and Rastall, 2012). This is a particularly relevant technological aspect since prebiotic substances differ markedly in this respect.

TABLE 2.3 Physiological and Health-Related Effects of Prebiotics

Effect	References
Improvement and stabilization of gut microbiota composition	Steer et al. (2000)
Improvement of intestinal functions (stool bulking, regularity, and consistency)	Costalos et al. (2008)
Increase in mineral absorption and improvement of bone health	Scholz-Ahrens et al. (2007)
Modulation of gastrointestinal peptide production	Cani et al. (2009)
Improvement of mucosal morphology by promoting mucin synthesis	Langlands et al. (2004)
Stimulation of antibiotic production	Liévin et al. (2000)
Alleviation of irritable bowel syndrome	Silk et al. (2009)
Control of appetite and body weight	Slavin (2013)
Control of intestinal infection and diarrhea	de Vrese and Marteau (2007)
Regulation and modulation of immune functions	Seifert and Watzl (2007) and Lomax and Calder (2009)
Reduction of pathogens by decoying their binding cellular receptors in the gut	Shoaf-Sweney and Hukins (2009)
Reduction of metabolic endotoxemia	Neyrinck et al. (2012)
Reduction of risk of intestinal infections	Bruzzese et al. (2009)
Reduction of risk of obesity	Delzenne et al. (2011) and Beserra et al. (2014)
Reduction of risk of type 2 diabetes and heart failure	Anderson et al. (2009)
Reduction of risk of colon cancer	Liong (2008)
Reduction of skin inflammation	Wickens et al. (2008)
Modulation of anxiety and depression	Luna and Foster (2015)

Selective fermentation in the large intestine is the most stringent criteria and usually the most difficult to meet (Rastall and Gibson, 2015). In principle, any foodstuff reaching the colon is a potential prebiotic; however, NDOs are those that more closely meet all requirements to be properly considered as such (Mussatto and Mancilha, 2007). Their fermentation in the colon produces mostly SCFAs (acetate, propionate, and butyrate), but also other metabolites including pyruvate, lactate, succinate, and ethanol, besides gases (CO_2, H_2, CH_4, and H_2S) (Sarbini and Rastall, 2011). Most health-promoting effects of NDOs are associated to SCFA production and the concomitant pH drop that contributes to reduce the pathogenic microbiota and increase the bifidobacteria population, in this way fulfilling the criterion of selective stimulation of beneficial bacteria (Cummings and Macfarlane, 2002). Propionate participates in the inhibition of cholesterol synthesis and the deposition of adipose tissue and has been proposed as a dietary-related factor to depress appetite and reduce obesity (Arora et al., 2011). Butyrate plays an important role in the maintenance of intestinal homeostasis and overall health status, and the effects it exerts are multiple, having been shown to be related to the prevention and inhibition of colorectal cancer and diarrhea, also acting at the extra-intestinal level (Berni Canani et al., 2011; Gonçalves and Martel, 2013). Cross-feeding among bacteria by the degradation products of NDO metabolism is likely to occur, as illustrated by the fact that lactate produced from fructo-oligosaccharides by a strain of *Bifidobacterium* was used as substrate by butyrate-producing bacteria (Belenguer et al., 2006), so that the metabolic fate of NDOs cannot be properly ascertained without considering the complex interaction of the products of carbohydrate metabolism with the intestinal microbiota (Sarbini and Rastall, 2011).

Each of one of the three fundamental conditions to be fulfilled by a prebiotic must be unambiguously proven, and several in vitro and in vivo tests have been performed to assess the prebiotic condition of NDOs both in human (Biedrzycka and Bielecka, 2004; Bouhnik et al., 2004; Gibson et al., 2004) and animal systems (Macfarlane and Macfarlane, 2007; Jung et al., 2008).

Nondegradation by gastric acid, intestinal enzymes and no gastrointestinal absorption can be demonstrated in vitro (resistance to pH conditions as the ones in the stomach and small intestine, resistance to the enzymes present in all of the digestive tract), and in vivo (determination of residual level in feces after an oral dose administration; in the case of test animals more invasive methods have been tested like gastrointestinal intubation). Indirect methods can also be used based on determining variation in glycemic and insulin levels after an oral dose administration.

Fermentation by the intestinal microbiota can also be tested in vitro or in vivo. Fermentation of NDOs, both in batch or continuous mode, can be done with pure cultures of selected bacteria, with mixed populations or with fecal samples (Gibson and Fuller, 2000; Macfarlane and Macfarlane, 2007). The former two, though simpler, are of limited value, and the latter should be preferred because the complex interactions among bacterial populations are taken into account (Rycroft et al., 2001). Multistage continuous culture systems also have been used to evaluate fermentation in an effort to simulate colon physiology (Gmeiner et al., 2000; Probert et al., 2004; Tzortis et al., 2005). In vivo studies of NDO fermentability have been done mostly with laboratory animals, where they are sacrificed after oral ingestion or intubation with the substance under study and then samples of feces and contents at several segments of the intestine are analyzed (Rodríguez-Cabezas et al., 2010), as well as with livestock and pets where the analysis is mostly based on fecal composition (Samanta et al., 2013). Studies with human volunteers have also been performed based on the effect of NDO administration on fecal microbiota (Depeint et al., 2008).

As noted previously, selective stimulation of growth and activity of intestinal bacteria is the ultimate criterion for determining a prebiotic condition. Studies on pure cultures, though frequently used for assessing a wide range of health-promoting *Bifidobacteria* and *Lactobacilli*, have the obvious disadvantage of not representing the colonic microbiota as a whole, disregarding interaction effects and the effects on undesirable species (Gibson et al., 2004), and being a methodology sometimes cumbersome and with poor resolution (Bouhnik et al., 2004); however, its simplicity makes it adequate for screening and comparative purposes. Prebiotic effect on nondesirable putrefactive bacteria, like *Bacteroides* and *Clostridia*, can also be assessed in this way (Roberfroid, 2001; Kolida et al., 2002). The use of fecal samples is clearly a better approximation although representing mostly the distal colon microbiota; more proximal areas may have a different composition but are hardly accessible with human volunteers so that complex gut models replicating the different intestinal sections represent an attempt to overcome this limitation (Payne et al., 2012). Identification of genera and species is a major hurdle when using a fecal inoculum. At first, use of selective culture media was used followed by biochemical identification, but this merely allowed identifying enrichment of a particular organism, being it desirable or undesirable, but did not reflect the population changes that occurred (Rossi et al., 2005).

A preferred strategy now is the use of molecular methods for bacterial identification, being more reliable and significant (Ben Amor et al., 2007). One such method that has been applied to identify representative colonic bacteria is fluorescence in situ hybridization (FISH), based on oligonucleotide probes that target discrete discriminatory (highly conserved) regions of the ribosomal RNA molecule (Franks et al., 1998; Bezirtzoglou et al., 2011). Another method for evaluating the genetic diversity of the colonic microbiota is denaturing gradient gel electrophoresis (DGGE) or temperature gradient gel electrophoresis (TGGE), which consist in the separation of short double-stranded DNA fragments according to their point of denaturation provoked by heat or chemicals (chaotropic compounds, like urea and formamide), linear gradients of denaturant (temperature in the case of TGGE or reagent concentration in the case of DGGE) being used along the gel (Muyzer and Smalla, 1998; Rehman et al., 2008). Separation is produced by the decreased electrophoretic mobility of the partially melted double-stranded DNA; identification is then carried out by sequencing excised fragments of the gel or else by comparing their mobility with that of known control sequences in databases (Gibson et al., 2004). The advantage of DGGE (or TGGE) over FISH is its speed and simplicity, no probes being required (Zoetendal et al., 1998; Roberfroid, 2008; Roger and McCartney, 2010).

In order to have a more quantitative approach to prebiotic effectiveness, different prebiotic indexes have been proposed, based on calculating and weighing the changes in microbial populations of selected species by the addition of the prebiotic. A prebiotic index was proposed by Palframan et al. (2003), considering variations in *Bifidobacteria* and *Lactobacilli* (beneficial), *Bacteroides* and *Clostridia* (potentially adverse), obtaining a quantitative score that has been used to evaluate different NDOs (Sanz et al., 2005; Depeint et al., 2008). A similar analysis was done by using representative probiotic (*Lactobacillus paracasei*) and enteric (*E. coli*) bacteria measuring the ratio of the increase in bacterial counts using a prebiotic and a glucose-based medium (Huebner et al., 2008). Another prebiotic index, based on the effect on growth rates of major gut bacterial groups, called measure of prebiotic effect, also has been proposed as a quantitative index for determining prebiotic potential (Vulevic et al., 2004). These indexes can be further refined by weighing the different bacteria according to their relevance. A specific prebiotic index has been proposed in which population variations are divided by the prebiotic dose and argued to be a better parameter to quantitatively describe the prebiotic effect of a substance (Roberfroid, 2007). Other parameters indirectly associated with prebiotic effect are the production of short-chain fatty acids and gas evolution (Rycroft et al., 2001); the former is considered advantageous by lowering colonic pH and so discouraging the growth of putrefactive bacteria, while the latter is considered detrimental by producing abdominal discomfort (Bouhnik et al., 2004; Tuohy et al., 2008). Acid and bile tolerance (Vernazza et al., 2006) and daily fecal output per unit amount of NDO ingested (Macfarlane et al., 2006) have also been considered as parameters determining prebiotic effectiveness.

In vivo tests performed through appropriate nutritional interventions in the targeted species, whether human or animal, should be the ultimate proof for prebiotic effectiveness (Van Loo, 2006; Roberfroid et al., 2010). Tests for humans involve the direct recovery of undigested molecules in distal ileum and feces following ingestion, and indirect assessment tests based on the absence of increase in glycemia or insulinemia following oral administration (Bandyopadhyay and Mandal, 2014). A thorough revision of current alternatives for determining prebiotic effects on intestinal microbiota can be found in the book edited by Gibson (Van Loo, 2006). In vitro models have been developed that can be considered as representative of the in vivo situation, having the advantages of not presenting ethical constraints and having a higher reproducibility. A complete and updated revision of such experimental models has been recently published by Venema and van den Abbeele (2013).

Although mostly considered as food ingredients for human consumption, prebiotics also have been used in animal feeding for livestock (Van Loo and Vancraeynest, 2008; Hajati and Rezaei, 2010) and companion animals (Verdonk et al., 2005) and also in aquaculture (Ringø et al., 2010; Song et al., 2014).

Regulatory issues, as in the case of probiotics, are a major concern for prebiotics (Tymczyszyn et al., 2014). Requirement of premarket approval for prebiotics is nonexisting in many countries lacking an established system for health claims; however, scientific substantiation on their physiological effects may be required by the regulatory authorities. Claims for prebiotic marketing refer to their structure and function rather than their health effects. In the United States, the Food and Drug Administration does not allow health claims for prebiotics and nutrient content claims have to be done based on dietary fiber intake values; in Japan, prebiotics have FOSHU status, so they are allowed to claim their physiological effects in humans (Brownawell et al., 2012). Several countries, like the United States, France, the Netherlands, Japan, Korea, Singapore, Malaysia, Thailand, Brazil, Chile, Colombia, and South Africa, have established guidelines for prebiotics. In Brazil and Chile inulin and FOS have been approved as prebiotics with daily intakes recommended of 2.5 and 3 g/day, respectively; in the case of Chile, approval has been strongly promoted by the installation of a large-scale production plant by the company BENEO Orafti. With respect to safety, at the intake doses recommended no risk to consumer health has been demonstrated for fructans and galactans where proper prebiotics belong (Pascal, 2008).

There is some controversy as to what NDOs can be properly considered as prebiotics, since the fulfillment of all of the discussed established criteria is difficult to prove conclusively. Those more properly considered as prebiotics are fructans—inulin and fructo-oligosaccharides (FOS)—and galactans—mostly GOSs (Sarbini and Rastall, 2011) and lactulose (Schuster-Wolff-Bühring et al., 2010; Mao et al., 2014). Other candidate NDO prebiotics fulfilling some but not all of the requirements are xylo-oligosaccharides (XOS), isomalto-oligosaccharides (IMO), soybean oligosaccharides (SOS), gluco-oligosaccharides, and resistant starch (RS). Performance of human trials is a condition to be met to be considered as prebiotic but not all NDOs have been tested in humans (Cummings et al., 2001).

The concept of prebiotic is not restricted to NDOs, even though most of the substances endowed with prebiotic properties are in fact NDOs. However, there is emerging information on novel compounds with prebiotic attributes, like steroid sapogenins (Huang et al., 2012) and also some novel oligosaccharides and polysaccharides extracted from different plants (Polari et al., 2012) and alga (O'Sullivan et al., 2010). A great amount of information is emerging from the Human Microbiome Project (Gevers et al., 2012), which may have a profound influence on prebiotic research by allowing a rational production of desired prebiotic compounds with specific functional properties (Patel and Goyal, 2012).

Prebiotics are produced both by enzymatic and chemical synthesis. The former has the advantages of higher specificity and mild reaction conditions that make the process simpler and less costly. One drawback of chemical synthesis is environmental threat, as waste management of such processes is a big issue, so that current chemical approaches for the synthesis of NDOs need to be reassessed and modified according to the principles of green chemistry (Arunrao Yadav et al., 2014).

A review of the concept, properties, health effects, and legal status of prebiotics has been recently published (Corzo et al., 2015).

2.4.1 Fructans

Inulin: Inulin is a polydisperse β(1–2) fructan composed by fructose units linked to a terminal glucose unit, with a chain length from 2 to 60 residues, its average degree of polymerization (DP) being between 10 and 12 (Niness, 1999a). A partially purified inulin, in which the shorter chains have been removed having a DP around 25, has been produced for special applications (Coussement, 1999). Inulin properly fulfills the criteria for being considered a prebiotic: it is not degraded or adsorbed during transit in the gastrointestinal tract; it is fermentable by the colonic microbiota, selectively stimulating the growth of health-promoting bacterial species; and human trials have been performed to assess its efficacy (Tuohy et al., 2001). Inulin is a natural product extracted from plants, including chicory (*Cichorium intybus*), Jerusalem artichoke (*Helianthus tuberosus*), yacon (*Smallanthus sonchifolius*), and agave (*Agave tequilana*) as the most relevant raw materials (Gómez et al., 2010; Ojansivu et al., 2011; Karimi et al., 2015). The production process closely resembles the one for beet sugar production, inulin being extracted with hot water and then purified and dried; the product is a white powder with less than 5% moisture, bland taste, poorly sweet (less than 10% of the sweetening score of sucrose), and moderately water soluble (solubility is 100 g/L at room temperature) (Frank, 2002), the latter depending on the degree of polymerization (Chawla and Patil, 2010). Some of the main producers of inulin are listed in Table 2.4. The leading producer is the Belgian company BENEO Orafti with production plants in Belgium, Chile, Germany, and Italy, using chicory roots as raw material, which have very high inulin content, from 15% to 20% on fresh-weight basis, and a crop yield of around 45 tons/hectare. Inulin is probably the most used NDO and applications go well beyond its prebiotic nature (Van Loo et al., 1995). In fact, inulin is used mostly as a fat replacer and texturizer in several foods. The former is based mostly on the gelling properties of inulin so that fat replacement is possible in water-containing food matrices, preferably those in which water is the continuous phase (Wouters, 2010). In this context, inulin has been used successfully as fat replacer in different types of low-fat cheese (Meyer et al., 2011; Miočinović et al., 2011). Inulin contributes significantly to the texture and rheology of the food matrix containing it, enhancing taste, mouthfeel, and shelf life without compromising the characteristics of the product as a functional food (Karimi et al., 2015). Inulin is therefore a multifunctional food ingredient so that despite not being the most prominent in prebiotic terms, it is probably the most appealing prebiotic ingredient to the food industry.

Fructo-oligosaccharides: FOSs, or oligofructose, are NDO fructans composed of a number of fructose units linked to a terminal glucose unit; glucose–fructose linkages are usually α-(1,2) and fructose–fructose linkages are usually

TABLE 2.4 Main Inulin Producers

Company	Trade Name or Catalog Number	Country
Alfa Aesar	A18425	United States
BENEO Orafti	Orafti GR	Belgium
Cargill	Oliggo-Fiber	United States
Cosucra Groupe Warcoing	Oliggo-Fiber	Belgium
Dr. Paul Lohmann	Calcium-Inulin-Granulate	Germany
Finetech Industry Ltd	FT-0627242	England
Gillco Products Inc.	Oliggo-Fiber	United States
IS Chemical Technology Ltd	I06-1119	Germany
Jinan Haohua Industry Co.	9005-80-5	China
Prebiotics Sigma–Aldrich	I2255	United States
Sensus	Frutafit	Holland
TCI America	I0041	United States
TIC Gums	LV 110	United States
The Iidea Company	Agave inulin	México

β-(2,1) (Barreteau et al., 2006). FOS can be produced by controlled hydrolysis of inulin with inulinases (Singh and Singh, 2010; Huitrón et al., 2013) or mineral acids (Ávila-Fernández et al., 2011), and also by enzymatic synthesis from sucrose with microbial β-D-fructofuranosidases (EC 3.2.1.26) or β-D-fructosyltransferases (EC 2.4.1.99) (Nemukula et al., 2009; Hang et al., 2012; Peña-Cardeña et al., 2015). The former are mixtures of oligosaccharides with variable chain lengths ranging from DP 2 to 9 (average DP 5), while the latter are shorter oligosaccharides with DP from 2 to 4. FOS can also be produced from sucrose with levansucrases (EC 2.4.1.10), which also belong to the glycosyltransferase family from sucrose (Santos-Moriano et al., 2015). Some authors use the term FOS to refer to the product of enzymatic synthesis from sucrose and oligofructose to refer to the products of inulin hydrolysis (Biedrzycka and Bielecka, 2004), but in most cases these terms are used interchangeably. Enzymatic synthesis of FOS involves several sequential reactions leading to the final products, which are mainly the trisaccharide 1-kestose (GF_2), the tetrasaccharide nystose (GF_3), and the pentasaccharide fructofuranosylnystose (GF_4). Several microbial β-D-fructofuranosidases (or β-D-fructosyltransferases) have been used for FOS synthesis, mostly from fungi like *Aureobasidium pullulans*,

Aspergillus flavus, *Aspergillus japonicus*, *Aspergillus niger*, *Aspergillus oryzae*, *Penicillium citrinum*, *Penicillium islandicum*, and some bacteria like *Bacillus macerans*, *Lactobacillus reuteri*, and *Zymomonas mobilis* (Dominguez et al., 2014). Enzymes vary considerably in terms of structural and functional properties, but all of them display both hydrolytic and transferase activities so that high substrate concentrations are required to depress the hydrolytic activity in favor of transfructosylation (Yoshikawa et al., 2008).

Kinetics of reaction of FOS synthesis is quite complex and several tentative models have been proposed to describe it, considering reaction under thermodynamic control and under kinetic control (Jung et al., 1989; Duan et al., 1994; Kim et al., 1996; Caicedo et al., 2009; Kovács et al., 2014).

Like inulin, FOSs fulfill all requirements to be considered prebiotics (Rastall, 2010); however, the marked differences in chain length confer them quite different physicochemical and functional properties (De Gennaro et al., 2000). FOSs are much more soluble than inulin (Franck, 2002), and the textural, rheological, and sensory properties are also different (Niness, 1999a; Morris and Morris, 2012); it has also been reported that DP of fructans may have an effect on their prebiotic action (Ito et al., 2011). As in the case of inulin, uses of FOS go well beyond their prebiotic condition, because of their appealing nutritional and functional properties that enhance shelf life, taste, and texture to the food matrices that contain them (Niness, 1999b; Renuka et al., 2009; Salazar et al., 2009; Sing and Singh, 2010).

Industrial production of FOS is done with soluble (Yoshikawa et al., 2008) or immobilized (Kurakake et al., 2010) β-D-fructofuranosidases or cells containing them (Jung et al., 2011) in different reactor configurations in batch and continuous processes (Sangeetha et al., 2005; Mussatto et al., 2009; Lorenzoni et al., 2015). Pectinex Ultra SP-L, a commercial pectinase preparation containing fructofuranosidase activity, has been used for FOS synthesis (Tanriseven and Aslan, 2005; Csanádi and Sisak, 2006) and also other commercial pectinase, protease, and invertase preparations have been evaluated (Vega-Paulino and Zúniga-Hansen, 2012). High-purity FOS can been obtained by removing glucose, fructose, and residual sucrose from the reaction medium by subjecting the raw FOS to membrane separation (nanofiltration) (Kuhn et al., 2010), selective adsorption (Kuhn and Maugeri Filho, 2010; Nobre et al., 2012), selective bioconversion with yeast cells (Yang et al., 2008), ion-exchange chromatography (Gramblička and Polakovič, 2007; Vaňková and Polakovič, 2010), and simulated moving bed chromatography (Vanneste et al., 2011). The design and economics of an industrial process for FOS production with immobilized fructosyltransferase was published by Vaňková et al. (2008); it is a very interesting work that contains valuable information on process design, which is not frequently found in the open scientific literature.

Most inulin producers also produce medium-chain FOS by controlled hydrolysis of inulin with endoinulinases (Singh and Singh, 2010). However, industrial production of FOS is done mostly by enzymatic transfructosylation of sucrose or

TABLE 2.5 Main Fructo-oligosaccharide Producers

Company	Trade Name or Catalog Number	Country
Akhil Healthcare Private Limited	Fructo-oligosaccharides (FOS)	India
Beghin Meiji	Actilight (human use) Profeed (animal use)	France
Cheil Foods and Chemicals Inc.	Oligo-Sugar	Korea
GTC Nutrition USA	NutraFlora	United States
Mitushi Pharma	FOS-G Syrup	India
Sensus	Frutalose	Holland
Triveni Aromatics and Perfumery Private Limited	FOS	India
Victory Biology Engineering Co.	scFOS	China

sucrose-containing raw materials (Ghazi et al., 2007; Dominguez et al., 2014). Most important industrial producers of FOS are listed in Table 2.5. Two updated reviews on the industrial aspects of FOS production and purification are recommended for the interested reader (Bali et al., 2015; Nobre et al., 2015).

2.4.2 Galactans

Galactans are galactose-containing NDOs that are mostly produced from lactose. Among them, GOS and lactulose can be properly considered as prebiotics, while lactitol, lactosucrose, lactobionic acid, and tagatose fulfill some but not all the requirements. Since this subject is the core of this book, it will be discussed in detail in the following chapters.

2.4.3 Other NDOs With Prebiotic Potential

Within NDOs there is a wide variety of compounds that comply with some or many, but not all, the qualifications to be considered prebiotics. They include: xylo-oligosaccharides, isomalto-oligosaccharides, soybean-oligosaccharides, gentio-oligosaccharides, gluco-oligosaccharides, and resistant starch.

2.4.3.1 Xylo-oligosaccharides

XOSs are NDOs composed of xylose units linked by β1-4 bonds and consist mainly of xylobiose (dimer), xylotriose (trimer) and xylotetraose (tetramer), although higher oligosaccharides can also be formed, their proportion being determined by the method of production (Aachari and Prapulla, 2011). XOSs comply with some but not all of the requirements of a prebiotic: there is no conclusive evidence of the selective growth stimulation of the beneficial bacterial microbiota, and evidence of their resistance to gastric digestion is still lacking. So, it cannot be properly considered as a prebiotic, but there is well-documented proof of their health-promoting effects (Madhukumar and Muralikrishna, 2010; Samanta et al., 2012) and they have been used as health-promoting agents in dairy products (Mumtaz et al., 2008).

XOSs are produced from lignocellulosic biomass (LCB), either by chemical or enzymatic hydrolysis. Among chemical methods, autohydrolysis stands out. In autohydrolysis, hemicellulose (mostly xylan and arabinoxylan) in the LCB is progressively broken down by hydronium ions produced by the generated organic acids in the aqueous medium, yielding soluble oligosaccharides, while the cellulose and lignin portions of the LCB remain insoluble; this process is quite appealing within a biorefinery concept (Liu et al., 2011). However, the hydrolysates obtained contain a variety of undesirable components, such as soluble lignin, large amounts of monosaccharides and their degradation products so that extensive purification is required. XOSs produced in this way have a significant proportion of high-molecular-weight oligosaccharides, which are less desirable. These drawbacks can be arrested if LCB is pretreated to extract xylan and then subjected to mild acid hydrolysis (Sun et al., 2002). The main soluble products from acid hydrolysis are oligosaccharides of variable chain length, XOS being usually in the range of 2 to 10 monomeric units. In this process, XOSs are intermediary products, molecular size distribution depending on the LCB and the reaction conditions (Moure et al., 2006). Controlling conditions to achieve a desired XOS distribution is a hard task, but this is important since health-promoting effects will vary according to it. Moreover, if reaction is not properly tuned, XOS formed will be further hydrolyzed to xylose and decomposed into furfural and hydroxymethylfurfural, which is highly undesirable. Many LCB feedstocks have been used for the production of XOS, including hardwoods and softwoods, bagasse, corncobs, wheat straw, and barley hulls, among others (Parajó et al., 2004; Driss et al., 2014).

XOS can also be produced by enzymatic hydrolysis with endoxylanase-containing preparations, and many LCBs have been tested as substrates (Kallel et al., 2015). Microbial xylanases are industrial enzymes currently used in detergents, in the production of pulp and paper and also in waste treatment (Juturu and Wu, 2012). For XOS production the enzyme preparations should have low exoxylanase and β-xylosidase activities to avoid xylose production. Such preparations are clearly advantageous and avoid the complexities of process control in the chemical synthesis. Desirably, endoxylanase preparations should be cellulase-free and stable at moderately high temperatures and high pH (Kumar and Satyanarayana, 2011). Significant advances in the production of improved xylanases have been reported by recombinant DNA technology (Jiang et al., 2004; Ruanglek et al., 2007) and protein engineering (Ruller et al., 2008; Zhang et al., 2010) to match such requirements. Enzyme cost is still a major hurdle for the enzymatic process, so strain improvement is still a pursued goal. Immobilization of xylanases will also contribute to reduce enzyme costs by increasing the efficiency of catalyst use (Maalej-Achouri et al., 2009; Driss et al., 2014).

Enzymatic hydrolysis of untreated LCB can be quite inefficient, except in the case of highly susceptible substrates, so that pretreatment is required (Otieno and Ahring, 2012). In fact, industrial production of XOS is conducted on xylan extracted from the LCB (Aachari and Prapulla, 2011). Enzymatic production of XOS has significant advantages over the chemical process: it does not produce undesirable by-products or significant amounts of monosaccharides and operation is conducted at mild conditions not requiring costly equipment (Akpinar et al., 2009a). In principle, the enzymatic process is more compliant with sustainable chemistry, but the requirement of chemical pretreatment is contradictory. As said before, enzyme cost is a major constraint for the enzymatic process and quality of the product is highly dependent on the enzyme source and substrate so that validation is problematic (Akpinar et al., 2009b). The enzymatic synthesis of XOS has also been reported by the reverse hydrolysis reaction catalyzed by β-xylosidase at very high xylose concentrations (Guerfali et al., 2009), which is a quite promising strategy to be further developed.

Purification of XOS for food use is a major issue. Degree of purification will vary according to the production process. Intense purification will be required in the case of using raw LCB, where monosaccharides, acetic acid and other extractives, acid-soluble lignin fractions, furfural and hydroxymethylfurfural will contaminate the product. Purification will be less stringent when hydrolysis is done on extracted xylan and even less in the case of enzymatic hydrolysis, which is more specific than acid hydrolysis. Several operations have been studied for XOS purification: solvent extraction (Vázquez et al., 2005), adsorption (Montané et al., 2006), membrane fractionation (Vegas et al., 2008), and simulated moving bed chromatography (Aachari and Prapulla, 2011).

Pros and cons of enzymatic and chemical hydrolysis make them competing technologies for XOS production that need to be confronted on a cost analysis basis, but no such analysis yet has been published.

2.4.3.2 Isomalto-oligosaccharides (IMOSs)

IMOSs are NDOs composed of glucose units linked by α1-6 bonds. Commercial IMOS preparations are a mixture of oligosaccharides containing from two to six sugar moieties; isomaltose (IG2) and isomaltotriose (IG3) are the main components. However, preparations with higher degree of polymerization are also produced. IMOSs have been considered potential prebiotics because of their well-documented stimulation of *Bifidobacteria* in the large intestine (Rycroft et al., 2001), but they not fulfill all requirements to be considered as such. IMOSs are partially digested and absorbed in the small intestine reducing their prebiotic effect; however, digestion and absorption are diminished with more highly polymerized IMOS (Goulas et al., 2004).

There are different processes for the production of IMOS. The one mostly used employs starch as starting material, which is first hydrolyzed by α-amylase (EC 3.2.1.1) and the debranching enzyme pullulanase (EC 3.2.1.41) to yield a mixture of malto-oligosaccharides, which are then the substrate of a glycosyl transfer reaction catalyzed by α-glucosidase (EC 2.4.1.20) that converts the α1-4 linked malto-oligosaccharides into α1-6 linked IMOS (Kohmoto et al., 1988). The first step is a conventional operation in the production of glucose and fructose syrups from cornstarch (van der Maarel et al., 2002); the second step is catalyzed by α-glucosidases, mostly from *Aspergillus* strains (Kato et al., 2002). In a variant of such process, starch is hydrolyzed by a mixture of α- and β-amylase yielding a maltose-rich syrup that is then transglycosylated with α-glucosidase to produce IMOS, maltose acting both as glycosyl donor and acceptor; alternatively, transglycosylation has been done with immobilized permeabilized cells expressing such activity (Yun et al., 1994).

A quite different approach is based on dextran as starting material. Dextran is produced by fermentation with *Leuconostoc mesenteroides* using sucrose as carbon source, being a high-molecular-weight polymer (over a million daltons) so that intense hydrolysis is required to produce oligosaccharides (Naessens et al., 2005). Smaller dextrans of 1000 daltons average molecular size can be produced by acid hydrolysis, but the production of IMOS requires the controlled hydrolysis of dextran with endodextranase (EC 3.2.1.11), which has been conducted in consecutive batch reactions with immobilized enzyme (Aslan and Tanriseven, 2007) and in continuous operation in enzyme membrane reactors (Mountzouris et al., 2001). In this case, IMOS of high molecular weight are obtained; IMOS of lower molecular weight and a more narrow distribution can be produced from sucrose by the combined action of dextransucrase (EC 2.4.1.5) and endodextranase. In principle, this strategy will lead to a complex mixture of products, mostly because dextransucrase synthesizes a mixture of oligosaccharides that then dextranase will hydrolyze randomly, so that optimization and careful control of reaction conditions is necessary to drive the reaction to a product with the composition appropriate to be used as health promoter (Goulas et al., 2004). IMOSs have been also produced by an acceptor reaction between sucrose and maltose catalyzed by dextransucrase (Rabelo et al., 2009). An interesting proposal for the production of controlled low-molecular-weight IMOS is the one-step synthesis from sucrose using glucose as acceptor of the transglycosylation reaction catalyzed by mutant dextransucrase variants obtained by protein-engineering techniques (Hellmuth et al., 2008; Moulis et al., 2008). Another promising strategy is the construction of a fusion enzyme of dextransucrase and dextranase that allows producing low-molecular-weight linear IMOS from sucrose in a one-step reaction (Kim et al., 2009).

Purification of raw IMOS is a critical issue whose complexity depends on the production strategy. Common by-products contaminating IMOS are glucose, maltose, and fructose, whose removal can be made by usual procedures of oligosaccharides purification, namely adsorption, solvent extraction, and membrane fractionation (Pinelo et al., 2009; Rabelo et al., 2013). Removal by selective yeast fermentation of contaminant monosaccharides has also been proposed (Pan and Lee, 2005).

Even though IMOS are not properly considered prebiotics, their consumption as a health-promoting food component is quite extended in Japan, where a market of 15,000 tons per year was estimated in 2002 (Nakakuki, 2002). The challenge of its introduction into the European and US markets has been analyzed by Goffin et al. (2011).

Other glucose-derived oligosaccharides, like malto-oligosaccharides and panose, have been enzymatically synthesized from sucrose by dextransucrase (Gibson et al., 2004) and from glucosidases by maltose transglycosylation (Mangas-Sánchez and Adlercreutz, 2015), being also prebiotic candidates.

2.4.3.3 Soybean Oligosaccharides (SOSs)

Numerous food crops contain potentially prebiotic NDOs. Cereals and legumes are considered rich sources of dietary fiber, mostly α-galacto-oligosaccharides containing galactose, glucose, and fructose units, like the trisaccharide raffinose (gal-glu-fru), the tetrasaccharide stachyose (gal-gal-glu-fru), and the pentasaccharide verbascose (gal-gal-gal-glu-fru) (Dwivedi et al., 2014). SOSs are mostly raffinose and stachyose that are NDOs reaching the colon intact where they can be fermented by *Lactobacilli* and *Bifidobacteria*. SOSs have demonstrated bifidogenic effect and compared favorably with other NDOs in terms of stimulation of *Bifidobacteria*, lactate production, low gas evolution, and increased SCFA production (Rycroft et al., 2001; Bouhnik et al., 2004; Bang et al., 2007). However, there is not enough scientific evidence of compliance of all requirements to be considered prebiotic, and well-designed studies in human volunteers are lacking (Rastall, 2010). Methods for SOS extraction from soybeans were originally developed to remove them in the production of soybean meal and are now the basis for the production of SOS (Parsons et al., 2000; Kim et al., 2003). Production of SOS considers solvent extraction of soybeans and precipitation, and further refinement by ultrafiltration and adsorption in diatomaceous earth and charcoal (Kim et al., 2003; Espinosa-Martos and Rupérez Antón, 2006). Despite not being properly prebiotics, SOSs are considered health-promoting food components that have GRAS status and are industrially produced in Japan and China.

2.4.3.4 Resistant Starch (RS)

Although originally considered as the fraction of starch resistant to exhaustive hydrolysis by α-amylase and pullulanase, the term resistant starch has been coined to refer to the recalcitrant fraction of starch that remains undigested after it passes through the gastrointestinal tract, reaching the colon where it is fermented by the indigenous microbiota (Sajilata et al., 2006). In this sense, RS behaves as soluble fiber, sharing its beneficial properties. Different from the NDOs described earlier, RS is a highly heterogeneous material that varies considerably in composition and properties according to its source and the processing conditions. RS is present in varying amounts in both raw and cooked vegetables, cereals, and legumes (Ranhotra et al., 1999; de Almeida Costa et al., 2006). Resistance of starch is related to its amylose content, high-amylose starches being the more resistant. Commercial products rich in RS, like Hi-maize 1043, Novelose 240, ActiStar, and CrystalLean have RS contents in the range from 40% to 60% (Nugent, 2005). RS has been classified into four subtypes identified as RS1 to RS4 (Leszczyński, 2004). RS refers to physically inaccessible starch surrounded by cell wall material that prevents amylolysis. RS2 refers to native starch granules containing raw uncooked starch or poorly gelatinized starch. RS3 refers to retrograded starch in the form of nongranular semicrystalline structures that are formed after cooking; retrogradation correlates with the amylose content of the starch. RS4 refers to purposely modified starch by physical or chemical treatment making it more indigestible. A fifth category, RS5, has been added to refer to amylose-lipid complexed starch that is formed in high-amylose starches requiring high-gelatinization temperatures (Fuentes-Zaragoza et al., 2010).

There is considerable interest both in the nutritional and functional properties of RS, as their physiological effects are similar to those attributed to dietary fiber. Most health effects of RS are the consequence of its fermentation in the colon, sharing most of the physiological effects that prebiotics promote (Fuentes-Zaragoza et al., 2011). As said before, RSs are naturally present in many vegetable foodstuffs, but their content can be significantly increased by processing conditions and/or by supplementation with RS-enriched products. RS has be used in synergistic mixtures with FOS and inulin (Rodríguez-Cabezas et al., 2010), increasing the individual prebiotic effects. Production and applications of RS have been recently reviewed by Homayouni et al. (2014).

2.5 SYNBIOTICS: CONCEPT, EVALUATION, PRESENT STATUS, TECHNOLOGICAL GOALS, AND CHALLENGES

Probiotics and prebiotics may act synergistically, and there is an obvious potential for their use in combination, their mixtures being referred to as synbiotics (Tuohy et al., 2008). Synbiotics can be simply defined as a combination of a probiotic and a prebiotic in one food matrix (Rastall and Maitin, 2002). A more sound and accepted definition of a synbiotic is "a mixture of probiotics and prebiotics that beneficially affect the host by improving the survival and implantation of live microbial dietary supplements in the gastrointestinal tract" (Andersson et al., 2001). The concept of synergism is implicit so this term has been suggested to be used when the prebiotic component of the mixture selectively favors the probiotic component (Schrezenmeir and de Vrese, 2001), since prebiotics per se can stimulate indigenous beneficial bacterial species in the gut, so in essence there is a synbiotic action when a prebiotic alone is administered (Itsaranuwat et al., 2003).

Based on this, the specification of two types of synbiotics has been proposed, designating them as "complementary" or "synergistic." The former are those in which the probiotic is chosen to exert its beneficial effect on the host and the prebiotic is chosen independently to stimulate the growth or activity of the beneficial components of the intestinal microbiota. The latter refers to the case when the prebiotic is chosen to selectively stimulate the growth or activity of the selected probiotic in the synbiotic mixture (Kolida and Gibson, 2011). The latter are nowadays considered more relevant, and according to the synergistic effect the term synbiotic implies; the term "integrated synbiotics" has been proposed to identify them (Bandyopadhyay and Mandal, 2014). These different approaches to construct a synbiotic product have practical consequences. In the case of complementary mixtures, each component is targeted separately so relatively high doses of each will be required. In the synergistic approach the prebiotic is targeted primarily to improve the survival and implantation of the added probiotic so in principle a smaller dose of the probiotic will be required.

Synbiotic efficacy should ideally be determined by in vivo studies with human volunteers. However, most studies have been conducted in vitro (Saulnier et al., 2008) or in simulated in vivo tests (van Zanten et al., 2012), and data relating their efficacy in human disease are still scarce (Quigley, 2010). In strict terms a symbiotic must prove that its effect exceeds those of its separate prebiotic and prebiotic components. However, these controls are not always done when evaluating a synbiotic and in some cases its effect may not be better than the one exerted by the prebiotic alone, so the ultimate proof of synbiotic effect should come from in vivo studies comparing the effect of the synbiotic mixture with respect to its separate components (Figueroa-González et al., 2011). In very few instances this has been proved (de Vrese and Schrezenmeir, 2008).

One of the main concerns of probiotic-containing foods is the delivery of an adequate amount of live bacteria to their site of action, so viability and stability during production and storage of the food product is a major issue. In a synbiotic mixture, the prebiotic component should contribute to improve the viability and stability of the probiotic component, allowing delivered doses higher than 10^8 cfu/mL of product (Roberfroid, 2000b; Champagne and Fustier, 2007).

Most synbiotic foods are fermented milk and dairy products already containing starter cultures to which, besides the prebiotic, probiotic bacteria (*Lactobacilli* and *Bifidobacteria*) are added. In this case the prebiotic should be selective for growth of the probiotic, and prebiotic fermentation by the starter microorganisms has to be taken into account as this may alter the organoleptic properties of the food and reduce the fermentable sugars available to the probiotic. Synbiotics combining probiotics and RS as the prebiotic counterpart have been developed, and RS is considered a promising synbiotic partner of many probiotic strains (Topping et al., 2003). RS can be advantageously used for microencapsulation in probiotic strains delivery, which is a nice way of administering a synbiotic product (Mirzaei et al., 2012). Galacto-oligosaccharides have been used successfully as protecting agent for probiotic strains during freeze-drying, the resulting product being in essence a synbiotic (Golowczyc et al., 2013).

Dosing is a complex task for synbiotics, and no boundaries between effectiveness in terms of health effects and unwanted side effects studies have yet been established (Kolida and Gibson, 2011).

Synbiotics have also been applied to animal feeding with promising results being obtained in in vivo trials (Awad et al., 2009; Gaggìa et al., 2010). Use of synbiotics as an alternative to the preventive use of antibiotics in livestock feeding has also been claimed (Scholz-Ahrens et al., 2007).

As in the case of probiotics and prebiotics, there are scientific, economic, and regulatory barriers that challenge widespread use of these products as active components in health foods (Rastall et al., 2005). Scientific barriers refer mostly to the limited knowledge about microbial communities in our body and their complex interactions. Economic barriers are related to the extra cost they represent when added to a food product that may lead to a negative cost-benefit analysis; prebiotic components have been in many cases produced from cheap by-products or wastes from agriculture and agroindustry. Probiotics are mostly *Lactobacilli* and *Bifidobacteria* strains, which are difficult to grow and maintain viable, so this may challenge the supply of validated and stable product certifying that the proper dose is delivered. Regulatory barriers are perhaps the most severe since regulatory processes are still complex in most countries, legislation is not always well developed, and guidelines based on scientific grounds do not exist in most cases. In addition, meaningful claims are difficult to sustain by scientific evidence and the cost of health claim applications (between €250,000 and 1 million) may certainly discourage many potential producers. Last but not least, there are educational barriers, the general public being unaware of the benefits and limitations of these products, which easily leads to misconception, hampering their widespread acceptance; educational campaigns and communication by authorities is a major contribution to improve public perception. However, scientific knowledge is advancing at a sustained pace and accordingly regulations are continuously under revision in the United States, Japan, Canada, and European countries (Brownawell et al., 2012). On the other hand, reduction of production costs by using waste biomass from agriculture and food processing is also underway (Gullon et al., 2009; Figueroa-González et al., 2011).

2.6 MARKET OUTLOOK

The functional foods market has experienced substantial development in the last decade with an increased rate of about 10% per year, much higher than the food market as a whole whose increase is estimated in about 2% per year. Functional foods represent therefore a growing fraction of the overall food market (El Sohaimy, 2012). It is, however, difficult to establish market figures for functional foods because boundaries between them and conventional foods are not clear-cut. In fact, global market sizes for functional foods ranging from US$30 to 60 billion were determined a decade ago (Siró et al., 2008). A figure of US$74 billion was estimated in 2005 with a forecast of a 10% increase per year (Granato et al., 2010), which, if sustained in the last decade, gives a projected market close to US$200 billion for 2015. A more conservative estimate of 7.5% increase per year has been forecast, according to which the present global market should be around US$150 billion. Estimates of market sizes in 2005 were US$20 billion for the United States, US$15 billion for the European Union, and US$12 billion for Japan (Matella et al., 2006), which would roughly represent 65% of the world market. The European market has been moving particularly fast: an increase of 60% was observed between 1998 and 2003 and an increase of 40% between 2003 and 2008. Japan and the United States are the leaders in per capita functional food consumption, which can be roughly estimated at US$190 and US$130 per year, respectively. Per capita consumption estimate for Europe is US$45/year. Japan has been a pioneer in health food development, and health eating habits of Japanese is part of their culture, which explains its leadership in functional food consumption. Functional foods are becoming increasingly important in Latin America but still its consumption is well below the levels of the leading developed countries. As an example, per capita consumption in Brazil is estimated around US$5/year. However, growing concern about the impact of eating habits on health is clearly perceived; besides, Latin America is an important producer of raw materials for functional foods, so its potential for developing a vigorous industrial platform for producing functional foods certainly exists (Lajolo, 2002).

Despite the impressive advantages in the functional food sector, it represents a small share of the total food market, estimated between 1% and 2%; however, it is significantly higher in Europe and Japan (Granato et al., 2010). The dairy industry is by far the most important sector contributing to functional food production.

In principle, due to the complexities of the regulations applying to functional foods, big companies and multinational consortia are the best positioned to overcome the challenges for launching new functional foods, as they have the required R&D resources, know-how, and market accessibility (Menrad, 2003).

Probiotics and prebiotics are important players in the functional food business and are considered as most promising components in health foods. This is more so in Europe and Japan, while in the United States focus on functional foods is more related to fortification with vitamins and minerals.

Probiotics may account for a significant share of the functional food market, with an estimated figure in Europe of €1.4 billion in 2008, most of it being represented by dairy products (Saxelin, 2008). A global estimate of the probiotic market in 2004 was US$17.5 billion, which might have increased to around US$35 billion after a decade (Siró et al., 2008). A more conservative estimate of US$16 billion in 2008 has also been reported with a projected value of US$26 for 2015 (Granato et al., 2010). About 60% of probiotics are based on *Lactobacillus* strains and the food applications are mostly in the dairy sector with yogurts and fermented milks representing close to 40% of probiotic-containing foods. Emerging applications of probiotics are ice cream (Pandiyan et al., 2012), nutrition bars (Ouwehand et al., 2004), cheese (Gomes da Cruz et al., 2009), and infant formula (Braegger et al., 2011).

The market for prebiotics in functional foods has been growing at an accelerated pace in the last decade. In 2007 there were 400 prebiotic foods registered in the market and more than 20 companies producing them. These figures may well have doubled by now. According to a survey by Frost and Sullivan in 2008 (http.//www.reuters.com/article/pressRelease), the prebiotic market in the United States was US$70 million in 2007 and may have reached US$200 million by 2015 (Panesar et al., 2013), which represents an increase rate of about 15% per year. The European prebiotic market was €87 million in 2007 with an impressive growth to €180 million in 2010 (Pineiro et al., 2008). This market is expected to keep on growing and an estimate of €1 billion was forecast for 2015 (Panesar et al., 2013). Japan was the first country incorporating NDOs in foods and is a world leader in the use of prebiotics. The concept of FOSHU was coined in 1991 in Japan and since then an explosive increase in the number of marketed FOSHU products has occurred. Today about 50% of FOSHU products in Japan contain NDOs. By 2001 the NDO market had reached US$125 million, a figure that might well have doubled (Taniguchi, 2005). As in the case of probiotics, dairy products are the most important food sector for prebiotic addition—yogurt and special milks are the most relevant. However, prebiotics are also being included in nondairy foods, like cereal bars, bakery products, fruit juices, and spreads (Charalampopoulos and Rastall, 2012).

REFERENCES

Aachary, A.A., Prapulla, S.G., 2011. Xylooligosaccharides (XOS) as an emerging prebiotic: microbial synthesis, utilization, structural characterization, bioactive properties, and applications. Comprehensive Reviews in Food Science and Food Safety 10, 2–16.

Akpinar, O., Erdogan, K., Bostanci, S., 2009a. Enzymatic production of xylooligosaccharide from selected agricultural wastes. Food and Bioproducts Processing 87, 145–151.

Akpinar, O., Erdogan, K., Bostanci, S., 2009b. Production of xylooligosaccharides by controlled acid hydrolysis of lignocellulosic materials. Carbohydrate Research 344, 660–666.

Alissa, E.M., Ferns, G.A., 2012. Functional foods and nutraceuticals in the primary prevention of cardiovascular diseases. Journal of Nutrition and Metabolism. http://dx.doi.org/10.1155/2012/569486 Article ID 569486, 16 pages.

Anal, A.K., Singh, H., 2007. Recent advances in microencapsulation of probiotics for industrial applications and targeted delivery. Trends in Food Science and Technology 18, 240–251.

Anderson, J.W., Baird, P., Davis, R.H., Ferreri, S., Knudtson, M., Koraym, A., Waters, V., Williams, C.L., 2009. Health benefits of dietary fiber. Nutrition Reviews 67, 188–205.

Andersson, H., Asp, N.G., Bruce, A., Roos, S., Wadström, T., Wold, A.E., 2001. Health effects of probiotics and prebiotics. A literature review on human studies. Scandinavian Journal of Nutrition 45, 58–75.

Annunziata, A., Vecchio, R., 2011. Functional foods development in the European market: a consumer perspective. Journal of Functional Foods 3, 223–228.

Arora, T., Sharma, R., Frost, G., 2011. Propionate. Anti-obesity and satiety enhancing factor? Appetite 56, 511–515.

Arunrao Yadav, S., Agte, V.V., Nilegaonkar, S.S., 2014. Enrichment of prebiotics in foods using green chemistry approach. Current Organic Chemistry 18, 2961–2971.

Aslan, Y., Tanriseven, A., 2007. Immobilization of *Penicillium lilacinum* dextranase to produce isomaltooligosaccharides from dextran. Biochemical Engineering Journal 34, 8–12.

Ávila-Fernández, A., Galicia-Lagunas, N., Rodríguez-Alegría, M.E., Olvera, C., López-Munguía, A., 2011. Production of functional oligosaccharides through limited acid hydrolysis of agave fructans. Food Chemistry 129, 380–386.

Awad, W.A., Ghareeb, K., Abdel-Raheem, S., Böhm, J., 2009. Effects of dietary inclusion of probiotic and synbiotic on growth performance, organ weights, and intestinal histomorphology of broiler chickens. Poultry Science 88, 49–55.

Bali, V., Panesar, P.S., Bera, M.B., Panesar, R., 2015. Fructo-oligosaccharides: production, purification and potential applications. Critical Reviews in Food Science and Nutrition 55, 1474–1490.

Bandyopadhyay, B., Mandal, N.C., 2014. Probiotics, prebiotics and synbiotics in health improvement by modulating gut microbiota: the concept revisited. International Journal of Current Microbiology and Applied Sciences 3, 410–420.

Bang, M.H., Chio, O.S., Kim, W.K., 2007. Soyoligosaccharide increases fecal *bifidobacteria* counts, short-chain fatty acids, and fecal lipid concentrations in young Korean women. Journal of Medicinal Food 10, 366–370.

Barreteau, H., Delattre, C., Michaud, P., 2006. Production of oligosaccharides as promising new food additive generation. Food Technology and Biotechnology 44, 323–333.

Belenguer, A., Duncan, S.H., Graham Calder, A., Holtrop, G., Louis, P., Lobley, G.E., Flint, H.J., 2006. Two routes of metabolic cross-feeding between *Bifidobacterium adolescentis* and butyrate-producing anaerobes from the human gut. Applied and Environmental Microbiology 72, 3593–3599.

Bellei, G., Haslberger, A., 2012. Dietary fibre and prebiotics. Journal für Ernährungsmedizin 14, 12–15.

Ben Amor, K., Vaughan, E.E., de Vos, W.M., 2007. Advanced molecular tools for the identification of lactic acid bacteria. The Journal of Nutrition 137, 741S–747S.

Berni Canani, R., Di Costanzo, M., Leone, L., Pedata, M., Meli, R., Calignano, A., 2011. Potential beneficial effects of butyrate in intestinal and extraintestinal diseases. World Journal of Gastroenterology 28, 1519–1528.

Beserra, B.T.S., Fernandes, R., do Rosario, V.A., Mocellin, M.C., Kuntz, M.G.F., Trindade, E.B.S.M., 2014. A systematic review and meta-analysis of the prebiotics and synbiotics effects on glycaemia, insulin concentrations and lipid parameters in adult patients with overweight or obesity. Clinical Nutrition. http://dx.doi.org/10.1016/j.clnu.2014.10.004 Available online.

Bezirtzoglou, E., Tsiotsias, A., Welling, G.W., 2011. Microbiota profile in feces of breast- and formula-fed newborns by using fluorescence in situ hybridization (FISH). Anaerobe 17, 478e–482e.

Biedrzycka, E., Bielecka, M., 2004. Prebiotic effectiveness of fructans of different degrees of polymerization. Trends in Food Science and Technology 15, 170–175.

Bigliardi, B., Galati, F., 2013. Innovation trends in the food industry: the case of functional foods. Trends in Food Science and Technology 31, 118–129.

Blatchhford, P., Ansell, J., de Godoy, M.R.C., Fahey, G., Garcia-Mazcorro, J.F., Gibson, G.R., Goh, Y.J., Hotchkiss, A.T., Hutkins, R., La Croix, C., Rastall, R.A., Reimer, R.A., Schotterman, M., Van Sinderen, D., Venema, K., Whelan, K., 2013. Prebiotic mechanisms, functions and applications – a review. International Journal of Probiotics and Prebiotics 8, 109–132.

Blaut, M., 2002. Relationship of prebiotics and food to intestinal microflora. European Journal of Nutrition 41, I11–I16.

Bocquier, F., González-García, E., 2010. Sustainability of ruminant agriculture in the new context: feeding strategies and features of animal adaptability into the necessary holistic approach. Animal 4, 1258–1273.

Boehm, G., Wind, R., Knol, J., 2010. Prebiotics and probiotics in infant formulae. In: Cho, S.S., Finocchiaro, E.T. (Eds.), Handbook of Prebiotics and Probiotics Ingredients. Health Benefits and Food Applications. CRC Press, Boca Raton, pp. 293–339.

Bouhnik, Y., Raskine, L., Simoneau, G., Vicaut, E., Neut, C., Flourié, B., Brouns, F., Bornet, F.R., 2004. The capacity of nondigestible carbohydrates to stimulate fecal bifidobacteria in healthy humans: a double-blind, randomized, placebo-controlled, parallel-group, dose-response relation study. American Journal of Clinical Nutrition 80, 1658–1664.

Boye, J.I., 2015. Nutraceutical and Functional Food Processing Technology. Wiley Blackwell, Oxford.

Braegger, C., Chmielewska, A., Decsi, T., Kolacek, S., Mihatsch, W., Moreno, L., Piescik, M., Puntis, J., Shamir, R., Szajewska, H., Turck, D., van Goudoever, J., 2011. Supplementation of infant formula with probiotics and/or prebiotics: a systematic review and comment by the ESPGHAN Committee on Nutrition. Journal of Pediatric Gastroenterology and Nutrition 52, 238–250.

Brownawell, A.M., Caers, W., Gibson, G.R., Kendall, C.W.C., Lewis, K.D., Ringel, Y., Slavin, J.L., 2012. Prebiotics and the health benefits of fiber: current regulatory status, future research, and goals. Journal of Nutrition 142, 962–974.

Bruzzese, E., Volpicelli, M., Squeglia, V., Bruzzese, D., Salvini, F., Bisceglia, M., Lionetti, P., Cinquetti, M., Iacono, G., Amarri, S., Guarino, A., 2009. A formula containing galacto- and fructo-oligosaccharides prevents intestinal and extra-intestinal infections: an observational study. Clinical Nutrition 156–161.

Caicedo, L., Silva, E., Sánchez, O., 2009. Semibatch and continuous fructooligosaccharides production by *Aspergillus* sp. N74 in a mechanically agitated airlift reactor. Journal of Chemical Technology and Biotechnology 84, 650–656.

Cani, P.D., Lecourt, E., Dewulf, E.M., Sohet, F.M., Pachikian, B.D., Naslain, D., De Backer, F., Neyrinck, A.M., Delzenne, N.M., 2009. Gut microbiota fermentation of prebiotics increases satietogenic and incretin gut peptide production with consequences for appetite sensation and glucose response after a meal. American Journal of Clinical Nutrition 90, 1236–1243.

Caplice, E., Fitzgerald, G.F., 1999. Food fermentations: role of microorganisms in food production and preservation. International Journal of Food Microbiology 50, 131–149.

Celec, P., Kukučková, M., Renczésová, V., Natarajan, S., Pálffy, R., Gardlík, R., Hodosy, J., Behuliak, M., Vlková, B., Minárik, G., Szemes, T., Stuchlík, S., Turňa, J., 2005. Biological and biomedical aspects of genetically modified food. Biomedicine and Pharmacotherapy 59, 531–540.

Cha, M.H., Lee, J., Song, M.J., 2010. Dieticians' intentions to recommend functional foods: the mediating role of consumption frequency of functional foods. Nutrition Research and Practice 4, 75–81.

Champagne, C.P., Fustier, P., 2007. Microencapsulation for the improved delivery of bioactive compounds into foods. Current Opinion in Biotechnology 18, 184–190.

Chan, E.S., Zhang, Z., 2002. Encapsulation of probiotic bacteria *Lactobacillus acidophilus* by direct compression. Food and Bioproducts 80, 78–82.

Charalampopoulos, D., Rastall, R.A., 2012. Prebiotics in foods. Current Opinion in Biotechnology 23, 187–191.

Chaucheyras-Durand, F., Durand, H., 2010. Probiotics in animal nutrition and health. Beneficial Microbes 1, 3–9.

Chawla, R., Patil, G.R., 2010. Soluble dietary fiber. Comprehensive Reviews in Food Science and Food Safety 9, 178–196.

Chen, M.J., Chen, K.N., Kuo, Y.T., 2007. Optimal thermotolerance of *Bifidobacterium bifidum* in gellan–alginate microparticles. Biotechnology and Bioengineering 98, 411–419.

Cho, S.S., Finocchiaro, E.T., 2010. Handbook of Prebiotics and Probiotics Ingredients. Health Benefits and Food Applications. CRC Press, Boca Raton.

Choudhari, A., Shinde, S., Ramteke, B.N., 2008. Prebiotics and probiotics as health promoter. Veterinary World 1, 59–61.

Coman, M.M., Cecchini, C., Verdenelli, M.C., Silvi, S., Orpianesi, C., Cresci, A., 2012. Functional foods as carriers for SYNBIO®, a probiotic bacteria combination. International Journal of Food Microbiology 157, 346–352.

Corzo, N., Alonso, J.L., Azpiroz, F., Calvo, M.A., Cirici, M., Leis, R., Lombó, F., Mateos-Aparicio, I., Plou, F.J., Ruas-Madiedo, P., Rúperez, P., Redondo-Cuenca, A., Sanz, M.L., Clemente, A., 2015. Prebióticos; concepto, propiedades y efectos beneficiosos. Nutrición Hospitalaria 31, 99–118.

Costalos, C., Kapiki, A., Apostolou, M., Papathoma, E., 2008. The effect of a prebiotic supplemented formula on growth and stool microbiology of term infants. Early Human Development 84, 45–49.

Coussement, P.A.A., 1999. Inulin and oligofructose: safe intakes and legal status. The Journal of Nutrition 129, 1412S–1417S.

Crittenden, R., Playne, M.J., 2009. Prebiotics. In: Lee, Y.K., Salminen, S. (Eds.), Handbook of Probiotics and Prebiotics, second ed. Wiley, New Jersey, pp. 535–581.

Csanádi, Z., Sisak, C., 2006. Immobilization of Pectinex Ultra SP-l pectinase and its application to production of fructooligosaccharides. Acta Alimentaria 35, 205–212.

Cummings, J.H., Macfarlane, G.T., 2002. Gastrointestinal effects of prebiotics. British Journal of Nutrition 87, S145–S151.

Cummings, J.H., Macfarlane, G.T., Englyst, H.N., 2001. Prebiotic digestion and fermentation. American Journal of Clinical Nutrition 73, 415S–420S.

Cutting, S.M., 2011. *Bacillus* probiotics. Food Microbiology 28, 214–220.

Czerucka, D., Piche, T., Rampal, P., 2007. Yeast as probiotics – *Saccharomyces boulardii*. Alimentary Pharmacology and Therapeutics 26, 767–778.

de Almeida Costa, G.E., da Silva Queiroz-Monici, K., Pissini Machado Reis, S.M., Costa de Oliveira, A., 2006. Chemical composition, dietary fibre and resistant starch contents of raw and cooked pea, common bean, chickpea and lentil legumes. Food Chemistry 94, 327–330.

De Gennaro, S., Birch, G., Parke, S., Stancher, B., 2000. Studies on the physicochemical properties of inulin and inulin oligomers. Food Chemistry 68, 179–183.

de Vrese, M., Marteau, P.R., 2007. Probiotics and prebiotics: effects on diarrhea. Journal of Nutrition 137, 803S–811S.

de Vrese, M., Schrezenmeir, J., 2008. Probiotics, prebiotics, and synbiotics. Advances in Biochemical Engineering/Biotechnology 111, 1–66.

Dean, M., Lampila, P., Shepherd, R., Arvola, A., Saba, A., Vassallo, M., Claupein, E., Winkelmann, M., Lähteenmäki, L., 2012. Perceived relevance and foods with health-related claims. Food Quality and Preference 24, 129–135.

Delzenne, N., Williams, C.M., 2002. Prebiotics and lipid metabolism. Current Opinion in Lipidology 13, 61–67.

Delzenne, N.M., Neyrinck, A.M., Bäckhed, F., Cani, P.D., 2011. Targeting gut microbiota in obesity: effects of prebiotics and probiotics. Nature Reviews Endocrinology 7, 639–646.

Delzenne, N.M., 2003. Oligosaccharides: state of the art. Proceedings of the Nutrition Society 62, 177–182.

Depeint, F., Tzortzis, G., Vulevic, J., I'Anson, K., Gibson, G.R., 2008. Prebiotic evaluation of a novel galactooligosaccharide mixture produced by the enzymatic activity of *Bifidobacterium bifidum* NCIMB 41171, in healthy humans: a randomized, double-blind, crossover, placebo-controlled intervention study. American Journal of Clinical Nutrition 87, 785–791.

Dominguez, A.L., Rodrigues, L.R., Lima, N.M., Teixeira, J.A., 2014. An overview of the recent developments on fructooligosaccharide production and applications. Food and Bioprocess Technology 7, 324–337.

Douglas, L.C., Sanders, M.E., 2008. Probiotics and prebiotics in dietetics practice. Journal of the American Dietetic Association 108, 510–521.

Doyon, M., Labrecque, J.A., 2008. Functional foods: a conceptual definition. British Food Journal 110, 1133–1149.

Driss, D., Zouari-Ellouzi, S., Chaari, F., Kallel, F., Ghazala, I., Bouaziz, F., Ghorbela, R., Chaabouni, S.E., 2014. Production and in vitro evaluation of xylooligosaccharides generated from corncobs using immobilized *Penicillium occitanis* xylanase. Journal of Molecular Catalysis B: Enzymatic 102, 146–153.

Duan, K.J., Chen, J.S., Sheu, D.C., 1994. Kinetic studies and mathematical model for enzymatic production of fructooligosaccharides from sucrose. Enzyme and Microbial Technology 16, 334–339.

Duc, L.H., Hong, H.A., Barbosa, T.M., Henriques, A.O., Cutting, S.M., 2004. Characterization of *Bacillus* probiotics available for human use. Applied and Environmental Microbiology 70, 2161–2171.

Dwivedi, S., Sahrawat, K., Puppala, N., Ortiz, R., 2014. Plant prebiotics and human health: biotechnology to breed prebiotic-rich nutritious food crops. Electronic Journal of Biotechnology 17, 238–245.

El Sohaimy, S.A., 2012. Functional foods and nutraceuticals- modern approach to food science. World Applied Sciences Journal 20, 691–708.

Espinosa Martos, I., Rupérez Antón, P., 2006. Soybean oligosaccharides. Potential as new ingredients in functional food. Nutrición Hospitalaria 21, 92–96.

Ezzati, M., Van der Hoorn, S., Lawes, C.M.M., Leach, R., James, W.P.T., Lopez, A.D., Rodgers, A., Murray, C.J.L., 2005. Rethinking the "diseases of affluence" paradigm: global patterns of nutritional risks in relation to economic development. PLoS Medicine 2, 404–412.

FAO/WHO, 2002. Joint FAO/WHO Working Group Guidelines for the Evaluation of Probiotics in Food. Food and Agriculture Organization, United Nations.

Figueroa-González, I., Quijano, G., Ramírez, G., Cruz-Guerrero, A., 2011. Probiotics and prebiotics – perspectives and challenges. Journal of the Science of Food and Agriculture 91, 1341–1348.

Foligné, B., Daniel, C., Pot, B., 2013. Probiotics from research to market: the possibilities, risks and challenges. Current Opinion in Microbiology 16, 284–292.

Franck, A., 2002. Technological functionality of inulin and oligofructose. British Journal of Nutrition 87, S287–S291.

Franks, A.H., Harmsed, H.J.M., Raangs, G.C., Jansen, G.J., Schut, F., Welling, G.W., 1998. Variations of bacterial populations in human faeces measured by fluorescent in situ hybridization with group-specific 16S rRNA-targeted oligonucleotide probes. Applied and Environmental Microbiology 64, 3336–3345.

Fu, L., Wang, Y., 2013. Sources and production of prebiotics. In: Ötleş, S. (Ed.), Probiotics and Prebiotics in Food. Nutrition and Health CRC Press, Boca Raton, pp. 50–64.

Fuentes-Zaragoza, E., Riquelme-Navarrete, M.J., Sánchez-Zapata, E., Pérez-Álvarez, J.A., 2010. Resistant starch as functional ingredient: a review. Food Research International 43, 931–942.

Fuentes-Zaragoza, E., Sánchez-Zapata, E., Sendra, E., Sayas, E., Navarro, C., Fernández-López, J., Pérez-Alvarez, J.A., 2011. Resistant starch as prebiotic: a review. Starch/Stärke 2011 (63), 406–415.

Fuller, R., 1989. Probiotics in man and animals. The Journal of Applied Bacteriology 66, 365–378.

Gaggìa, F., Mattarelli, P., Biavati, B., 2010. Probiotics and prebiotics in animal feeding for safe food production. International Journal of Food Microbiology 141, S15–S28.

Gevers, D., Knight, R., Petrosino, J.F., Huang, K., McGuire, A.L., Birren, B.W., Nelson, K.E., White, O., Methé, B.A., Huttenhower, C., 2012. The human microbiome project: a community resource for the healthy human microbiome. PLoS Biology 10 (8), e1001377. http://dx.doi.org/10.1371/journal.pbio.1001377.

Ghazi, I., Fernandez-Arrojo, L., Garcia-Arellano, H., Ferrer, M., Ballesteros, A., Plou, F.J., 2007. Purification and kinetic characterization of a fructosyltransferase from *Aspergillus aculeatus*. Journal of Biotechnology 128, 204–211.

Gibson, G.R., Fuller, R., 2000. Aspects of *in vitro* and *in vivo* research approaches directed toward identifying probiotics and prebiotics for human use. Journal of Nutrition 130, 391S–395S.

Gibson, G.R., Roberfroid, M.B., 1995. Dietary modulation of the human colonic microbiota: introducing the concept of prebiotics. Journal of Nutrition 125, 1401–1412.

Gibson, G.R., Probert, H.M., Van Loo, J., Rastall, R.A., Roberfroid, M.B., 2004. Dietary modulation of the human colonic microbiota: updating the concept of prebiotics. Nutrition Research Reviews 17, 259–275.

Gmeiner, M., Kneifel, W., Kulbe, K.D., Wouters, R., De Boever, P., Nollet, L., Verstraete, W., 2000. Influence of a synbiotic mixture consisting on *Lactobacillus acidophilus* 74-2 and a fructooligosaccharide preparation on the microbial ecology sustained in a simulation of the human intestinal ecosystem (SHIME reactor). Applied Microbiology and Biotechnology 53, 219–223.

Goffin, D., Delzenne, N., Blecker, C., Hanon, E., Deroanne, C., Paquot, M., 2011. Will isomalto-oligosaccharides, a well-established functional food in Asia, break through the European and American market? The status of knowledge on these prebiotics. Critical Reviews in Food Science and Nutrition 51, 394–409.

Golowczyc, M.A., Gerez, C.L., Silva, J., Abraham, A.G., De Antoni, G.L., Teixeira, P., 2011. Survival of spray-dried *Lactobacillus kefir* is affected by different protectants and storage conditions. Biotechnology Letters 33, 681–686.

Golowczyc, M., Vera, C., Santos, M., Guerrero, C., Carasi, P., Illanes, A., Gómez-Zavaglia, A., Tymczyszyn, E.E., 2013. Use of whey permeate containing in situ synthesised galacto-oligosaccharides for the growth and preservation of *Lactobacillus plantarum*. Journal of Dairy Research 80, 374–381.

Gomes da Cruz, A., Buriti, F.C.A., Batista de Souza, C.H., Fonseca Faria, J.A., Isay Saad, S.M., 2009. Probiotic cheese: health benefits, technological and stability aspects. Trends in Food Science and Technology 20, 344–354.

Gomez, D.E., Tuohy, K.M., Gibson, G.R., Klinder, A., Costabile, A., 2010. *In vitro* evaluation of the fermentation properties and potential prebiotic activity of Agave fructans. Journal of Applied Microbiology 108, 2114–2121.

Gonçalves, P., Martel, F., 2013. Butyrate and colorectal cancer: the role of butyrate transport. Current Drug Metabolism 14, 994–1008.

Gosálbez, L., Ramón, D., 2015. Probiotics in transition: novel strategies. Trends in Biotechnology 33, 195–196.

Goulas, A.K., Fisher, D.A., Grimble, G.K., Grandison, A.S., Rastall, R.A., 2004. Synthesis of isomaltooligosaccharides and oligodextrans by the combined use of dextransucrase and dextranase. Enzyme and Microbial Technology 35, 327–338.
Gourbeyre, P., Denery, S., Bodinier, M., 2011. Probiotics, prebiotics, and synbiotics: impact on the gut immune system and allergic reactions. Journal of Leukocyte Biology 89, 685–695.
Gramblička, M., Polakovič, M., 2007. Adsorption equilibria of glucose, fructose, sucrose, and fructooligosaccharides. Journal of Chemical and Engineering Data 52, 345–350.
Granato, D., Branco, G.F., Nazzaro, F., Cruz, A.G., Faria, J.A.F., 2010. Functional foods and nondairy probiotic food development: trends, concepts, and products. Comprehensive Reviews in Food Science and Food Safety 9, 292–302.
Guerfali, M., Maalej, I., Gargouri, A., Belghith, H., 2009. Catalytic properties of the immobilized *Talaromyces thermophilus* β-xylosidase and its use for xylose and xylooligosaccharides production. Journal of Molecular Catalysis B: Enzymatic 57, 242–249.
Gullon, P., Gullon, B., Moure, A., Alonso, J.L., Dominguez, H., Parajo, J.C., 2009. Manufacture of prebiotics from biomass sources. In: Charalampopoulos, D., Rastall, R.A. (Eds.), Prebiotics and Probiotics Science and Technology. Springer, New York, pp. 535–589.
Gupta, C., Prakash, D., Gupta, S., 2014. Genetically engineered probiotics. African Journal of Basic and Applied Sciences 6, 57–64.
Haas, G., Wetterich, F., Köpke, U., 2001. Comparing intensive, extensified and organic grassland farming in southern Germany by process life cycle assessment. Agriculture, Ecosystems and Environment 83, 43–53.
Hajati, H., Rezaei, M., 2010. The application of prebiotics in poultry production. International Journal of Poultry Science 9, 298–304.
Hang, H., Mu, W., Jiang, B., Zhao, M., Zhou, L., Zhang, T., Miao, M., 2012. Enzymatic hydrolysis of inulin in a bioreactor coupled with an ultrafiltration membrane. Desalination 284, 309–315.
Hardy, G., 2000. Nutraceuticals and functional foods: introduction and meaning. Nutrition 16, 688–697.
Heasman, M., Mellentin, J., 2001. The Functional Foods Revolution: Healthy People, Healthy Profits? Earthscan Publications Ltd, London.
Hellmuth, H., Wittrock, S., Kralj, S., Dijkhuizen, L., Hofer, B., Seibel, J., 2008. Engineering the glucansucrase GTFR enzyme reaction and glycosidic bond specificity: toward tailor-made polymer and oligosaccharide products. Biochemistry 47, 6678–6684.
Henker, J., Laass, M., Blokhin, B.M., Bolbot, Y.K., Maydannik, V.G., Elze, M., Wolff, C., Schulze, J., 2007. The probiotic *Escherichia coli* strain Nissle 1917 (EcN) stops acute diarrhoea in infants and toddlers. European Journal of Pediatrics 166, 311–318.
Homayouni, A., Amini, A., Keshtiban, A.K., Mortazavian, A.M., Esazadeh, K., Pourmoradian, S., 2014. Resistant starch in food industry: a changing outlook for consumer and producer. Starch/Stärke 66, 102–114.
Hoppe, C., Nexmann Larsen, C., 2009. Commercially available human probiotic microorganisms. In: Lee, Y.K., Salminen, S. (Eds.), Handbook of Probiotics and Prebiotics, second ed. Wiley, New Jersey, pp. 441–532.
Huang, C.H., Cheng, J.Y., Deng, M.C., Chou, C.H., Jan, T.R., 2012. Prebiotic effect of diosgenin, an immunoactive steroidal sapogenin of the Chinese yam. Food Chemistry 132, 428–432.
Huebner, J., Wehling, R.L., Parkhurst, A., Hutkins, R.W., 2008. Effect of processing conditions on the prebiotic activity of commercial prebiotics. International Dairy Journal 18, 287–293.
Huitrón, C., Pérez, R., Gutiérrez, L., Lappe, P., Petrosyan, P., Villegas, J., Aguilar, C., Rocha-Zavaleta, L., Blancas, A., 2013. Bioconversion of *Agave tequilana* fructans by exo-inulinases from indigenous *Aspergillus niger* CH-A-2010 enhances ethanol production from raw *Agave tequilana* juice. Journal of Industrial Microbiology and Biotechnology 40, 123–132.

Ishibashi, N., Yamazaki, S., 2001. Probiotics and safety. American Journal of Clinical Nutrition 73, 465S–470S.

Ito, H., Takemura, N., Sonoyama, K., Kawagishi, H., Topping, D.L., Conlon, M.A., Morita, T., 2011. Degree of polymerization of inulin-type fructans differentially affects number of lactic acid bacteria, intestinal immune functions, and immunoglobulin A secretion in the rat cecum. Journal of Agricultural and Food Chemistry 59, 5771–5778.

Itsaranuwat, P., Al-Haddad, K.S.H., Robinson, R.K., 2003. The potential therapeutic benefits of consuming 'health-promoting' fermented dairy products: a brief update. Journal of Dairy Technology 56, 203–210.

Jiang, Z.Q., Deng, W., Zhu, Y.P., Li, L.T., Sheng, Y.J., Hayashi, K., 2004. The recombinant xylanase B of *Thermotoga maritima* is highly xylan specific and produces exclusively xylobiose from xylans, a unique character for industrial applications. Journal of Molecular Catalysis B: Enzymatic 27, 207–213.

Jung, K.H., Yun, J.W., Kang, K.R., Lim, J.Y., Lee, J.H., 1989. Mathematical model for enzymatic production of fructooligosaccharides from sucrose. Enzyme and Microbial Technology 11, 491–494.

Jung, S.J., Houde, R., Baurhoo, B., Zhao, X., Lee, B.H., 2008. Effect of galacto-oligosaccharides and a *Bifidobacteria lactis*-based probiotic strain on the growth performance and fecal microflora of broiler chickens. Poultry Science 87, 1694–1699.

Jung, K.H., Bang, S.H., Oh, T.K., Park, H.J., 2011. Industrial production of fructooligosaccharides by immobilized cells of *Aureobasidium pullulans* in a packed bed reactor. Biotechnology Letters 33, 1621–1624.

Juturu, V., Wu, J.C., 2012. Microbial xylanases: engineering, production and industrial applications. Biotechnology Advances 30, 1219–1227.

Kallel, F., Driss, D., Bouaziz, F., Neifer, M., Ghorbel, R., Chaabouni, S.E., 2015. Production of xylo-oligosaccharides from garlic straw xylan by purified xylanase from *Bacillus mojavensis UEB-FK* and their *in vitro* evaluation as prebiotics. Food and Biproducts Processing 94, 536–546.

Karimi, R., Azziz, M.H., Ghasemlou, M., Vaziri, M., 2015. Application of inulin in cheese as prebiotic, fat replacer and texturizer: a review. Carbohydrate Polymers 119, 85–100.

Kato, N., Suyama, S., Shirokane, M., Kato, M., Kobayashi, T., Tsukagoshi, N., 2002. Novel α-glucosidase from *Aspergillus nidulans* with strong transglycosylation activity. Applied and Environmental Microbiology 68, 1250–1256.

Kaur, S., Das, M., 2011. Functional foods: an overview. Food Science and Biotechnology 20, 861–875.

Kaur, I.P., Kuhad, A., Garg, A., Chopra, K., 2010. In: Cho, S.S., Finocchiaro, E.T. (Eds.), Handbook of Prebiotics and Probiotics Ingredients. Health Benefits and Food Applications. CRC Press, Boca Raton, pp. 381–415.

Kellow, N.J., Coughlan, M.T., Reid, C.M., 2014. Metabolic benefits of dietary prebiotics in human subjects: a systematic review of randomised controlled trials. British Journal of Nutrition 111, 1147–1161.

Kim, M.H., In, M.J., Cha, H.J., Yoo, Y.J., 1996. An empirical rate equation for the fructooligosaccharide-producing reaction catalyzed by β-fructofuranosidase. Journal of Fermentation and Bioengineering 82, 458–463.

Kim, S., Kim, W., Hwang, I.K., 2003. Optimization of the extraction and purification of oligosaccharides from defatted soybean meal. International Journal of Food Science and Technology 38, 337–342.

Kim, Y.M., Seo, M.Y., Kang, H.K., Atsuo, K., Kim, D., 2009. Construction of a fusion enzyme of dextransucrase and dextranase: application for one-step synthesis of isomalto-oligosaccharides. Enzyme and Microbial Technology 44, 159–164.

Kloosterman, J., Fransen, H.P., de Stoppelaar, J., Verhagen, H., Rompelberg, C., 2007. Safe addition of vitamins and minerals to foods: setting maximum levels for fortification in the Netherlands. European Journal of Nutrition 46, 220–229.

Kohmoto, T., Fukui, F., Takaku, H., Machida, Y., Arai, M., Mitsuoka, T., 1988. Effect of isomalto-oligosaccharides on human fecal flora. Bifidobacteria Microflora 7, 61–69.

Kolida, S., Gibson, G.R., 2011. Synbiotics in health and disease. Annual Review of Food Science and Technology 2, 373–393.

Kolida, S., Tuohy, K., Gibson, G.R., 2002. Prebiotic effects of inulin and oligofructose. British Journal of Nutrition 87, S193–S197.

Kotilainen, L., Rajalahti, R., Ragasa, C., Pehu, E., 2006. Health enhancing foods opportunities for strengthening the sector in developing countries. In: Agriculture and Rural Development Discussion Paper 30. World Bank, Washington D.C.

Kovács, Z., Benjamins, E., Grau, K., Rehman, A.U., Ebrahimi, M., Czermak, P., 2014. Recent developments in manufacturing oligosaccharides with prebiotic functions. Advances in Biochemical Engineering/Biotechnology 143, 257–295.

Kuhn, R.C., Maugeri Filho, F., 2010. Purification of fructooligosaccharides in an activated charcoal fixed bed column. New Biotechnology 27, 862–869.

Kuhn, R.C., Maugeri Filho, F., Silva, V., Palacio, L., Hernández, A., Prádanos, P., 2010. Mass transfer and transport during purification of fructooligosaccharides by nanofiltration. Journal of Membrane Science 365, 356–365.

Kumar, V., Satyanarayana, T., 2011. Applicability of thermo-alkali-stable and cellulase-free xylanase from a novel thermo-halo-alkaliphilic *Bacillus halodurans* in producing xylooligosaccharides. Biotechnology Letters 33, 2279–2285.

Kumar, H., Salminen, S., Verhagen, H., Rownd, I., Heimbach, J., Bañares, S., Young, T., Nomoto, K., Lalonde, M., 2015. Novel probiotics and prebiotics: road to the market. Current Opinion in Biotechnology 32, 99–103.

Kurakake, M., Masumoto, R., Maguma, K., Kamata, A., Saito, E., Ukita, N., Komaki, T., 2010. Production of fructooligosaccharides by β-fructofuranosidases from *Aspergillus oryzae* KB. Journal of Agricultural and Food Chemistry 58, 488–492.

Lähteenmäki, L., Lampila, P., Grunert, K., Boztug, Y., Ueland, Ø., Åström, A., Martinsdóttir, E., 2010. Impact of health-related claims on the perception of other product attributes. Food Policy 35, 230–239.

Lajolo, F.M., 2002. Functional foods: Latin American perspectives. British Journal of Nutrition 88, S145–S150.

Langlands, S.J., Hopkins, M.J., Coleman, N., Cummings, J.H., 2004. Prebiotic carbohydrates modify the mucosa associated microflora of the human large bowel. Gut 53, 1610–1616.

Laparra, J.M., Sanz, Y., 2010. Interactions of gut microbiota with functional food components and nutraceuticals. Pharmacological Research 61, 219–225.

Lee, Y.K., 2009. Effective dosage for probiotic effects. In: Lee, Y.K., Salminen, S. (Eds.), Handbook of Probiotics and Prebiotics, second ed. Wiley, New Jersey, pp. 52–58.

Lee, Y.K., Salminen, S. (Eds.), 2009. Handbook of Probiotics and Prebiotics, second ed. Wiley, New Jersey.

Lenoir-Wijnkoop, I., Sanders, M.E., Cabana, M.D., Caglar, E., Corthier, G., Rayes, N., Sherman, P.M., Timmerman, H.M., Vaneechoutte, M., Van Loo, J., Wolvers, D.A., 2007. Probiotic and prebiotic influence beyond the intestinal tract. Nutrition Reviews 65, 469–489.

Leszczyński, W., 2004. Resistant starch - classification, structure, production. Polish Journal of Food and Nutrition Sciences 13, 37–50.

Liévin, V., Peiffer, I., Hudault, S., Rochat, F., Brassart, D., Neeser, J.R., Servin, A.L., 2000. Bifidobacterium strains from resident infant human gastrointestinal microflora exert antimicrobial activity. Gut 47, 646–652.

Lilly, D.M., Stillwell, R.H., 1965. Probiotics. Growth promoting factors produced by microorganisms. Science 147, 747–748.

Lim, A., Tan, H.M., 2009. Effects on farm animals. In: Lee, Y.K., Salminen, S. (Eds.), Handbook of Probiotics and Prebiotics, second ed. Wiley, New Jersey, pp. 321–350.

Liong, M.T., 2008. Roles of probiotics and prebiotics in colon cancer prevention: postulated mechanisms and in-vivo evidence. International Journal of Molecular Sciences 9, 854–863.

Liu, Z., Fatehi, P., Jahan, M.S., Ni, Y., 2011. Separation of lignocellulosic materials by combined processes of pre-hydrolysis and ethanol extraction. Bioresource Technology 102, 1264–1269.

Lomax, A.R., Calder, P.C., 2009. Prebiotics, immune function, infection and inflammation: a review of the evidence. British Journal of Nutrition 101, 633–658.

Lorenzoni, A.S.G., Aydos, L.F., Klein, M.P., Ayub, M.A.Z., Rodrigues, R.C., Hertz, P.F., 2015. Continuous production of fructooligosaccharides and invert sugar by chitosan immobilized enzymes: comparison between fluidized and packed bed reactors. Journal of Molecular Catalysis B: Enzymatic 111, 51–55.

Louis, P., Flint, H.J., 2009. Diversity, metabolism and microbial ecology of butyrate-producing bacteria from the human large intestine. FEMS Microbiology Letters 294, 1–8.

Lourens-Hattingh, A., Viljoen, B.C., 2001. Yogurt as probiotic carrier food. International Dairy Journal 11, 1–17.

Luna, R.A., Foster, J.A., 2015. Gut brain axis: diet microbiota interactions and implications for modulation of anxiety and depression. Current Opinion in Biotechnology 32, 35–41.

Lunn, J., Buttriss, J.L., 2007. Carbohydrates and dietary fibre. Nutrition Bulletin 32, 21–64.

Lupton, J.R., 2009. Scientific substantiation of claims in the USA: focus on functional foods. European Journal of Nutrition 48, S27–S31.

Lutz, W., Sanderson, W., Scherbov, S., 2008. The coming acceleration of global population ageing. Nature 451, 716–719.

Maalej-Achouri, I., Guerfali, M., Gargouri, A., Belghith, H., 2009. Production of xylo-oligosaccharides from agro-industrial residues using immobilized *Talaromyces thermophilus* xylanase. Journal of Molecular Catalysis B: Enzymatic 59, 145–152.

Macfarlane, G.T., Macfarlane, S., 2007. Models for intestinal fermentation: association between food components, delivery systems, bioavailability and functional interactions in the gut. Current Opinion in Biotechnology 18, 156–162.

Macfarlane, S., Macfarlane, G.T., Cummings, J.H., 2006. Review article: prebiotics in the gastrointestinal tract. Alimentary Pharmacology and Therapeutics 24, 701–714.

Macfarlane, G.T., Steed, H., Macfarlane, S., 2008. Bacterial metabolism and health-related effects of galacto-oligosaccharides and other prebiotics. Journal of Applied Microbiology 104, 305–344.

Madhukumar, M.S., Muralikrishna, G., 2010. Structural characterisation and determination of prebiotic activity of purified xylo-oligosaccharides obtained from Bengal gram husk (*Cicer arietinum* L.) and wheat bran (*Triticum aestivum*). Food Chemistry 118, 215–223.

Mangas-Sánchez, J., Adlercreutz, P., 2015. Enzymatic preparation of oligosaccharides by transglycosylation: a comparative study of glucosidases. Journal of Molecular Catalysis B: Enzymatic 122, 51–55.

Mao, B., Li, D., Zhao, J., Liu, X., Gu, Z., Chen, Y.Q., Zhang, H., Chen, W., 2014. In vitro fermentation of lactulose by human gut bacteria. Journal of Agricultural and Food Chemistry 62, 10970–10977.

Marshall, B.M., Levy, S.B., 2011. Food animals and antimicrobials: impacts on human health. Clinical Microbiology Reviews 24, 718–733.

Matella, N.J., Dolan, K.D., Lee, Y.S., 2006. Comparison of galactooligosaccharide production in free-enzyme ultrafiltration and in immobilized-enzyme systems. Journal of Food Science 71, C363–C368.

Menrad, K., 2003. Market and marketing of functional food in Europe. Journal of Food Engineering 56, 181–188.

Meyer, D., Bayarri, S., Tárrega, A., Costell, E., 2011. Inulin as texture modifier in dairy products. Food Hydrocolloids 25, 1881–1890.

Miočinović, J., Puđa, P., Radulović, Z., Pavlović, V., Miloradović, Z., Radovanović, M., Paunović, D., 2011. Development of low fat UF cheese technology. Mljekarstvo 61, 33–44.

Mirzaei, H., Pourjafar, H., Homayouni, A., 2012. Effect of calcium alginate and resistant starch microencapsulation on the survival rate of *Lactobacillus acidophilus* La5 and sensory properties in Iranian white brined cheese. Food Chemistry 132, 1966–1970.

Montané, D., Nabarlatz, D., Martorell, A., Torné-Fernández, V., Fierro, V., 2006. Removal of lignin and associated impurities from xylo-oligosaccharides by activated carbon adsorption. Industrial and Engineering Chemistry Research 45, 2294–2302.

Morelli, L., 2007. *In vitro* assessment of probiotic bacteria: from survival to functionality. International Dairy Journal 17, 1278–1283.

Morris, C., Morris, G.A., 2012. The effect of inulin and fructo-oligosaccharide supplementation on the textural, rheological and sensory properties of bread and their role in weight management: a review. Food Chemistry 133, 237–248.

Moulis, C., Vaca Medina, G., Suwannarangsee, S., Monsan, P., Remaud-Simeon, M., Potocki-Veronese, G., 2008. One-step synthesis of isomalto-oligosaccharide syrups and dextrans of controlled size using engineered dextransucrase. Biocatalysis and Biotransformation 26, 141–151.

Mountzouris, K.C., Gilmour, S.G., Rastall, R.A., 2001. Continuous production of oligodextrans via controlled hydrolysis of dextran in an enzyme membrane reactor. Journal of Food Science 675, 1767–1771.

Moure, A., Gullón, P., Domínguez, H., Parajó, J.C., 2006. Advances in the manufacture, purification and applications of xylo-oligosaccharides as food additives and nutraceuticals. Process Biochemistry 41, 1913–1923.

Mumtaz, S., Rehman, S.U., Huma, N., Jamil, A., Nawaz, H., 2008. Xylooligosaccharide enriched yoghurt: physicochemical and sensory evaluation. Pakistan Journal of Nutrition 7, 566–569.

Muñoz, P., Bouza, E., Cuenca-Estrella, M., Eiros, J.M., Pérez, M.J., Sánchez-Somolinos, M., Rincón, C., Hortal, J., Peláez, T., 2005. *Saccharomyces cerevisiae* fungemia: an emerging infectious disease. Clinical Infectious Diseases 40, 1625–1634.

Mussatto, S.I., Mancilha, I.M., 2007. Non-digestible oligosaccharides: a review. Carbohydrate Polymers 68, 587–597.

Mussatto, S.I., Rodrigues, L.R., Teixeira, J.A., 2009. Fructofuranosidase production by repeated batch fermentation with immobilized *Aspergillus japonicas*. Journal of Industrial Microbiology and Biotechnology 36, 923–928.

Muyzer, G., Smalla, K., 1998. Application of denaturing gradient gel electrophoresis (DGGE) and temperature gradient gel electrophoresis (TGGE) in microbial ecology. Antonie van Leeuwenhoek 73, 127–141.

Naessens, M., Cerdobbel, A., Soetaert, W., Vandamme, E.J., 2005. *Leuconostoc* dextransucrase and dextran: production, properties and applications. Journal of Chemical Technology and Biotechnology 80, 845–860.

Nakakuki, T., 2002. Present status and future of functional oligosaccharide development in Japan. Pure and Applied Chemistry 74, 1245–1251.

Nemukula, A., Mutanda, T., Wilhelmi, B.S., Whiteley, C.G., 2009. Response surface methodology: synthesis of short chain fructooligosaccharides with a fructosyltransferase from *Aspergillus aculeatus*. Bioresource Technology 100, 2040–2045.

Neyrinck, A.M., Van Hee, V.F., Piront, N., De Backer, F., Toussaint, O., Cani, P.D., Delzenne, N.M., 2012. Wheat-derived arabinoxylan oligosaccharides with prebiotic effect increase satietogenic gut peptides and reduce metabolic endotoxemia in diet-induced obese mice. Nutrition and Diabetes 2, e28. http://dx.doi.org/10.1038/nutd.2011.24.

Niness, K.R., 1999a. Inulin and oligofructose: what are they? The Journal of Nutrition 129, 1402S–1406S.

Niness, K.R., 1999b. Breakfast foods and the health benefits of inulin and oligofructose. Cereal Foods World 44, 79–81.

Niva, M., 2007. All foods affect health: understandings of functional foods and healthy eating among health-oriented Finns. Appetite 48, 384–393.

Nobre, C., Teixeira, J.A., Rodrigues, R.L., 2012. Fructo-oligosaccharides purification from a fermentative broth using an activated charcoal column. New Biotechnology 29, 395–401.

Nobre, C., Teixeira, J.,A., Rodrigues, L.R., 2015. New trends and technological challenges in the industrial production and purification of fructo-oligosaccharides. Critical Reviews in Food Science and Nutrition 55, 1444–1455.

Nocella, G., Kennedy, O., 2012. Food health claims – what consumers understand. Food Policy 37, 571–580.

Nugent, A.P., 2005. Health properties of resistant starch. British Nutrition Foundation Nutrition Bulletin 30, 27–54.

Ogueke, C.C., Owuamanam, C.I., Ihediohanma, N.C., Iwouno, J.O., 2010. Probiotics and prebiotics: unfolding prospects for better human health. Pakistan Journal of Nutrition 9, 833–843.

Ohama, H., Ikeda, H., Moriyama, H., 2006. Health foods and foods with health claims in Japan. Toxicology 221, 95–111.

Ojansivu, I., Ferreira, C.L., Salminen, S., 2011. Yacon, a new source of prebiotic oligosaccharides with a history of safe use. Trends in Food Science and Technology 22, 40–46.

Ooi, L.G., Liong, M.T., 2010. Cholesterol-lowering effects of probiotics and prebiotics: a review of *in vivo* and *in vitro* findings. International Journal of Molecular Sciences 11, 2499–2522.

O'Sullivan, D.J., 2001. Screening of intestinal microflora for effective probiotic bacteria. Journal of Agricultural and Food Chemistry 49, 1751–1760.

O'Sullivan, L., Murphy, B., McLoughlin, P., Duggan, P., Lawlor, P.G., Hughes, H., Gardiner, G.E., 2010. Prebiotics from marine macroalgae for human and animal health applications. Marine Drugs 8, 2038–2064.

Otieno, D.O., Ahring, B.K., 2012. The potential for oligosaccharide production from the hemicellulose fraction of biomasses through pretreatment processes: xylooligosaccharides (XOS), arabinooligosaccharides (AOS), and mannooligosaccharides (MOS). Carbohydrate Research 360, 84–92.

Ouwehand, F.H., Lahtinen, A., 2009. Legal status and regulatory issues in Europe. In: Lee, Y.K., Salminen, S. (Eds.), Handbook of Probiotics and Prebiotics, second ed. Wiley, New Jersey, pp. 106–110.

Ouwehand, A.C., Salminen, S., 2003. *In vitro* adhesion assays for probiotics and their *in vivo* relevance: a review. Microbial Ecology in Health and Disease 15, 175–184.

Ouwehand, A.C., Kurvinen, T., Rissanen, P., 2004. Use of a probiotic *Bifidobacterium* in a dry food matrix, an in vivo study. International Journal of Food Microbiology 95, 103–106.

Palframan, R., Gibson, G.R., Rastall, R.A., 2003. Development of a quantitative tool for the comparison of the prebiotic effect of dietary oligosaccharides. Letters in Applied Microbiology 37, 281–284.

Pan, Y.C., Lee, W.C., 2005. Production of high-purity isomalto-oligosaccharides syrup by the enzymatic conversion of transglucosidase and fermentation of yeast cells. Biotechnology and Bioengineering 89, 797–804.

Pandiyan, C., Annal Villi, R., Kumaresan, G., Murugan, B., Gopalakrishnamurthy, T.R., 2012. Development of synbiotic ice cream incorporating *Lactobacillus acidophilus* and *Saccharomyces boulardii*. International Food Research Journal 19, 1233–1239.

Panesar, P.S., Kumari, S., Panesar, R., 2013. Biotechnological approaches for the production of prebiotics and their potential applications. Critical Reviews in Biotechnology 33, 345–364.

Parajó, J.C., Garrote, G., Cruz, J.M., Dominguez, H., 2004. Production of xylooligosaccharides by autohydrolysis of lignocellulosic materials. Trends in Food Science and Technology 15, 115–120.

Parsons, C.M., Zhang, Y., Araba, M., 2000. Nutritional evaluation of soybean meals varying in oligosaccharide content. Poultry Science 79, 1127–1131.

Pascal, G., 2008. Prebiotics and food safety. In: Gibson, G.R., Roberfroid, M.B. (Eds.), Handbook of Prebiotics. Taylor & Francis, Boca Raton, pp. 449–470.

Patel, S., Goyal, A., 2012. The current trends and future perspectives of prebiotics research: a review. 3 Biotechnology 2, 115–125.

Payne, A.N., Zihler, A., Chassard, C., Lacroix, C., 2012. Advances and perspectives in *in vitro* human gut fermentation modeling. Trends in Biotechnology 30, 17–25.

Peña-Cardeña, A., Rodríguez-Alegría, M.E., Olvera, C., López Munguía, A., 2015. Synthesis of fructooligosaccharides by IslA4, a truncated inulosucrase from *Leuconostoc citreum*. BMC Biotechnology 15, 2. http://dx.doi.org/10.1186/s12896-015-0116-1.

Pieters, N., Brunt, J., Austin, B., Lyndon, A.R., 2008. Efficacy of in-feed probiotics against *Aeromonas bestiarum* and *Ichthyophthirius multifiliis* skin infections in rainbow trout (*Oncorhynchus mykiss*, Walbaum). Journal of Applied Microbiology 105, 723–732.

Pineiro, M., Stanton, C., 2007. Probiotic bacteria: legislative framework-requirements to evidence basis. Journal of Nutrition 137, 850S–853S.

Pineiro, M., Asp, N.G., Reid, G., Macfarlane, S., Morelli, L., Brunser, O., Tuohy, K., 2008. FAO technical meeting on prebiotics. Journal of Clinical Gastroenterology 42, S156–S159.

Pinelo, M., Jonsson, G., Meyer, A.S., 2009. Membrane technology for purification of enzymatically produced oligosaccharides: molecular and operational features affecting performance. Separation and Purification Technology 70, 1–11.

Polari, L., Ojansivu, P., Ma¨kela¨, S., Eckerman, C., Holmbom, B., Salminen, S., 2012. Galactoglucomannan extracted from spruce (*Picea abies*) as a carbohydrate source for probiotic bacteria. Journal of Agricultural and Food Chemistry 60, 11037–11043.

Prakash, S., Tomaro-Duchesneau, C., Saha, S., Cantor, A., 2011. The gut microbiota and human health with an emphasis on the use of microencapsulated bacterial cells. Journal of Biomedicine and Biotechnology. http://dx.doi.org/10.1155/2011/981214 Article ID 981214, 12 pages.

Probert, H.M., Apajalahti, J.H.A., Rautonen, N., Stowell, J., Gibson, G.R., 2004. Polydextrose, lactitol, and fructo-oligosaccharide fermentation by colonic bacteria in a three-stage continuous culture system. Applied and Environmental Microbiology 70, 4505–4511.

Qiang, X., Lie, C.Y., Bing, W.Q., 2009. Health benefit application of functional oligosaccharides. Carbohydrate Polymers 77, 435–441.

Quigley, E.M., 2010. Prebiotics and probiotics; modifying and mining the microbiota. Pharmacological Research 61, 213–218.

Rabelo, M.C., Honorato, T.L., Gonçalves, L.R.B., Pinto, G.A.S., Rodrigues, S., 2009. Optimization of enzymatic synthesis of isomalto-oligosaccharides production. Journal of Food Biochemistry 33, 342–354.

Rabelo, M.C., Pereira, C.S.M., Rodrigues, S., Rodrigues, A.E., Azevedo, D.C.S., 2013. Chromatographic separation of isomaltooligosaccharides on ion-exchange resins: effect of the cationic form. Adsorption Science and Technology 30, 773–784.

Ranhotra, G.S., Gelroth, J.A., Leinen, S.D., 1999. Resistant starch in selected grain-based foods. Cereal Foods World 44, 357–359.

Rastall, R.A., Gibson, G.R., 2015. Recent developments in prebiotics to selectively impact beneficial microbes and promote intestinal health. Current Opinion in Biotechnology 32, 42–46.

Rastall, R.A., Maitin, V., 2002. Prebiotics and synbiotics: towards the next generation. Current Opinion in Biotechnology 13, 490–496.

Rastall, R.A., Gibson, G.R., Gill, H.S., Guarner, F., Klaenhammer, T.R., Pot, B., Reid, G., Rowland, I.R., Sanders, M.E., 2005. Modulation of the microbial ecology of the human colon by probiotics, prebiotics and synbiotics to enhance human health: an overview of enabling science and potential applications. FEMS Microbiology Ecology 52, 145–152.

Rastall, R.A., 2010. Functional oligosaccharides: application and manufacture. Annual Review of Food Science and Technology 1, 305–339.

Rehman, H., Hellweg, P., Taras, D., Zentek, J., 2008. Effects of dietary inulin on the intestinal short chain fatty acids and microbial ecology in broiler chickens as revealed by denaturing gradient gel electrophoresis. Poultry Science 87, 783–789.

Reid, G., Gibson, G.R., Gill, H.S., Klaenhammer, T.R., Rastall, R.A., Rowland, I.R., Sanders, M.E., 2006. Use of genetically modified microbes for human health. Microbial Ecology in Health and Disease 18, 75–76.

Renuka, B., Kulkarni, S.G., Vijayanand, P., Prapulla, S.G., 2009. Fructooligosaccharide fortification of selected fruit juice beverages: effect on the quality characteristics. LWT - Food Science and Technology 42, 1031–1033.

Ringø, E., Olsen, R.E., Gifstad, T.Ø., Dalmo, R.A., Amlund, H., Hemre, G.I., Bakke, A.M., 2010. Prebiotics in aquaculture: a review. Aquaculture Nutrition 16, 117–136.

Roberfroid, M., Gibson, G.R., Hoyles, L., McCartney, A.L., Rastall, R., Rowland, I., Wolvers, D., Watzl, B., Szajewska, H., Stahl, B., Guarner, F., Respondek, F., Whelan, K., Coxam, V., Davicco, M.J., Léotoing, L., Wittrant, Y., Delzenne, N.M., Cani, P.D., Neyrinck, A.M., Meheust, A., 2010. Prebiotic effects: metabolic and health benefits. British Journal of Nutrition 104, S1–S63.

Roberfroid, M.B., 2000a. Concepts and strategy of functional food science: the European perspective. American Journal of Clinical Nutrition 71, 1660S–1664S.

Roberfroid, M.B., 2000b. Prebiotics and probiotics: are they functional foods? American Journal of Clinical Nutrition 71, 1682S–1687S.

Roberfroid, M.B., 2001. Prebiotics: preferential substrates for specific germs? American Journal of Clinical Nutrition 73, 406s–409s.

Roberfroid, M., 2002. Functional food concept and its application to prebiotics. Digestive and Liver Disease 34, S105–S110.

Roberfroid, M., 2007. Prebiotics: the concept revisited. The Journal of Nutrition 137, 830S–837S.

Roberfroid, M.B., 2008. Prebiotics: concepts, definition, criteria, methodologies and products. In: Gibson, G.R., Roberfroid, M. (Eds.), Handbook of Prebiotics. Taylor & Francis, Boca Raton, pp. 39–68.

Rodríguez-Cabezas, M.E., Camuesco, D., Arribas, B., Garrido-Mesa, N., Comalada, M., Bailón, E., Cueto-Sola, M., Utrilla, P., Guerra-Hernández, E., Pérez-Roca, C., Gálvez, J., Zarzuelo, A., 2010. The combination of fructooligosaccharides and resistant starch shows prebiotic additive effects in rats. Clinical Nutrition 29, 832e–839e.

Roger, L.C., McCartney, A.L., 2010. Longitudinal investigation of the faecal microbiota of healthy full-term infants using fluorescence *in situ* hybridization and denaturing gradient gel electrophoresis. Microbiology 156, 3317–3328.

Roselli, M., Finamore, A., Britti, M.S., Mengheri, E., 2006. Probiotic bacteria *Bifidobacterium animalis* MB5 and *Lactobacillus rhamnosus* GG protect intestinal Caco-2 cells from the inflammation-associated response induced by enterotoxigenic *Escherichia coli* K88. British Journal of Nutrition 95, 1177–1184.

Rossi, M., Corradini, C., Amaretti, A., Nicolini, M., Pompei, A., Zanoni, S., Matteuzzi, D., 2005. Fermentation of fructooligosaccharides and inulin by Bifidobacteria: a comparative study of pure and fecal cultures. Applied and Environmental Microbiology 71, 6150–6158.

Ruanglek, V., Sripranga, R., Ratanaphan, N., Tirawongsaroj, P., Chantasigh, D., Tanapongpipat, S., Pootanakit, K., Eurwilaichitr, L., 2007. Cloning, expression, characterization, and high cell-density production of recombinant endo-1,4-xylanase from *Aspergillus niger* in *Pichia pastoris*. Enzyme and Microbial Technology 41, 19–25.

Ruller, R., Deliberto, L., Lopes Ferreira, T., Ward, R.J., 2008. Thermostable variants of the recombinant xylanase A from *Bacillus subtilis* produced by directed evolution show reduced heat capacity changes. Proteins: Structure, Function, and Bioinformatics 70, 1280–1293.

Rycroft, C.E., Jones, M.R., Gibson, G.R., Rastall, R.A., 2001. A comparative *in vitro* evaluation of the fermentation properties of prebiotic oligosaccharides. Journal of Applied Microbiology 91, 878–887.

Saad, N., Delattre, C., Urdaci, M., Schmitter, J.M., Bressollier, P., 2013. An overview of the last advances in probiotic and prebiotic field. LWT - Food Science and Technology 50, 1–16.

Saarela, M., Virkajärvi, I., Alakomi, H.L., Sigvart-Mattila, P., Mättö, J., 2006. Stability and functionality of freeze-dried probiotic *Bifidobacterium* cells during storage in juice and milk. International Dairy Journal 16, 1477–1482.

Saavedra, J.M., Degnan, 2009. Legal status and regulatory issues in the United States of America. In: Lee, Y.K., Salminen, S. (Eds.), Handbook of Probiotics and Prebiotics, second ed. Wiley, New Jersey, pp. 111–123.

Sajilata, M.G., Singhal, R.S., Kulkarni, P.R., 2006. Resistant starch – a review. Comprehensive Reviews in Food Science and Food Safety 5, 1–17.

Salazar, P., García, M.L., Selgas, M.D., 2009. Short-chain fructooligosaccharides as potential functional ingredient in dry fermented sausages with different fat levels. International Journal of Food Science and Technology 44, 1100–1107.

Salminen, S., von Wright, A., Morelli, L., Marteau, P., Brassart, D., de Vos, W.M., Fondén, R., Saxelin, M., Collins, K., Mogensen, G., Birkeland, S.E., Mattila-Sandholm, T., 1998. Demonstration of safety of probiotics – a review. International Journal of Food Microbiology 4, 93–106.

Salminen, S., Ouwehand, A., Benno, Y., Lee, Y.K., 1999. Probiotics: how should they be defined? Trends in Food Science and Technology 10, 107–110.

Samanta, A.K., Senani, S., Kolte, A.P., Sridhar, M., Sampath, K.T., Jayapal, N., Devi, A., 2012. Production and *in vitro* evaluation of xylooligosaccharides generated from corn cobs. Food and Bioproducts Processing 90, 466–474.

Samanta, A.K., Jayapal, N., Senani, S., Kolte, A.P., Sridhar, M., 2013. Prebiotic inulin: useful dietary adjuncts to manipulate the livestock gut microflora. Brazilian Journal of Microbiology 44, 1–14.

Sanders, M.E., 2003. Probiotics: considerations for human health. Nutrition Reviews 61, 91–99.

Sanders, M.E., 2008. Probiotics: definition, sources, selection and uses. Clinical Infectious Diseases 46, S58–S61.

Sangeetha, P.T., Ramesh, M.N., Prapulla, S.G., 2005. Recent trends in the microbial production, analysis and application of fructooligosaccharides. Trends in Food Science and Technology 16, 442–457.

Santos-Moriano, P., Fernández-Arrojo, L., Rodríguez-Colinas, B., Ballesteros, A., Plou, F.J., 2015. Enzymatic synthesis of fructooligosaccharides that stimulate the gut microbiota. Anales de la Real Academia Nacional de Farmacia 81, 48–62.

Sanz, M.L., Polemis, N., Morales, V., Corzo, N., Drakoularakou, A., Gibson, G.R., Rastall, R.A., 2005. In vitro investigation into the potential prebiotic activity of honey oligosaccharides. Journal of Agricultural and Food Chemistry 53, 2914–2921.

Sarbini, S.R., Rastall, R.A., 2011. Prebiotics: metabolism, structure, and function. Functional Food Reviews 3, 93–106.

Satokari, R., Rantanen, R., Pitkälä, K., Salminen, S., 2010. Prebiotics and probiotics in infant formulae. In: Cho, S.S., Finocchiaro, E.T. (Eds.), Handbook of Prebiotics and Probiotics Ingredients. Health Benefits and Food Applications. CRC Press, Boca Raton, pp. 293–339.

Saulnier, D.M.A., Gibson, G.R., Kolida, S., 2008. *In vitro* effects of selected synbiotics on the human faecal microbiota composition. FEMS Microbiology Ecology 66, 516–527.

Saulnier, D.M.A., Spinler, J.K., Gibson, G.R., Versalovic, J., 2009. Mechanisms of probiosis and prebiosis: considerations for enhanced functional foods. Current Opinion in Biotechnology 20, 135–141.

Savini, M., Cecchini, C., Verdenelli, M.C., Silvi, S., Orpianesi, C., Cresci, A., 2010. Pilot-scale production and viability analysis of freeze-dried probiotic bacteria using different protective agents. Nutrients 2, 330–339.

Saxelin, M., 2008. Probiotic formulations and applications, the current probiotics market, and changes in the marketplace: a European perspective. Clinical Infectious Diseases 46, S76–S79.

Scantlebury Manning, T., Gibson, G.R., 2004. Prebiotics. Best Practice and Research in Clinical Gastroenterology 18, 287–298.

Scholz-Ahrens, K.E., Ade, P., Marten, B., Weber, P., Timm, W., Asil, Y., Glüer, C.C., Schrezenmeir, J., 2007. Prebiotics, probiotics, and synbiotics affect mineral absorption, bone mineral content, and bone structure. The Journal of Nutrition 137, 838S–846S.

Schrezenmeir, J., de Vrese, M., 2001. Probiotics, prebiotics, and synbiotics – approaching a definition. The American Journal of Clinical Nutrition 73, 361S–364S.

Schuster-Wolff-Bühring, R., Fischer, L., Hinrichs, J., 2010. Production and physiological action of the disaccharide lactulose. International Dairy Journal 20, 731–741.

Seifert, S., Watzl, B., 2007. Inulin and oligofructose: review of experimental data on immune modulation. The Journal of Nutrition 137, 2563S–2567S.

Sharma, A., Trivedi, S., 2015. Evaluation of in vitro probiotic potential of phytase-producing bacterial strain as a new probiotic candidate. International Journal of Food Science and Technology 50, 507–514.

Shimizu, M., 2012. Functional food in Japan: current status and future of gut-modulating food. Journal of Food and Drug Analysis 20, 213–216.

Shoaf-Sweney, K.D., Hukins, R.W., 2009. Adherence, anti-adherence and oligosaccharides preventing pathogens from sticking to the host. Advances in Food and Nutrition Research 55, 101–161.

Silk, D.B.A., Davis, A., Vulevic, J., Tzortzis, G., Gibson, G.R., 2009. Clinical trial: the effects of a trans-galactooligosaccharide prebiotic on faecal microbiota and symptoms in irritable bowel syndrome. Alimentary Pharmacology and Therapeutics 29, 508–518.

Simon, O., 2005. Micro-organisms as feed additives – probiotics advances in pork production. Advances in Pork Production 16, 161–167.
Singh, R.S., Singh, R.P., 2010. Production of fructooligosaccharides from inulin by endoinulinases and their prebiotic potential. Food Technology and Biotechnology 48, 435–450.
Siró, I., Kápolna, E., Kápolna, B., Lugasi, A., 2008. Functional food. Product development, marketing and consumer acceptance – a review. Appetite 51, 456–467.
Slavin, J., 2013. Fiber and prebiotics: mechanisms and health benefits. Nutrients 5, 1417–1435.
Snydman, D.R., 2008. The safety of probiotics. Clinical Infectious Diseases 46, S104–S111.
Song, S.K., Beck, B.R., Kim, D., Park, J., Kim, J., Kim, H.D., Ringø, E., 2014. Prebiotics as immunostimulants in aquaculture: a review. Fish and Shellfish Immunology 40, 40–48.
Sorokulova, I.B., Pinchuk, I.V., Denayrolles, M., Osipova, I.G., Huang, J.M., Cutting, S.M., Urdaci, M.C., 2008. The safety of two *Bacillus* probiotic strains for human use. Digestive Diseases and Sciences 53, 954–963.
Sreeja, V., Prajapati, J.B., 2013. Probiotic formulations: application and status as pharmaceuticals – a review. Probiotics and Antimicrobial Proteins 5, 81–91.
Stanton, C., Desmond, C., Coakley, M., Collins, J.K., Fitzgerald, G., Ross, R.P., 2003. Challenges facing development of probiotic-containing functional foods. In: Farnworth, E.R. (Ed.), Handbook of Fermented Functional Foods. CRC Press, Boca Ratón, pp. 28–58.
Steer, T., Carpenter, H., Tuohy, K., Gibson, G.R., 2000. Perspectives on the role of the human gut microbiota and its modulation by pro- and prebiotics. Nutrition Research Reviews 13, 229–254.
Sun, H.J., Yoshida, S., Park, N.H., Kusakabe, I., 2002. Preparation of (1-4)-β-D-xylooligosaccharides from an acid hydrolysate of cotton-seed xylan: suitability of cotton-seed xylan as a starting material for the preparation of (1-4)-β-D-xylooligosaccharides. Carbohydrate Research 337, 657–661.
Tamine, A., 2005. Probiotic Dairy Products. Blackwell Publishing, Oxford.
Taniguchi, H., 2005. Carbohydrate active enzymes for the production of oligosaccharides. In: Hou, C.T. (Ed.), Handbook of Industrial Biocatalysts, Chapter 20. CRC Press, Boca Raton, pp. 1–15.
Tanriseven, A., Aslan, Y., 2005. Immobilization of Pectinex Ultra SP-L to produce fructooligosaccharides. Enzyme and Microbial Technology 36, 550–554.
Taper, H.S., Roberfroid, M.B., 2002. Inulin/oligofructose and anticancer therapy. British Journal of Nutrition 87, S283–S286.
Teitelbaum, J.E., Walker, W.A., 2002. Nutritional impact of pre- and probiotics as protective gastrointestinal organisms. Annual Review of Nutrition 22, 107–138.
Tien, W., Chang, H., 2009. Legal status and regulatory issues in Asia. In: Lee, Y.K., Salminen, S. (Eds.), Handbook of Probiotics and Prebiotics, second ed. Wiley, New Jersey, pp. 95–105.
Topping, D.L., Fukushima, M., Bird, A.R., 2003. Resistant starch as a prebiotic and synbiotic: state of the art. Proceedings of the Nutrition Society 62, 171–176.
Tuohy, K.M., Finlay, R.K., Wynne, A.G., Gibson, G.R., 2001. A human volunteer study on the prebiotic effects of HP-inulin-faecal bacteria enumerated using fluorescent in situ hybridisation (FISH). Anaerobe 7, 113–118.
Tuohy, K.M., Probert, H.M., Smejkal, C.W., Gibson, G.R., 2008. Using probiotics and prebiotics to improve gut health. Drug Discovery Today 8, 692–700.
Tymczyszyn, E.E., Gerbino, E., Illanes, A., Gómez-Zavaglia, A., 2011. Galacto-oligosaccharides as protective molecules in the preservation of *Lactobacillus delbrueckii* subsp. *bulgaricus*. Cryobiology 62, 123–129.
Tymczyszyn, E.E., Santos, M.I., Costa, M.C., Illanes, A., Gómez-Zavaglia, A., 2014. History, synthesis, properties, applications and regulatory issues of prebiotic oligosaccharides. In: Gil, M.H. (Ed.), Carbohydrates Applications in Medicine. Research Signpost, Kerala, pp. 127–154.

Tzortzis, G., Goulas, A.K., Gee, J.M., Gibson, G.R., 2005. A novel galactooligosaccharide mixture increases the bifidobacterial population numbers in a continuous *in vitro* fermentation system and in the proximal colonic contents of pigs *in vivo*. Journal of Nutrition 135, 1726–1731.

Uauy, R., Monteiro, C.A., 2004. The challenge of improving food and nutrition in Latin America. Food and Nutrition Bulletin 25, 175–182.

Ungar, P.S., Teaford, M.F., 2002. Human Diet: Its Origin and Evolution. Greenwood Publishing, Westport Connecticut.

van der Maarel, M.J.E.C., van der Veen, B., Uitdehaag, J.C.M., Leemhuis, H., Dijkhuizen, L., 2002. Properties and applications of starch-converting enzymes of the α-amylase family. Journal of Biotechnology 94, 137–155.

Van Loo, J., Vancraeynest, D., 2008. Prebiotics and animal nutrition. In: Gibson, G.R., Roberfroid, M. (Eds.), Handbook of Prebiotics. Taylor & Francis, Boca Raton, pp. 421–436.

Van Loo, J., Coussement, P., de Leenheer, L., Hoebregs, H., Smits, G., 1995. On the presence of inulin and oligofructose as natural ingredients in the Western diet. Critical Reviews in Food Science and Nutrition 35, 525–552.

Van Loo, J., 2006. Prebiotics: a nutritional concept gaining momentum in modern nutrition. In: Gibson, G.R. (Ed.), Food Science and Technology Bulletin Functional Foods, vol. 2. IFIS Publishing, Reading, pp. 83–100.

van Zanten, G.C., Knudsen, A., Röytiö, H., Forssten, S., Lawther, K., Blennow, A., Lahtinen, S.J., Jakobsen, M., Svensson, B., Jespersen, L., 2012. The effect of selected synbiotics on microbial composition and short-chain fatty acid production in a model system of the human colon. PLoS One 7 (10), e47212. http://dx.doi.org/10.1371/journal.pone.0047212.

Vaňková, K., Polakovič, M., 2010. Optimization of single-column chromatographic separation of fructooligosaccharides. Process Biochemistry 45, 1325–1329.

Vaňková, K., Onderková, Z., Antošová, M., Polakovič, M., 2008. Design and economics of industrial production of fructooligosaccharides. Chemical Papers 62, 375–381.

Vanneste, J., De Ron, S., Vandecruys, S., Soare, S.A., Darvishmanesh, S., Van der Bruggen, B., 2011. Techno-economic evaluation of membrane cascades relative to simulated moving bed chromatography for the purification of mono- and oligosaccharides. Separation and Purification Technology 80, 600–609.

Vasiljevic, T., Shah, N.P., 2008. Probiotics – from Metchnikoff to bioactives. International Dairy Journal 18, 714–728.

Vázquez, M.J., Garrote, G., Alonso, J.L., Domínguez, H., Parajó, J.C., 2005. Refining of autohydrolysis liquors for manufacturing xylooligosaccharides: evaluation of operational strategies Bioresource Technology 96, 889–896.

Vega-Paulino, R.J., Zúñiga-Hansen, M.E., 2012. Potential application of commercial enzyme preparations for industrial production of short-chain fructooligosaccharides. Journal of Molecular Catalysis B: Enzymatic 76, 44–51.

Vegas, R., Moure, A., Domínguez, H., Parajó, J.C., Alvarez, J.R., Luque, S., 2008. Evaluation of ultra- and nanofiltration for refining soluble products from rice husk xylan. Bioresource Technology 99, 5341–5351.

Venema, K., van den Abbeele, P., 2013. Experimental models of the gut microbiome. Best Practice and Research Clinical Gastroenterology 27, 115–126.

Verdonk, J.M.A.J., Shim, S.B., van Leeuwen, P., Verstegen, M.W.A., 2005. Application of inulin-type fructans in animal feed and pet food. British Journal of Nutrition 93, S125–S138.

Vernazza, C.L., Gibson, G.R., Rastall, R.A., 2006. Carbohydrate preference, acid tolerance and bile tolerance in five strains of *Bifidobacterium*. Journal of Applied Microbiology 100, 846–853.

Vester, B., Fahey, G.C., 2010. Prebioics and probiotics in companion animal nutrition. In: Cho, S.S., Finocchiaro, E.T. (Eds.), Handbook of Prebiotics and Probiotics Ingredients. Health Benefits and Food Applications. CRC Press, Boca Raton, pp. 355–380.

Vulevic, J., Rastall, R.A., Gibson, G.R., 2004. Developing a quantitative approach for determining the in vitro prebiotic potential of dietary oligosaccharides. FEMS Microbiology Letters 236, 153–159.

Wang, Y., 2009. Prebiotics: present and future in food science and technology. Food Research International 42, 8–12.

Weststrate, J.A., van Poppel, G., Verschuren, P.M., 2002. Functional foods, trends and future. British Journal of Nutrition 88, S233–S235.

Wickens, K., Black, P.N., Stanley, T.V., Mitchell, E., Fitzharris, P., Tannock, G.W., Purdie, G., Crane, J., 2008. A differential effect of 2 probiotics in the prevention of eczema and atopy: a double-blind, randomized, placebo-controlled trial. Journal of Allergy and Clinical Immunology 122, 788–794.

Wouters, R., 2010. Inulin. In: Imeson, A. (Ed.), Food Stabilisers, Thickeners and Gelling Agents. Wiley-Blackwell, Oxford, pp. 180–197.

Xu, X., Wang, Z., Zhang, W., 2015. The human microbiota associated with overall health. Critical Reviews in Biotechnology 35, 129–140.

Yan, F., Polk, D., 2006. Probiotics as functional food in the treatment of diarrhea. Current Opinion in Clinical Nutrition and Metabolic Care 9, 717–721.

Yang, Y.L., Wang, J.H., Teng, D., Zhang, F., 2008. Preparation of high-purity fructo-oligosaccharides by *Aspergillus japonicus* β-fructofuranosidase and successive cultivation with yeast. Journal of Agricultural and Food Chemistry 56, 2805–2809.

Yoshikawa, J., Amachi, S., Shinoyama, H., Fujii, T., 2008. Production of fructooligosaccharides by crude enzyme preparations of β-fructofuranosidase from *Aureobasidium pullulans*. Biotechnology Letters 30, 535–539.

Yun, J.W., Lee, M.G., Song, S.K., 1994. Continuous production of isomaltooligosaccharides from maltose syrup by immobilized cells of permeabilized *Aureobasidium pullulans*. Biotechnology Letters 16, 1145–1150.

Zhang, Z.G., Yi, Z.L., Pei, X.Q., Wu, Z.L., 2010. Improving the thermostability of *Geobacillus stearothermophilus* xylanase XT6 by directed evolution and site-directed mutagenesis. Bioresource Technology 101, 9272–9278.

Zoetendal, E.G., Akkermans, A.D.L., De Vos, W.M., 1998. Temperature gradient gel electrophoresis analysis of 16S rRNA from human fecal samples reveals stable and host-specific communities of active bacteria. Applied and Environmental Microbiology 64, 3834–3859.

Chapter 3

Lactose-Derived Nondigestible Oligosaccharides and Other High Added-Value Products

C. Vera and A. Illanes

3.1 PRODUCTION OF LACTOSE-DERIVED NONDIGESTIBLE OLIGOSACCHARIDES: CHEMICAL OR ENZYMATIC?

Oligosaccharides have been the subject of considerable research in the last two decades because of their valuable functional properties (Bucke, 1996). Health benefits associated with their consumption is now well documented, which, associated with their good physicochemical properties, has placed them in a privileged position within the functional food business (Grajek et al., 2005; Siró et al., 2008; Boye, 2015).

Oligosaccharides can be produced either by chemical or enzymatic-catalyzed reactions. Chemical synthesis of oligosaccharides is a complex task that has been pursued for a long time with limited success, despite the fact that significant advances in the field have occurred with the development of automated solid-phase carbohydrate synthesis (Plante et al., 2001; Hsu et al., 2011) and computer-controlled design of monosaccharide sequences (Seeberger and Werz, 2007). A major constraint is that six-carbon monosaccharides have five hydroxyl groups that can form links with different stereochemistry so the number of possible combinations is exceedingly high, even for short-chain oligosaccharides. Therefore, protection and deprotection reactions, reaction conditions, and donor leaving groups and acceptors have to be carefully selected for obtaining the desired regio- and stereochemistry of the glycosidic bond to be formed (Hanson et al., 2004). For instance, in the case of galacto-oligosaccharide (GOS), the anomeric configuration and the type of glycosidic bond are important for the prebiotic activity (Hernández–Hernández et al., 2012). The anomeric configuration on carbon atom 1 leads to a glycosidic bond resistant to the attack by digestive enzymes in the upper intestinal tract of mammals (Vester and Fahey, 2012). GOS preparations with proven prebiotic effect are the result of transgalactosylation reactions catalyzed by microbial β-galactosidases, and therefore most glycoside bonds are β 1–4 and β 1–6, and also β 1–2 and β 1–3 but

Lactose-Derived Prebiotics. http://dx.doi.org/10.1016/B978-0-12-802724-0.00003-2

to a much lesser extent (Torres et al., 2010). Chemical synthesis of oligosaccharides with such glycosidic bonds can be formed by two main routes: Fischer transgalactosylation and Koenigs–Knorr glycosylation. As stated earlier, both methods are complex and laborious requiring several protection/deprotection steps, and exhaustive purification of the resulting product. Fischer transgalactosylation is simpler and less expensive than Koenigs–Knorr reaction, but the use of strong mineral acids complicates product purification (Rather and Mishra, 2013). Besides, in Fischer transgalactosylation different bonds with α and β configurations and anhydrous sugars are formed, with the consequent reduction in product yield (Huh et al., 1990, 1991). Koenigs–Knorr reaction has the advantage of producing mostly glycosidic bonds in the β configuration, but its use at productive scale is limited because of using heavy metal salts as promoters, making it unsuitable for food applications (Rather and Mishra, 2013).

Enzymatic synthesis appears then as a preferred alternative owing to its high selectivity, mild reaction conditions, and process simplicity because no protection reactions are required (Planas and Faijes, 2002; Bojarová and Křen, 2009). Interestingly, this is a quite different situation than occurring in the synthesis of peptides, where chemical solid-phase synthesis is a mature, well-established technology with which enzymatic synthesis with proteases hardly competes despite its potential advantages (Guzmán et al., 2007).

Enzymatic synthesis of oligosaccharides can be performed either with glycosyl transferases (E.C. 2.4.) or glycosidases (E.C. 3.2.1.). Glycosyl transferases are the enzymes that catalyze glycosidic bond formation in cell metabolism. According to the nature of the donor molecule, they are divided into three subgroups: Leloir-type, non-Leloir type, and transglycosidases. The former two require activated sugars: sugar nucleotides, like UDP, in the case of Leloir-type, and sugar phosphates in the case of non-Leloir type; the latter do not require activated sugars but can also display hydrolytic activity (Plou et al., 2007). Glycosyl transferases are intracellular complex coenzyme-requiring enzymes of limited availability, therefore in practice, they are poor candidates for the synthesis of oligosaccharides at production level. Much more interesting is the use of glycosidases, which are robust, frequently extracellular proteins, not requiring coenzymes, and readily available because of their conventional use as hydrolases in several industrial applications (Monsan and Paul, 1995).

Glycosidases can be used for oligosaccharide synthesis in thermodynamically or kinetically controlled reactions. The former is based simply on the reversion of the hydrolytic reaction, while the latter implies the use of activated glycosyl donors (Faijes and Planas, 2007). In both cases, low water activity in the reaction medium will be required to either favor the displacement of the equilibrium in favor of glycosidic bond formation or to selectively depress the nucleophilic attack to the enzyme–glycoside complex by water, respectively. In practice, this can be hard to achieve for oligosaccharide synthesis due to the limited solubility of sugars in nonconventional (nonaqueous) media. However, the use of (very) high sugar concentrations will reduce water activity as well and depress the hydrolytic potential of the enzyme favoring glycosidic bond formation (Vera et al., 2012). Kinetically controlled synthesis is a preferred option because higher conversions and productivities can be obtained. However, maximum conversion is limited since hydrolysis will inevitably take place, so product yields will rarely be over 40%, corresponding to limiting substrate conversions not exceeding 60%.

An interesting approach to reduce the hydrolytic potential of glycosidases is their molecular redesign, which led to the concept of glycosynthases as a novel family of engineered glycosidases with attenuated hydrolytic potential; this strategy involves site-directed mutagenesis in which the catalytic nucleophile in the active center is replaced by a noncatalytic residue to arrest the hydrolytic activity (Hancock et al., 2006; Strohmeier et al., 2011). However, the resulting mutant enzymes are usually poorly active so that high amounts of enzyme or extended reaction times are required, precluding for the moment their use as process catalysts (Perugino et al., 2004).

Lactose is a plentiful substance even though its presence in nature is restricted to mammal's milk (Martinez-Ferez et al., 2006). Lactose is the energy-supplying component in milk, and, as such, it is a key nutrient. However, in order to be metabolized it must be hydrolyzed to its monosaccharide components, which occurs by the enzyme-catalyzed hydrolysis with β-galactosidase at the intestinal brush border membrane (Van Beers et al., 1995). Activity of intestinal β-galactosidase (lactase-phlorizin hydrolase, or simply lactase) varies considerably according to ethnicity (Vesa et al., 2000) and decreases with age (Heitlinger et al., 1991) so that an important fraction of the world population suffers from lactose intolerance or maldigestion. As analyzed in Chapter 1, Lactose: Production and Upgrading, lactose-containing whey is the major byproduct of cheese production, its output exceeding current uses. Therefore, copious amounts of whey are disposed to treatment plants or underutilized. Among the many alternatives for lactose upgrading (see Section 1.2), its use as substrate for producing different types of oligosaccharides is quite appealing because of the significant added value of such products, which is much higher than those attainable by more conventional production strategies. Even though lactose and whey permeate are still the most significant outputs of the cheese industry, high added-value lactose-derived products have an already-established market with an annual growth rate around 10% (Lifran et al., 2009). Therefore, production of nondigestible oligosaccharide (NDO) and other high added-value products from lactose are powerful technological options for its upgrading.

3.2 PRESENT STATUS AND TECHNOLOGICAL PERSPECTIVES

Lactosucrose, lactitol, lactobionic acid, tagatose, lactulose, and GOS are lactose-derived compounds considered as health promoters. Among them, GOS (Torres et al., 2010) and lactulose (Mayer et al., 2010) are properly considered as prebiotics, while the rest fulfill some but not all the requirements for being considered as such, despite their use as health-promoting food components

(Saarela et al., 2003; Schaafsma, 2008) and also as nonnutritive and noncariogenic sweeteners (Gänzle, 2012; Shankar et al., 2013).

3.2.1 Galacto-Oligosaccharides and Lactulose

GOS and lactulose are undoubtedly the most significant lactose-derived NDOs. Having different properties and fields of applications, they share their scientifically sustained prebiotic condition. However, their production technologies are quite different: GOS is exclusively produced by enzyme biocatalysis and no chemical counterpart exists; while in the case of lactulose, the current technology of production is chemical catalysis, biocatalysis not being yet competitive despite evident advantages in terms of sustainable chemistry. The production of GOS and lactulose will be thoroughly reviewed in Chapter 4, Enzymatic Production of Galacto-Oligosaccharides and Chapter 5, Enzymatic Production of Lactulose, respectively.

3.2.2 Lactosucrose

Lactosucrose (O-β-D-galactopyranosyl-(1–4)-O-α-D-glucopyranosyl-(1–2)-β-D-fructo-furanoside) (see Fig. 3.1), is a trisaccharide sweetener that is also endowed with some prebiotic properties: it is poorly digestible and stimulates beneficial gut microbiota (Rycroft et al., 2001; Gänzle et al., 2008); some other health-related effects, like the increase in calcium absorption, have also been claimed (Teramoto et al., 2006). It has been used as an ingredient in low-calorie foods and claimed to be effective against irritable bowel syndrome (Teramoto et al., 1996). More insight on lactosucrose can be found in Chapter 6, Enzymatic Production of Other Lactose-Derived Prebiotic Candidates, Section 6.1.

3.2.3 Lactitol

Lactitol (4-O-β-D-galactopyranosyl-D-glucitol) (see Fig. 3.1) is a nonnatural sugar alcohol obtained from lactose that shares some of the properties of prebiotics since it is not metabolized nor absorbed in the small intestine and stimulates colonic *Bifidobacteria* and *Lactobacilli* (Finney et al., 2007), although to a lesser extent than other NDOs (Kneifel, 2000). A proteomic analysis of *Lactobacillus acidophilus* has revealed the enzymes involved in lactitol metabolism that may explain its prebiotic action (Majumder et al., 2011). Lactitol is mildly sweet (about 35% of the sweetening power of sucrose) with a bland taste and poorly caloric (about 2 kcal/g); it has been used as a sugar substitute in several diet food products (Kummel and Brokx, 2001). Aside from the food sector, it has been used also as moistener, plasticizer, and adhesive (Audic et al., 2003). Lactitol, as lactulose, has been used also as a laxative and in the treatment of hepatic encephalopathy (Patil et al., 1987; Clausen and Mortensen, 1997). More insight on lactitol can be found in Chapter 6, Enzymatic Production of Other Lactose-Derived Prebiotic Candidates, Section 6.2.

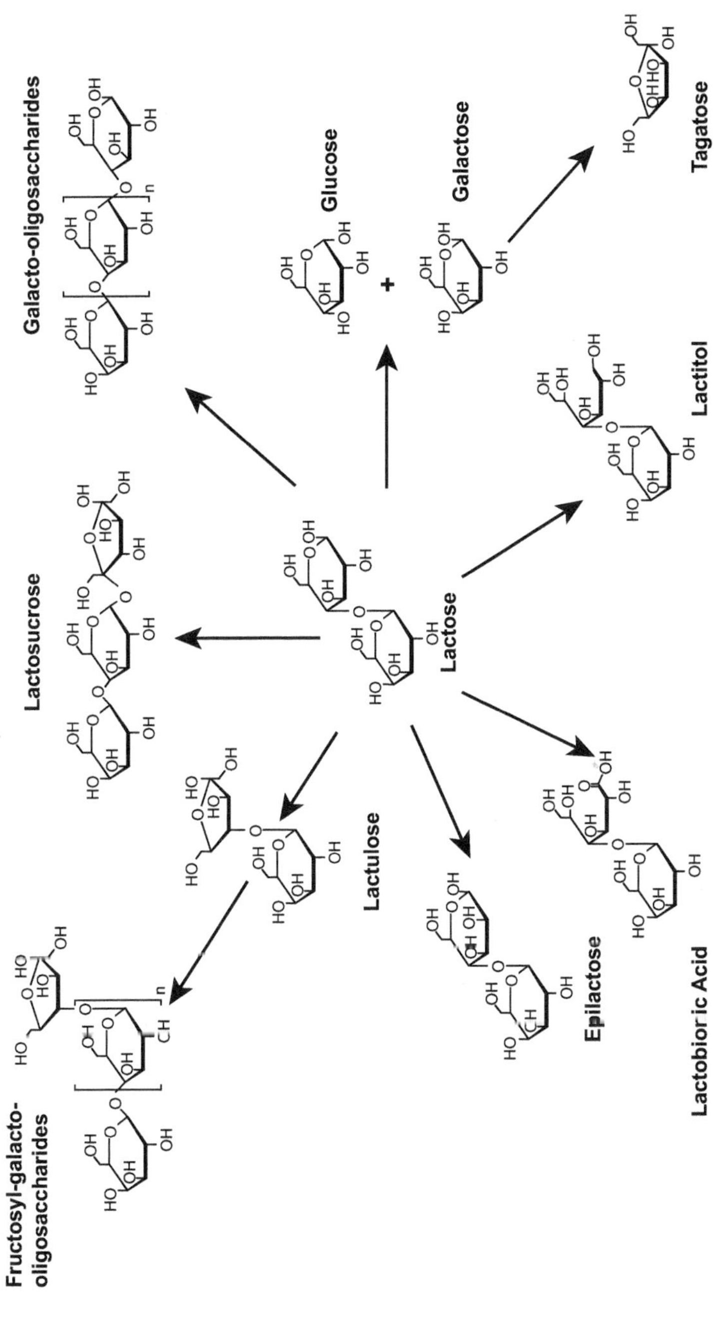

FIGURE 3.1 Chemical substructure of lactose-derived nondigestible oligosaccharides.

3.2.4 Lactobionic Acid

Lactobionic acid (4-O-β-galactopyranosyl-D-gluconate) (see Fig. 3.1) is a polyhydroxy acid having several applications of industrial relevance (Alonso et al., 2013). It is an aldonic acid industrially produced by the chemical oxidation of the aldehyde group in the glucose residue of lactose in an energy-intensive process requiring expensive catalysts and extensive downstreaming to remove unwanted side products (Kuusisto et al., 2007); dehydrogenation of lactose at high pH catalyzed by noble metals is used for the commercial production of lactobionic acid (Chia et al., 2008). Applications of lactobionic acid are many and in most cases unrelated to the field of functional foods; its most important use is for the preservation of organs for transplant where it acts as a chelating agent and impermeant, reducing oxidative damage and cell swelling (Karam et al., 2005). More insight on lactobionic acid can be found in Chapter 6, Enzymatic Production of Other Lactose-Derived Prebiotic Candidates, Section 6.3.

3.2.5 Tagatose

D-Tagatose (see Fig. 3.1) is a galactose isomer whose chemical structure is close to D-fructose, being its epimer (Yoshida et al., 2007). It is found only in trace amounts in some fruits and exudates from plants like *Sterculia* (tropical chestnut) and produced in very small quantities during the heat processing of dairy products (Mendoza et al., 2005). Therefore, tagatose production must follow a synthetic route from readily available raw materials. Initially, tagatose was produced by the chemical isomerization of galactose using calcium catalysts (Beadle et al., 1992). Low specificity of the chemical reaction made necessary a complex purification process and waste treatment of offensive compounds, which was also a problem (Oh, 2007). Therefore, more specific biotransformations were sought as technological alternatives. Tagatose is mostly produced as a low-calorie sweetener since its achievement of generally recognized as safe (GRAS) status in 2000 (Vastenavond and François, 2013). It has a sweetening power similar to sucrose but is poorly absorbable, and its caloric content is 70% lower (Levin, 2002). Tagatose has a very low glycemic index (Donner et al., 1999; Saunders et al., 1999), is acceptable for diabetics (Lu et al., 2008), and does not contribute to dental plaque formation (Lu and Levin, 2002; US FDA, 2003), which allows its use as a sweetener in toothpaste, mouthwash, and lipstick. Many other uses for tagatose have been claimed, and certainly its health-promoting effects have captured attention. Being poorly absorbable in the gastrointestinal tract and fermented by *Lactobacillus* producing short-chain fatty acids (Lærke and Jensen, 1999), it is a candidate prebiotic despite not fulfilling all requirements for being considered as such (Fujimaru et al., 2012); actually it has very low effect on *Bifidobacteria* (Bertelsen et al., 2001) and its storage stability may be not enough for effective administration (Dobbs and Bell, 2010). However, its health-promoting effects will certainly contribute to a better appreciation of its use as a noncaloric sweetener. More insight on tagatose production can be found in Chapter 6, Enzymatic Production of Other Lactose-Derived Prebiotic Candidates, Section 6.4.

3.2.6 Epilactose and Fructosyl-Galacto-Oligosaccharides

Epilactose (see Fig. 3.1) and fructosyl-galacto-oligosaccharides are also high added-value products derived from lactose, as analyzed in Chapter 6, Enzymatic Production of Other Lactose-Derived Prebiotic Candidates, Sections 6.5 and 6.6, respectively.

3.3 MARKET OUTLOOK: PAST, PRESENT, AND FUTURE PERSPECTIVES

First reports on the use of oligosaccharides as bifidogenic factors were published in Japan in the 1970s and referred to inulin and dextrans (Yazawa et al., 1978).

3.3.1 Galacto-Oligosaccharides

The use of GOS was mostly driven by the necessity of developing formulas favoring the implantation of *Bifidobacteria* in the gut of newborn babies, so mimicking the effect of human milk oligosaccharides in mother's milk (Yazawa et al., 1978; Mitsuhashi et al., 1982; Minami et al., 1983, 1985). Large-scale production of GOS began in Japan around 1985 (Playne and Crittenden, 2009), spreading then to Europe (United Kingdom and The Netherlands), Australia, New Zealand, and the United States (Torres et al., 2010; Nguyen and Haltrich, 2013; Austin et al., 2014). Massive incorporation of GOS into foods started in Japan in 1990, which very much coincided with the approval in 1991 by the Japanese government of a special category of foods "food for specified health use" (FOSHU) claiming a special health benefit (Taniguchi, 2005). GOS was one of the first NDOs recognized as FOSHU (Sako et al., 1999), and by 1995 several food products containing GOS were already sold in Japan. In Europe, the first product containing GOS, a dessert called Umer, was launched in 1997 in The Netherlands (Spherix Consulting, 2010). In 1996 the Dutch government approved the use of Vivinal GOS in foods and qualified it as a common (non-novel) food ingredient, so GOS was exempted from the Novel Food Regulation 258/907 EC that was adopted, and could be used as a common ingredient in foods both in the European Union and the United Kingdom (Spherix Consulting, 2010; van Leusen et al., 2014). A big impulse for GOS use in the European Union occurred in 2001 when Danone introduced it in their products for infants, claiming a health-related benefit in the immune system; this was a powerful driving force for other Western companies to include GOS in their product lines for infants and toddlers (van Leusen et al., 2014). GOS was introduced in the United States in 1998, when Vivinal under the trade brand Elix'or was approved by the US Food and Drug Administration (FDA) as a dietary supplement ingredient (Spherix Consulting, 2010). Almost a decade later, in

2007, Friesland Foods Domo (Vivinal) and later on in 2010 Yakult (Oligomate) applied for GRAS status for their products (Environ International Corporation, 2007; Spherix Consulting, 2010); since then, four other companies have applied for GRAS status. Since GOS adoption in the United States was a later event than in the European Union and in the Asia–Pacific Region, the US market is still incipient, but it is expected to grow at a higher pace in forthcoming years, representing an important market for prebiotics in general.

Market evolution assessment for GOS-containing products was recently done (van Leusen et al., 2014), clearly showing an actively growing market. From 1998 to 2011 the number of companies including GOS in their product lines increased from 1 to 111. Application of GOS in new product lines has been led by companies having used them and gained experience in their handling, but the number of newcomers is gradually increasing. Most GOS-containing products have been launched in the Asia–Pacific Region (57%) and the European Union (37%), and correspond mostly to infant formulas and baby foods.

World market size for GOS is difficult to estimate since in some important producing countries, like Australia, most of the production is domestic and used in closed industrial operations (Playne and Crittenden, 2009). Table 3.1 presents an estimate of global GOS production showing a moderate increase in the period from 1995 to 2007. By 2007 the market for GOS was estimated between US$135 and 170 million (Paterson and Kellam, 2009). Since then, production has increased considerably reaching an estimate of 94,000 tons in 2013, which represents 16% of total prebiotic market, and is secondly most demanded after inulin, which represents 42% of such market. Market size for GOS is expected to double by the end of the present decade as a result of the increasing global demand for functional foods, mostly in India and China, with an estimate of US$1010 million, corresponding to 17.5% of the forecasted total prebiotic market (Grand View Research, 2014, 2015).

TABLE 3.1 Global Estimated Production of Galacto-Oligosaccharides

Year	Tons/Year	References
1995	15,000	Sako et al. (1999)
2004	12–14,000	Playne and Crittenden (2009)
2007	25,000	Paterson and Kellam (2009)
2009	21,000	Playne and Crittenden (2009)
2010[a]	33–44,000	Paterson and Kellam (2009)
2013	94,100	Grand View Research (2015)
2020[a]	175,700	Grand View Research (2015)

[a]*Forecasted value.*

The main GOS-producing companies are listed in Table 3.2, identifying their products and the minimum declared GOS content. As can be seen, there are only a few companies producing and trading GOS and some producing them for their own use (Torres et al., 2010).

GOS preparations are presented in the form of concentrated syrups (70–75 °Brix) and in powder form. Minimum declared GOS content varies between 55% and 70% of total carbohydrates. High-purity products containing 90% or more of GOS, like Purimune (GTC Nutrition) and TOS-100 (Yakult-Honsha) have been recently discontinued.

In recent years new GOS manufacturers have entered in the market. In 2009 the Great Ocean Ingredients plant (Allansford, Australia) was opened, having

TABLE 3.2 Galacto-Oligosaccharides (GOS) Manufacturers

Company	Product(s)	Minimum Declared GOS Content
Friesland Foods Domo (Netherlands)	Vivinal GOS Syrup	59% on dry matter
	Vivinal GOS Easy Drying Syrup	72% on dry matter
	Vivinal GOS Powder	69% (w/w)
	Vivinal GOS Powder WPC	29% (w/w)
	Vivinal GOS Powder Maltodextrin	29% (w/w)
Yakult Honsha (Japan)	Oligomate 55N (syrup)	55% on dry matter
	Oligomate 55NP (Powder)	55% (w/w)
Nissin Sugar Manufacturing Company (Japan)	Cup Oligo H-70	70% on dry matter
	Cup Oligo P	70% (w/w)
Clasado Bioscience (Malta)	Bimuno Powder	50% (w/w)
	Bimuno Travelaid	32% (w/w)
	Bimuno Immunaid	35% (w/w)
	Bimuno Ibaid	35% (w/w)
Ingredion Inc. (South Korea)	Bioligo GL 5700 IMF (syrup)	57% on dry matter
BaolinBao Biology Co., Ltd. (China)	Galacto-oligosaccharide (powder)	27% on dry matter
		57% on dry matter
		70% on dry matter

a production capacity of 15,000 tons of Vivinal GOS per year. The facility is a joint venture between FrieslandCampina Domo (FCD) and the Australia-based Warrnambool Cheese & Butter Factory (WCB). The project required a total investment of €30 million, shared by FCD (62%) and WCB (32%) (Foodprocessing-technology.com, 2009). On the other hand, in 2014 Dairy Crest (UK) announced a £20 million capital investment at its Davidstow site to manufacture 13,500 tons of GOS per year. Dairy Crest will enter into a joint venture to produce GOS with Fayrefield Foods Limited (UK) that will provide access to the required enzyme and other intellectual properties. Fonterra (New Zealand) will supply lactose and will also take responsibility for marketing and selling this product. Commercial production is expected to begin in 2016 or early in 2017 (Dairy Crest, 2015).

3.3.2 Lactulose

Lactulose has been known since the early work of Montgomery and Hudson (1930), but it was after the pioneering work of Friedrich Petuely when lactulose was for the first time considered an interesting compound to modulate the gut microbiota in infants and improve the human intestinal peristalsis (Petuely, 1957, 1963). According to Playne and Crittenden (2009), lactulose began to be marketed in infant's milk by Morinaga (Japan) in 1960. In these early days, lactulose was sold only in a syrup format because of its hygroscopic character and difficult crystallization; the powder obtained by spray-drying turned easily into a syrup (Nagasawa et al., 1971). Since then, the processes for lactulose production have been steadily improving by the development of new catalysts and enhanced purification processes (Carobbi et al., 1985; Deya et al., 1991; Krumbholz and Dorscheid, 1991; Shimamura et al., 1993). Currently, lactulose is marketed as a syrup and in the form of anhydrous crystals.

Lactulose market value, similar to the GOS market, was estimated between US$160 and 176 million in 2007 and between US$180 and 190 million in 2010 (Paterson and Kellam, 2009). Recently, lactulose production for medical and pharmaceutical uses was estimated at 12,000 tons/year (Westhoff et al., 2014), being more than three-quarters of the overall lactulose production destined for the food industry. Table 3.3 shows the evolution of the estimated global production of lactulose.

Table 3.4 shows the major lactulose manufacturers. For many years, the lactulose market was controlled by Solvay Pharmaceuticals and Morinaga Milk Industry, the former being mainly devoted to medical application and the later toward the food industry (Playne and Crittenden, 2009). Nowadays, the scenario has changed because Solvay Pharmaceuticals sold its Solactis nutrition division, which is now operating independently as Solactis (Starling, 2013).

TABLE 3.3 Estimated Global Production of Lactose

Year	Tons/Year	References
1995	20,000	Sako et al. (1999)
2004	40,000	Playne and Crittenden (2009)
2007	30–40,000	Paterson and Kellam (2009)
2009	45,000	Playne and Crittenden (2009)
2010[a]	36,000–42,000	Paterson and Kellam (2009)

[a]*Forecasted value.*

TABLE 3.4 Major Lactulose Manufacturers

Company	Product(s)	Minimum Declared GOS Content
Solactis (France)	Solactis Powder	74% w/w
	Solactis Syrup	52% on dry matter
	Vitaloa Syrup (animal feed)	66% w/v
	Vitaloa Powder (animal feed)	74% w/w
Fresenius Kabi (Austria and Italy)	Lactulose liquid standar	67% w/v
	Lactulose liquid special	up to 70% w/v
	Purified lactulose solution	636–703 g/L
	Lactulose crystals	97% (w/w)
Inalco (Italy)	Lactulose crystals	–
	Lactulose solution (USP)	50%
DaniPharm (Denmark)	Danilax	67% w/v
Dephasaar Chem.-Pharm. Fabrik GmbH (Germany)	Lactulose-MIP	65% w/v
Morinaga Milk Industry Co. (Japan and Germany)	MLS-50	50% w/v
	MLC-A	98% w/v (anhydride form)

3.3.3 Lactosucrose

Lactosucrose has been known since Avigard published his pioneering article dealing with the use of levosucrases in the synthesis of new sugars and analogs (Avigard, 1957). The production of this oligosaccharide started in 1990, manufactured by Ensuiko Sugar Refining Co. and Hayashibara Shoji Inc (Ensuiko Sugar Refining Co., 2015; Hayashibara Shoji Inc., 2015). Both processes are probably based on the use of the β-fructofuranosidase I from *Arthrobacter* sp. K-1, discovered by Fujita et al. (1990a,b), which is able to efficiently transfer a fructosyl residue of sucrose to a wide range of sugars and alcohols. Taniguchi (2005) described in detail the process for the production of lactosucrose: "5000 kg of a equimolar solution of lactose and sucrose at 40% w/w is contacted with 750,000 IU of β-fructofuranosidase from *Arthrobacter* sp. K-1 at 55°C for 10 h in the presence of a yeast without invertase activity. The reaction mixture is heated to stop the reaction, then it is subjected to decoloring, desalting and concentration or spray-dried to obtain a powder." The product obtained by this process has a purity ranging between 45% and 55%. The earliest study about the effect of lactosucrose in the intestinal microbiota appeared in 1992 (Iwagaki et al., 1991). Similar to lactulose, lactosucrose is a compound difficult to dry because of the hygroscopicity of the resulting powder. For this reason, the development of an efficient technique of crystallization of lactosucrose was sought by the industry. In 2005 Hayashibara Shoji Inc. launched a new process for the production of high-purity lactosucrose (Hayashibara Shoji Inc., 2015). According to Playne and Crittenden (2009), lactosucrose production has been developed in Japan in a cooperative manner by Ensuiko, Hayashibara, and Biofermin Pharmaceutical.

Table 3.5 shows the estimated production of lactosucrose, whose market value was US$28 million in 2007 and estimated between US$32 and 37 million in 2010 (Paterson and Kellam, 2009). It is mainly marketed in Japan, where it is included in the formulation of more than 30 food products. Lactosucrose is also marketed in the United States and in Europe, being considered an emerging prebiotic (Díez-Municio et al., 2014).

TABLE 3.5 Global Lactosucrose Estimated Production

Year	Tons/Year	References
1995	1600	Sako et al. (1999)
2004	3000	Playne and Crittenden (2009)
2007	3500	Paterson and Kellam (2009)
2010[a]	4000–4600	Paterson and Kellam (2009)

[a]*Forecasted value.*

TABLE 3.6 Major Lactosucrose Manufacturers

Company	Product	Minimum Declared Lactosucrose Content
Hayashibara Shoji Inc. (Japan)	Nyūka-origo (乳果オリゴ)	
	Nyūka-origo 700 (syrup)	68–73% on dry matter
	Nyūka-origo 550 (syrup)	55–60% on dry matter
	Nyūka-origo 400 (syrup)	42–47% on dry mater
	Nyūka-origo 700 Powder	68–73% on dry matter
Ensuiko Sugar Refining Co. (Japan)	Origonōkage (オリゴのおかげ)	
	Origonōkage Syrup[a]	30% and 40% on dry matter
	Origonōkage Powder[a]	30% and 40% on dry matter
	Origonōkage LS 40L (syrup)	42% on dry matter
	Origonōkage LS 55L (syrup)	55% on dry matter
	Origonōkage LS 55P (Powder)	55% on dry matter
	Origonōkage LS 90P	90% on dry matter

[a]*For personal consumption.*

The two major producers of lactosucrose are Hayashibara Shoji Inc. and Ensuiko Sugar Refining Co. Both companies produce syrups with a total solid content ranging between 72% and 78% w/w and a lactosucrose content from 40% to 70% on dry matter. These companies produce also a high-purity lactosucrose powder. Ensuiko Sugar Refining Co. has a product line of low-caloric sweeteners based on lactosucrose, which are sold in drugstores and grocery stores. The latter products are marketed advertising a bifidogenic effect and a promotion of calcium absorption. Major lactosucrose manufacturers are identified in Table 3.6.

3.3.4 Lactitol

Lactitol was obtained for the first time in 1920 by J.B. Senderens. The bases for lactitol industrial production were settled in 1937 by Wolfrom et al. (1938). In their developed process, a solution of lactose (30–40% w/w) is hydrogenated at temperatures ranging from 70 to 130°C and at pressures between 30 and 74 atm in the presence of nickel Raney catalyst. According to Playne and Crittenden (2009), lactitol was first marketed in 1987. Lactitol has been mainly employed as an alternative to lactulose in the treatment of constipation and hepatic encephalopathy, and as a sweetening agent (Faruqui and Joshi, 2012). In the latter application, lactitol is used as a bulk sweetener in total or partial replacement of sucrose, having a caloric content about half of sucrose and not causing increase in blood glucose content (Grembecka, 2015).

Currently lactitol is produced and marketed by DuPont Danisco as crystalline lactitol and milled-crystalline lactitol both in anhydrous and monohydrate form. Also, this company manufactures OsmoAid Lactitol, a gentle laxative, suitable for babies, pregnant women, and elderly people (Danisco, 2015). Lactitol is also marketed by B-Food Science (Japan) under the following brands: Lactitol LC-0 and Lactitol LC-1, supplied as white granular powders, and Lactitol LC 50 M, a white powder supplied in 20-kg bags. A few years ago, PURAC Biochem (The Netherlands) discontinued Lacty from its product portfolio, which was a lactitol monohydrate powder with several presentations. At present, Mitsubishi Shoji Food Tech Co. Ltd. is the only supplier of lactitol monohydrate powder sold under the brand Milchem. Global production of lactitol was estimated between 10,000 and 11,000 tons in 2007; a slight increase was expected for 2010 with production between 11,000 and 12,000 tons being forecasted. The market value corresponding to those years was estimated between US$40 to 50 million and between US$44 to 55 million, respectively (Paterson and Kellam, 2009). According to Playne and Crittenden (2009), lactitol had a price between US$2 and 2.5/kg, being much cheaper than GOS and lactosucrose but similar to lactulose syrup.

In terms of regulation, lactitol has been approved as a food additive (sweetening agent E966) in the European Union (Annex II of Regulation [EC] No 1333/2008) and its use was also approved in Israel, Japan, Norway, Sweden, and Switzerland; in the United States, lactitol has a self-affirmed GRAS status (Zacharis, 2012). In 2006 Danisco requested from the FDA an exemption from food allergen labeling for their products lactitol anhydrous and lactitol-monohydrate. The request was denied because Danisco did not properly demonstrate the absence of milk protein in these products (US FDA, 2006). However, the FDA has accepted the claims of low caloric content (Calorie Control Council, 2015) and noncariogenic carbohydrate sweetener (US FDA, 2015) for labeling purposes.

3.3.5 Lactobionic Acid

Lactobionic acid is formed by a galactose moiety linked to a gluconic acid molecule via an ether linkage. It was first obtained by Fisher and Meyer (1889) from the oxidation of lactose. Nowadays lactobionic acid is synthesized at industrial level by dehydrogenation of lactose at high pH over a noble metal catalyst (Playne and Crittenden, 2009). The major drawback of this process is the energy cost, which is high and represents a significant share of the operation costs (Quitmann et al., 2014). Several alternatives have been proposed to overcome this problem, such as biocatalytic oxidation, electrochemical oxidation, and heterogeneous catalysis (Gutierrez et al., 2012).

Lactobionic acid has numerous applications in food, pharmaceutical, and cosmetic industries due to its antioxidant, chelating, humectant, and emulsifying properties (Gutierrez et al., 2012). Two of the most important uses of lactobionic acid are the formulation of erythromycin lactobionate, which is 50–100 times more soluble than erythromycin, and its addition as chelating and antioxidant agent in the preservation of organs prior to transplantation (Nakano et al., 2010). Other interesting uses of this molecule are as calcium carrier and as a healing, humectant, antiaging, and keratinizing compound in skin care products (Nakano et al., 2010; Gutierrez et al., 2012). In addition, the FDA has approved the use of lactobionic acid as a firming agent in dry pudding mixes (US FDA Code of Federal Regulation, Title 21, 2015), owing to the protein coagulation ability of this acid.

Among the manufacturers of lactobionic acid, Global Lactobionic Acid Ltd. (India), Reliable Biopharmaceutical Corporation (USA), Bio-sugar (China), Hubei YuanCheng SaiChuang Technology Co. (China), and Hangzhou Dayangchem Co., Ltd. (China) stand out. Also, Sandoz Private Limited, the generic pharmaceutical division of Novartis, has been involved in the production of calcium lactobionate. In spite of the several applications of lactobionic acid, its market is still rather small and only significant in Asian countries. Its annual production was estimated at 1000 tons per year (Paterson and Kellam, 2009).

3.3.6 Tagatose

Tagatose is a six-carbon sugar, which corresponds to the keto form of galactose; it is usually defined as an epimer of D-fructose that only differs in the position of the hydroxyl group at carbon atom 4 (Bertelsen et al., 1999). Tagatose is a rare natural carbohydrate found in some fruits (apples, oranges, and pineapple) and sterilized milk at levels ranging from 1 to 3.5 g/kg (Petersen, 2006). As presented in Section 3.2.5, tagatose has several appealing attributes as a food additive where it can be used as sweetener, texturizer, stabilizer, humectant, formulation aid (NutraSource Inc., 2010), and, being a highly Maillard reactive reducing sugar, it can be used also as flavoring agent (Petersen, 2006). Health-promoting effect is also a bonus of tagatose as food additive.

The commercial production of tagatose has come a long way since the early 1990s, when the American company Biospherics Inc. patented a chemical method for tagatose manufacturing (Beadle et al., 1991, 1992). The patented process involved the production of galactose by hydrolysis of food grade lactose in a packed bed reactor with immobilized *Aspergillus oryzae* β-galactosidase, chromatographic separation of galactose from the hydrolysate, and chemical isomerization of galactose in the presence of calcium hydroxide. Impurities such as residual galactose, talose, aldol condensation products, and other minor components were removed by demineralization and chromatographic fractionation, followed by tagatose crystallization as the final step (McColl, 2001; Petersen, 2006).

In 1996 Arla Foods bought the rights to produce tagatose from Spherix and created a 50/50 joint venture with the German company Nordzücker, named SweetGredients, to manufacture it as a sweetener under the brand name Gaio Tagatose (Arla, 2003). Five years later, tagatose was given GRAS status by the FDA (McColl, 2001). The entrance of tagatose into the market was, however, troublesome. In 2002 Spherix carried out legal proceedings against Arla Food claiming that the Danish company had taken a long time to bring tagatose into the market, causing an economic detriment to Spherix because the royalties were linked to tagatose sales (Nutraingredients-USA.com, 2002). One year later, they reached a settlement and tagatose finally entered the US market (Donner et al., 2010). In 2005 tagatose was approved as a novel food ingredient in the European Union (Lu et al., 2008). Unfortunately, one year later SweetGredients decided to discontinue the manufacturing of D-tagatose because the chemical process of synthesis was too expensive (Vastenavond et al., 2012) and a market development justifying more investment was not envisaged. In 2007 Nutrilab NV, a subsidiary of the Belgian company Damhert, purchased the remaining supply of tagatose from Arla Food in order to settle down a market for it, as a previous step to manufacturing using an enzymatic process (Halliday, 2008). Five years later, once the market was set, Nutrilab NV opened a facility near Rome, Italy, that was able to manufacture 10,000 tons/year of tagatose. The plant entered into operation producing 800 tons/year of crystalline tagatose under the brand name Tagatesse. However, production of 2500 tons/year is soon expected to satisfy the product demand coming from PepsiCo and Yoplait (Flanderstoday, 2013). It is believed that Nutrilab NV is a two-step enzymatic process in which galactose is obtained by hydrolyzing whey lactose using a psychrophilic β-galactosidase from *Pseudoalteromonas haloplanktis* (Université de Liège, 2010), and then galactose is isomerized by galactose isomerase, which is a genetically modified *Geobacillus stearothermophilus* arabinose isomerase. The reaction product is then purified employing customary operations: active carbon adsorption, ion exchange chromatography, simulated moving bed chromatography, crystallization, and drying (Vastenavond and François, 2013).

The vicissitudes did not cease. In 2008 Spherix Inc. announced an agreement with Inalco SpA (Italy) to supply the tagatose used in the Spherix Phase 3 clinical trial for the treatment of type 2 diabetes. Three years later, Spherix Inc. presented a complaint to the US Securities and Exchange Commission against Inalco SpA arguing that they refused to supply the tagatose previously paid for by Spherix Inc. Inalco SpA refused to provide a promised bank guarantee and shut down their D-tagatose production facilities (Spherix Inc., 2011). A few months later, both companies reached a settlement whereby Inalco agreed to supply 8.5 tons of tagatose to Spherix plus a lump payment (Yahoo Finance, 2012).

In 2010 the South Korean company CJ CheilJedang obtained the GRAS status from the FDA for its tagatose-based product (Nutrasource, 2010). Since 2012, CJ CheilJedang has been manufacturing tagatose by an enzymatic process under the brand Beksul Tagatose. The Korea Food and Drug Administration

(KFDA) approved health claims for its antihyperglycemic effect and tagatose has been honored as one of the World Class Products of Korea by the Ministry of Knowledge Economy of that country (CheilJedang, 2013). CJ CheilJedang has announced its intention to export Beksul Tagatose to Europe, India, and the United States (CheilJedang, 2012). The manufacturing process employed by this company is the following: a solution of lactose (20% w/w) is subjected to acid hydrolysis (140°C and 0.1% w/w of sulfuric acid) by 1 h. Next, calcium carbonate is added to the hydrolyzed syrup to remove the precipitants, especially sulfate residues. The neutralized syrup is then mixed with manganese chloride and the resulting solution is contacted with nonviable cells of *Corynebacterium glutamicum* immobilized in calcium alginate gel beads harboring an L-arabinose isomerase derived from *Thermotoga neapolitana*. Isomerization of galactose to tagatose is conducted at 75°C during 4–8 h. Then, color and impurities are removed using activated carbon and ion-exchange chromatography, respectively. The purified solution is concentrated to 60 °Brix and then tagatose is chromatographically separated from the other carbohydrates. Tagatose raffinate is concentrated from 10 to 15 °Brix to 75 to 80 °Brix and then fed to a continuous crystallizer. Crystalline tagatose is separated by centrifugation, crystals are washed, and finally drum-dried (NutraSource, 2010). A schematic representation of the process is shown in Fig. 3.2.

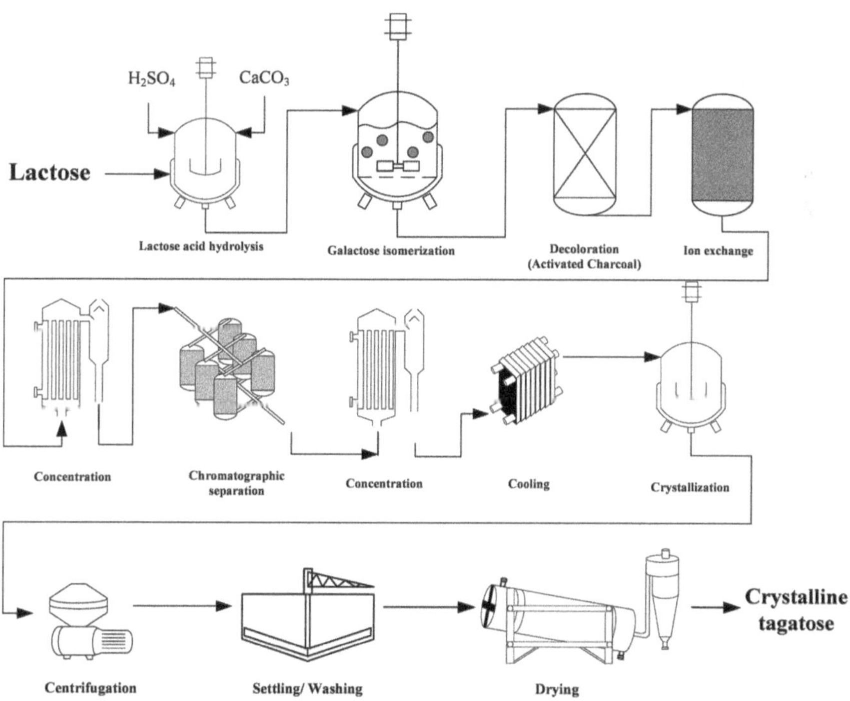

FIGURE 3.2 Process diagram of D-tagatose production.

The future of tagatose is promising due to its remarkable properties: high sweetness, low caloric content, low glycemic index, prebiotic potential, and desirable physicochemical properties for food product formulation. Paterson and Kellam (2009) estimated the tagatose market at 900 tons/year in 2009, with a market value of US$1.3 million. The same authors forecasted a significant growth in the tagatose market size for 2015, expecting an output of 20,000–30,000 tons/year with a market value of US$37.5 million.

REFERENCES

Alonso, S., Rendueles, M., Díaz, M., 2013. Bio-production of lactobionic acid: current status, applications and future prospects. Biotechnology Advances 31, 1275–1291.

Annex II of Regulation (EC) No 1333/2008, 2008. Community List of Food Additives Approved for Use in Foods and Conditions of Use. E 966Lactitol. Available online in: https://webgate.ec.europa.eu/sanco_foods/main/?event=substance.view&identifier=323.

Arla, 2003. The World's First Production of Gaio Tagatose Gets Off the Ground. Available online on: http://www.arla.com/about-us/news-and-press/2003/pressrelease/the-world-s-first-production-of-gaio-tagatose-gets-off-the-ground-760328/.

Audic, J.L., Chaufer, B., Dauphin, G., 2003. Non-food applications of milk components and dairy co-products: a review. Lait 83, 417–438.

Austin, S., Bénet, T., Michaud, J., Cuany, D., Rohfritsch, P., 2014. Determination of β-galactooligosaccharides by liquid chromatography. International Journal of Analytical Chemistry:768406 10 pp.

Avigard, G., 1957. Enzymatic synthesis and characterization of a new trisaccharide, α-lactosyl-β-fructofuranoside. Journal of Biological Chemistry 229, 121–129.

Beadle, J.R., Saunders, J.P., Wajda, T.J., 1991. Process for Manufacturing Tagatose. United States Patent 5,002,612.

Beadle, J.R., Saunders, J.P., Wajda, T.J., 1992. Process for Manufacturing Tagatose. United States Patent 5,078,796.

Bertelsen, H., Jensen, B.B., Buemann, B., 1999. D-Tagatose – a novel low-calorie bulk sweetener with prebiotic properties. In: Corti, A. (Ed.), IUFoST World Conference on Low-Calorie Sweeteners. Karger, Basel, pp. 98–109.

Bertelsen, H., Andersen, H., Tvede, M., 2001. Fermentation of D-tagatose by human intestinal bacteria and dairy lactic acid bacteria. Microbial Ecology in Health and Disease 13, 87–95.

Bojarová, P., Křen, V., 2009. Glycosidases: a key to tailored carbohydrates. Trends in Biotechnology 27, 199–209.

Boye, J.I., 2015. Nutraceutical and Functional Food Processing Technology. Wiley Blackwell, Oxford.

Bucke, C., 1996. Oligosaccharide synthesis using glycosidases. Journal of Chemical Technology and Biotechnology 67, 217–220.

Calorie Control Council, 2015. Lactitol. Available online in: http://polyol.org/facts-about-polyols/lactitol/.

Carobbi, R., Miletti, S., Franci, V., 1985. Process for Preparing Lactulose in the Form a Syrup or a Crystalline Product. United States Patent 4,536,221.

CheilJedang, C.J., 2012. Tagatose Is Now Ready to Enter European Market at Its Full Scale. Available online in: http://english.cj.net/company/news/press/press_view.asp?ps_idx=3927&NO=345.

CheilJedang, C.J., 2013. Sustainability Report 2013. Available online on: http://static.globalreporting.org/report-pdfs/2013/e0b216d6af19e117c8fe07601bedd6c4.pdf.

Chia, Y.N., Latusek, M.P., Holles, J.H., 2008. Catalytic wet oxidation of lactose. Industrial and Engineering Chemistry Research 47, 4049–4055.

Clausen, M.R., Mortensen, P.B., 1997. Lactulose, disaccharides and colonic flora. Drugs 53, 930–942.

Dairy Crest, 2015. Dairy Crest Update. Available online in: http://www.dairycrest.co.uk/~/media/Files/D/Dairy-Crest-Group/reports-and-presentations/2015/analyst-presentation. pdf.

Danisco, 2015. Lactitol. Available online in: http://www.danisco.com/product-range/sweeteners/lactitol/.

Deya, E., Takahashi, K., Ikeuchi, Y., 1991. Production Process of High-Purity Lactulose Syrup. United States Patent 5,034,064.

Diez-Municio, M., Herrero, M., Olano, O., Moreno, F.J., 2014. Synthesis of novel bioactive lactose-derived oligosaccharides by microbial glycoside hydrolases. Microbial Biotechnology 7, 315–331.

Dobbs, C.M., Bell, L.N., 2010. Storage stability of tagatose in buffer solutions of various compositions. Food Research International 43, 382–386.

Donner, T.W., Wilber, J.F., Ostrowski, D., 1999. D-Tagatose, a novel hexose: acute effects on carbohydrate tolerance in subjects with and without type 2 diabetes. Diabetes, Obesity and Metabolism 1, 285–291.

Donner, T.W., Magderb, L.S., Zarbalian, K., 2010. Dietary supplementation with D-tagatose in subjects with type 2 diabetes leads to weight loss and raises high-density lipoprotein cholesterol. Nutrition Research 30, 801–806.

Ensuiko Sugar Refining Co., 2015. オリゴのおかげの歩み. Available in: http://www.ensuiko.co.jp/product/oligostep.html.

Environ International Corporation, 2007. Determination for the Use of Galacto-Oligosaccharides (GOS) in Foods and in Foods for Term Infants. GRAS Notification for Galacto-Oligosaccharides (GOS). Presented: October 18.

Faijes, M., Planas, A., 2007. In vitro synthesis of artificial polysaccharides by glycosidases and glycosynthases. Carbohydrate Research 342, 1581–1594.

Faruqui, A.A., Joshi, C., 2012. Lactitol: a review of its use in the treatment of constipation. International Journal of Recent Advances in Pharmaceutical Research 2, 1–5.

Finney, M., Smullen, J., Foster, H.A., Brokx, S., Storey, D.M., 2007. Effects of low doses of lactitol on faecal microflora, pH, short chain fatty acids and gastrointestinal symptomology. European Journal of Nutrition 46, 307–314.

Fisher, E., Meyer, J., 1889. Oxydation des Milchzuckers. Berichte der Deutschen Chemischen Gesellschaft 22, 361–364.

Flanderstoday, 2013. The Sweet Taste of Success. Available online on: http://www.flanderstoday.eu/business/sweet-taste-success.

Foodprocessing-technology.com, 2009. Great Ocean Ingredients Plant, Australia. http://www.foodprocessing-technology.com/projects/greatoceanplant/.

Fujimaru, T., Park, J.H., Lim, J., 2012. Sensory characteristics and relative sweetness of tagatose and other sweeteners. Journal of Food Science 77, S323–S328.

Fujita, K., Hara, K., Hashimoto, H., Kitahata, S., 1990a. Purification and some properties of β-fructofuranosidase I from *Arthrobacter* sp. K-1. Agricultural and Biological Chemistry 54, 913–919.

Fujita, K., Hara, K., Hashimoto, H., Kitahata, S., 1990b. Transfructosylation catalyzed by β-fructofuranosidase I from *Arthrobacter* sp. K-1. Agricultural and Biological Chemistry 54, 2655–2661.

Gänzle, M.G., Haase, G., Jelen, P., 2008. Lactose: crystallization, hydrolysis and value-added derivatives. International Dairy Journal 18, 685–694.

Gänzle, M.G., 2012. Enzymatic synthesis of galacto-oligosaccharides and other lactose derivatives (hetero-oligosaccharides) from lactose. International Dairy Journal 22, 116–122.

Grajek, W., Olejnik, A., Sip, A., 2005. Probiotics, prebiotics and antioxidants as functional foods. Acta Biochimica Polonica 52, 665–671.

Grand View Research, 2014. Global Prebiotics Market by Ingredient (FOS, Inulin, GOS, MOS), by Application (Food & Beverages, Animal Feed, Dietary Supplements) Expected to Reach USD 5.75 Billion by 2020: Grand View Research, Inc. Available online: http://www.grandviewresearch.com/press-release/global-prebiotics-market Cited on: August 2015.

Grand View Research, 2015. Galacto-Oligosaccharides (GOS) Market Analysis by Application (Food & Beverage, Prebiotic Dietary Supplements) and Segment Forecasts to 2020. Available online: http://www.grandviewresearch.com/industry-analysis/galacto-oligosaccharides-gos-market Cited on: August, 2015.

Grembecka, M., 2015. Sugar alcohols – their role in the modern world of sweeteners: a review. European Food Research and Technology 241, 1–14.

Gutiérrez, L.F., Hamoudi, S., Belkacemi, K., 2012. Lactobionic acid: a high value-added lactose derivative for food and pharmaceutical applications. International Dairy Journal 26, 103–111.

Guzmán, F., Barberis, S., Illanes, A., 2007. Peptide synthesis: chemical or enzymatic. Electronic Journal of Biotechnology 10, 279–314.

Halliday, J., 2008. Nutrilab Works towards a Bright Future for Tagatose. Available online on: http://www.foodnavigator.com/MarketTrends/Nutrilabworkstowardsabrightfuturefortag-atose.

Hancock, S.M., Vaughan, M.D., Withers, S.G., 2006. Engineering of glycosidases and glycosyltransferases. Current Opinion in Chemical Biology 10, 509–519.

Hanson, S., Best, M., Bryan, M.C., Wong, C.H., 2004. Chemoenzymatic synthesis of oligosaccharides and glycoproteins. Trends in Biochemical Sciences 29, 656–663.

Hayashibara Shoji Inc., 2015. History. Available online: http://www.hayashibara.co.jp/groupinformation.php?jpml=information_en_3.

Heitlinger, L.A., Rossi, T.M., Lee, P.C., Lebenthal, E., 1991. Human intestinal disaccharidase activities: correlations with age, biopsy technique, and degree of villus atrophy. Journal of Pediatric Gastroenterology and Nutrition 12, 204–208.

Hernández-Hernández, O., Marín-Manzano, M.C., Rubio, L.A., Moreno, F.J., Sanz, M.L., Clement, A., 2012. Monomer and linkage type of galacto-oligosaccharides affect their resistance to ileal digestion and prebiotic properties in rats. Journal of Nutrition 142, 1232–1239.

Hsu, C.H., Hung, S.C., Wu, C.Y., Wong, C.H., 2011. Toward automated oligosaccharide synthesis. Angewandte Chemie International 50, 11872–11923.

Huh, K.T., Toba, T., Adachi, S., 1990. Oligosaccharide formation during the hydrolysis of lactose with hydrochloric-acid and cation-exchange resin. Food Chemistry 38, 305–314.

Huh, K.T., Toba, T., Adachi, S., 1991. Oligosaccharide structures formed during acid-hydrolysis of lactose. Food Chemistry 39, 39–49.

Iwagaki, H., Fuchimoto, S., Matsubara, N., Akazai, Y., Suzaki, K., Watanabe, T., Nonaka, Y., Kimura, S., Orita, K., Yoneyyama, M., Madai, T., Aga, H., Fujii, K., Sakai, S., 1991. The influence of lactosucrose (LS-98) on the intestinal microflora of patients with resected colon. Journal of the Japan Society of Colo-Proctology 44, 426–430.

Karam, G., Compagnon, P., Hourmant, M., Despins, P., Duveau, D., Noury, D., Boudjema, K., 2005. A single solution for multiple organ procurement and preservation. Transplant International 18, 657–663.

Kneifel, W., 2000. *In vitro* growth behaviour of probiotic bacteria in culture media with carbohydrates of prebiotic importance. Microbial Ecology in Health and Disease 12, 27–34.

Krumbholz, R.E., Dorscheid, M.G., 1991. Method of Manufacturing Lactulose. United States Patent 5,071,530.

Kummel, K.F., Brokx, S., 2001. Lactitol as a functional prebiotic. Cereal Foods World 46, 424–429.

Kuusisto, J., Tokarev, A.V., Murzina, E.V., Roslund, M.U., Mikkola, J.P., Murzin, D.Y., Salmi, T., 2007. From renewable raw materials to high value-added fine chemicals-catalytic hydrogenation and oxidation of D-lactose. Catalysis Today 121, 92–99.

Lærke, H.N., Jensen, B.B., 1999. D-Tagatose has low small intestinal digestibility but high large intestinal fermentability in pigs. The Journal of Nutrition 129, 1002–1009.

Levin, G.V., 2002. Tagatose, the new GRAS sweetener and health product. Journal of Medicinal Food 5, 23–36.

Lifran, E.V., Hourigan, J.A., Sleigh, R.W., 2009. Lactose derivatives: turning waste into functional foods. Australian Journal of Dairy Technology 64, 89–93.

Lu, Y., Levin, G.V., 2002. Removal and prevention of dental plaque with D-tagatose. International Journal of Cosmetic Science 24, 225–234.

Lu, Y., Levin, G.V., Donner, T.W., 2008. Tagatose, a new antidiabetic and obesity control drug. Diabetes, Obesity and Metabolism 10, 109–134.

Majumder, A., Sultan, A., Jersie-Christensen, R.R., Ejby, M., Gregers Schmidt, B., Lahtinen, S.J., Jacobsen, S., Svensson, B., 2011. Proteome reference map of *Lactobacillus acidophilus* NCFM and quantitative proteomics towards understanding the prebiotic action of lactitol. Proteomics 11, 3470–3481.

Martinez-Ferez, A., Rudloff, S., Guadix, A., Henkel, C.A., Pohlentz, G., Boza, J.J., Guadix, E.M., Kunz, C., 2006. Goats' milk as a natural source of lactose-derived oligosaccharides: isolation by membrane technology. International Dairy Journal 16, 173–181.

Mayer, J., Kranz, B., Fischer, L., 2010. Continuous production of lactulose by immobilized thermostable β-glycosidase from *Pyrococcus furiosus*. Journal of Biotechnology 145, 387–393.

McColl, D.B., 2001. Notice of a GRAS Exemption for D-Tagatose. Document Prepared by Hyman, Phelps & McNamara, on Behalf of Arla Foods Ingredients Amba. Available online on: http://www.fda.gov/ucm/groups/fdagov-public/@fdagov-foods-gen/documents/document/ucm261321.pdf.

Mendoza, M.R., Olano, A., Villamiel, M., 2005. Chemical indicators of heat treatment in fortified and special milks. Journal of Agricultural and Food Chemistry 53, 2995–2999.

Minami, Y., Yazawa, K., Tamura, Z., Tanaka, T., Yamamoto, T., 1983. Selectivity of utilization of galactosyl-oligosaccharides by bifidobacteria. Chemical and Pharmaceutical Bulletin 31, 1688–1691.

Minami, Y., Yazawa, K., Nakamura, K., Tamura, Z., 1985. Selectivity and efficiency of utilization of galactosyl-oligosaccharides by bifidobacteria. Chemical and Pharmaceutical Bulletin 33, 710–714.

Mitsuhashi, S., Yoshihama, M., Yahiro, M., Nishikawa, I., Deya, E., Ahiko, K., Mitsuoka, T., 1982. Effect of oligosaccharides on intestinal flora and fecal characteristics of neonates. In: Proceedings of the 111 IPCR Symposium on Intestinal Flora, Tokyo.

Monsan, P., Paul, F., 1995. Enzymatic synthesis of oligosaccharides. FEMS Microbiology Reviews 16, 187–192.

Montgomery, E., Hudson, C., 1930. Relations between rotary power and structure in sugar group XXVII. Synthesis of a new disaccharide ketose (lactulose) from lactose. Journal of the American Chemical Society 52, 2101–2106.

Nagasawa, T., Kawashima, T., Suzuki, K., Kawase, K., 1971. Process for Preparing a Lactulose Powder. United States Patent 3,716,403.

Nakano, H., Kiryu, T., Kiso, T., Murakami, H., 2010. Biocatalytic production of lactobionic acid. In: Hou, C.T., Shaw, J. (Eds.), Biocatalysis and Biomolecular Engineering. John Wiley & Sons Inc., New Jersey, pp. 391–404.

Nguyen, T.H., Haltrich, D., 2013. Microbial production of prebiotics oligosaccharides. In: McNeil, B., Archer, D., Giavasis, I., Harvey, L. (Eds.), Microbial Production of Food Ingredients, Enzymes and Nutraceuticals. Woodhead Publishing Limited, Cambridge, pp. 494–532.

Nutraingredients-USA.com, 2002. Spherix Begins Legal Proceedings Over Tagatose Delay. Available online on: http://www.nutraingredients-usa.com/Markets/Spherix-begins-legal-proceedings-over-tagatose-delay.

NutraSource Inc., 2010. GRAS Exemption Claim for D-Tagatose as an Ingredient in Foods. Submitted on behalf of CJ Cheiljedang on August 16.

Oh, D.K., 2007. Tagatose: properties, applications, and biotechnological processes. Applied Microbiology and Biotechnology 76, 1–8.

Paterson, A.H.J., Kellam, S.J., 2009. Transformation of lactose for value-added ingredients. In: Corredig, M. (Ed.), Dairy-Derived Ingredients: Food and Nutraceutical Uses. CRC, Boca Raton, pp. 625–643.

Patil, D.H., Westaby, D., Mahida, Y.R., Palmer, K., Rees, R., Clark, M.L., Dawson, A.M., Silk, D.B., 1987. Comparative modes of action of lactitol and lactulose in the treatment of hepatic encephalopathy. Gut 28, 255–259.

Perugino, G., Trincone, A., Rossi, M., Moracci, M., 2004. Oligosaccharide synthesis by glycosynthases. Trends in Biotechnology 22, 31–37.

Petersen, U., 2006. Tagatose. In: Mitchell, H. (Ed.), Sweeteners and Sugar Alternatives in Food Technology. Blackwell Publishing Ltd., Chennai, pp. 262–294.

Petuely, F., 1957. Biochemische Untersuchungen zur Regulation der Dickdarmflora des Säuglings: über den Bifidusfaktor. Notring der Wiss. Verbaende Österreichs, Wien. 90 pp.

Petuely, F., 1963. Laxative Composition and Method of Using Same. United States Patent 3,272,705.

Planas, A., Faijes, M., 2002. Glycosidases and glycosynthases in enzymatic synthesis of oligosaccharides. An Overview. Afinidad 59, 295–313.

Plante, O.J., Palmacci, E.R., Seeberger, P.H., 2001. Automated solid-phase synthesis of oligosaccharides. Science 291, 1523–1527.

Playne, M.J., Crittenden, R.G., 2009. Galacto-oligosaccharides and other products derived from lactose. In: Fox, P.F., McSweeney, P.L.H. (Eds.), Advanced Dairy Chemistry. Lactose, Water, Salts and Minor Constituents, vol. 3. Springer, New York, pp. 121–202.

Plou, F.J., Gómez de Segura, A., Ballesteros, A., 2007. Applications of glycosidases and transglycosidases in the synthesis of oligosaccharides. In: Polaina, J., MacCabe, A.P. (Eds.), Industrial Enzymes. Structure, Function and Applications. Springer, Dordrecht, pp. 141–157.

Quitmann, H., Fan, R., Czermak, P., 2014. Acid organic compound in beverage, food, and feed production. In: Zorn, H., Czermak, P. (Eds.), Biotechnology of Food and Feed Additives. Springer-Verlag Berlin Heidelberg, Berlin, pp. 91–142.

Rather, M.Y., Mishra, S., 2013. β-Glycosidases: an alternative enzyme based method for synthesis of alkyl-glycosides. Sustainable Chemical Processes 1, 7. http://dx.doi.org/10.1186/2043-7129-1-7.

Rycroft, C.E., Jones, M.R., Gibson, G.R., Rastall, R.A., 2001. A comparative *in vitro* evaluation of the fermentation properties of prebiotic oligosaccharides. Journal of Applied Microbiology 91, 878–887.

Saarela, M., Hallamaa, K., Mattila-Sandholm, T., Mättö, J., 2003. The effect of lactose derivatives lactulose, lactitol and lactobionic acid on the functional and technological properties of potentially probiotic *Lactobacillus* strains. International Dairy Journal 13, 291–302.

Sako, T., Matsumoto, K., Tanaka, R., 1999. Recent progress on research and applications of non-digestible galacto-oligosaccharides. International Dairy Journal 9, 69–80.

Saunders, J.P., Donne, T.W., Sadler, J.H., Levin, G.V., Makris, N.G., 1999. Effects of acute and repeated oral doses of D-tagatose on plasma uric acid in normal and diabetic humans. Regulatory Toxicology and Pharmacology 29, S57–S65.

Schaafsma, G., 2008. Lactose and lactose derivatives as bioactive ingredients in human nutrition. International Dairy Journal 18, 458–465.

Seeberger, P.H., Werz, D.B., 2007. Synthesis and medical applications of oligosaccharides. Nature 446, 1046–1051.

Shankar, P., Ahuja, S., Sriram, K., 2013. Non-nutritive sweeteners: review and update. Nutrition 29, 1293–1299.

Shimamura, S., Tamura, Y., Mizota, T., Nakano, S., Suzawa, I., 1993. Crystalline Lactulose Trihydrate and a Method for its Manufacture. European Patent Application 0532173 A2.

Siró, I., Kápolna, E., Kápolna, B., Lugasi, A., 2008. Functional food. Product development, marketing and consumer acceptance – a review. Appetite 51, 456–467.

Spherix Consulting, Inc., 2010. Generally Recognized as Safe (GRAS) Determination for the Use of Galacto-Oligosaccharides in Foods and Infants Formulas. Prepared for. Yakult Pharmaceutical Industry Co., Ltd.

Spherix Inc., 2011. Complaint for Injunction Relief and Damages. Available online on: http://globaldocuments.morningstar.com/documentlibrary/document/8119baa1b6f33d7fca781077cad1f8c9.msdoc/original.

Starling, S., 2013. Solvay Management Buys Out Solactis. Available in: http://www.nutraingredients.com/Suppliers2/Solvay-management-buys-out-Solactis.

Strohmeier, G.A., Pichler, H., May, O., Gruber-Khadjawi, M., 2011. Application of designed enzymes in organic synthesis. Chemical Reviews 111, 4141–4164.

Taniguchi, H., 2005. Carbohydrate active enzymes for the production of oligosaccharides. In: Hou, C.T. (Ed.), Handbook of Industrial Biocatalysis. CRC Taylor & Francis, Boca Raton. pp. 20–21, 20–25.

Teramoto, F., Rokutan, K., Kawakami, Y., Uchida, J., Oku, K., Oka, M., Yoneyama, M., 1996. Effect of 4G-β-D-galactosylsucrose (lactosucrose) on fecal microflora in patients with chronic inflammatory bowel disease. Journal of Gastroenterology 31, 33–39.

Teramoto, F., Rokutan, K., Sugano, Y., Oku, K., Kishino, E., Fujita, K., Hara, K., Kishi, K., Fukunaga, M., Morita, T., 2006. Long-term administration of 4G-beta-D-galactosylsucrose (lactosucrose) enhances intestinal calcium absorption in young women: a randomized, placebo-controlled 96-wk study. Journal of Nutritional Science and Vitaminology 52, 337–346.

Torres, D.P.M., Gonçalves, M.P.F., Teixeira, J.A., Rodrigues, L.T., 2010. Galacto-oligosaccharides production, properties, applications, and significance as prebiotics. Comprehensive Reviews in Food Science and Food Safety 9, 438–454.

Université de Liège, 2010. Reflexions, The University of Liège Website that Makes Knowledge Accessible. The Enzyme Which Produces Sugar. Available online on: http://reflexions.ulg.ac.be/cms/c_30824/en/the-enzyme-which-produces-sugar?portal=j_55 &printView=true.

U.S. FDA, 2003. Food Labeling: Health Claims; D-Tagatose and Dental Caries. Available online in: http://www.fda.gov/ohrms/dockets/98fr/02P-0177-NFR00001.pdf.

U.S. FDA, 2006. Agency Response Letter – Objection; Anhydrous Lactitol and Lactitol Monohydrate; FALN No. 005 (Docket No. 2006FL-120). Available online in: http://www.fda.gov/Food/GuidanceRegulation/GuidanceDocumentsRegulatoryInformation/Allergens/ucm076706.htm.

U.S. FDA, 2015. Code of Federal Regulations, Title 21 (Rev. April 2015). Available online in: http://www.accessdata.fda.gov/scripts/cdrh/cfdocs/cfcfr/CFRSearch.cfm?fr=101.80&SearchTerm=sugar.

Van Beers, E.H., Büller, H.A., Grand, R.J., Einerhand, A.W.C., Dekker, J., 1995. Intestinal brush border glycohydrolases: structure, function, and development. Critical Reviews in Biochemistry and Molecular Biology 30, 197–262.

van Leusen, E., Torringa, E., Groenink, P., Kortleve, P., Geene, R., Schoterman, M., Klarenbeek, B., 2014. Industrial applications of galactooligosaccharides. In: Moreno, F.J., Sanz, M.L. (Eds.), Food Oligosaccharides: Production, Analysis and Bioactivity. John Wiley & Sons Ltd., Chichester, pp. 470–491.

Vastenavond, C.M., François, 2013. Improved Galactose Isomerases and Use Thereof in the Production of Tagatose. European Patent EP 2834354.

Vastenavond, C.M., Bertelsen, H., Hansen, S.J., Laursen, R.S., Saunders, J., Eriknauer, K., 2012. Tagatose (D-tagatose). In: O'Brien Nabors, L. (Ed.), Alternative Sweeteners, fourth ed. CRC Press, Boca Raton, pp. 198–222.

Vera, C., Guerrero, C., Conejeros, R., Illanes, A., 2012. Synthesis of galacto-oligosaccharides by β-galactosidase from *Aspergillus oryzae* using partially dissolved and supersaturated solution of lactose. Enzyme and Microbial Technology 50, 188–194.

Vesa, T.H., Marteau, P., Korpela, R., 2000. Lactose intolerance. Journal of the American College of Nutrition 19, S165–S175.

Vester, B.M., Fahey, G., 2012. Prebiotics of plant and microbial origin. In: Callaway, T.R., Ricke, S.C. (Eds.), Direct-Fed Microbials and Prebiotics for Animals. Springer, NewYork, pp. 13–26.

Westhoff, G.M., Kuster, B.F.M., Heslinga, M.C., Pluim, H., Verhage, M., 2014. Lactose and Derivatives. Ullmann's Encyclopedia of Industrial Chemistry, pp. 1–9.

Wolfrom, M.L., Burke, W.J., Brown, K.R., Rose, R.S., 1938. Crystalline lactositol. Journal of the American Chemical Society 60, 571–573.

Yahoo Finance, 2012. Form 10-Q for Spherix Inc. Available online on: http://biz.yahoo.com/e/120813/spex10-q.html.

Yazawa, K., Imai, K., Tamura, Z., 1978. Oligosaccharides and polysaccharides specifically utilizable by bifidofacteria. Chemical and Pharmaceutical Bulletin 26, 3306–3311.

Yoshida, M., Yamada, M., Nishitani, T., Takada, G., Izumori, K., Kamitori, S., 2007. Crystal structures of D-tagatose 3-epimerase from *Pseudomonas cichorii* and its complexes with D-tagatose and D-fructose. Journal of Molecular Biology 374, 443–453.

Zacharis, C., 2012. Lactitol. In: O'Donnell, K., Kearsley, M.W. (Eds.), Sweeteners and Sugar Alternatives in Food Technology. Blackwell Publishing Ltd., New Delhi, pp. 275–294.

Chapter 4

Enzymatic Production of Galacto-Oligosaccharides

A. Illanes, C. Vera and L. Wilson

4.1 GALACTO-OLIGOSACCHARIDES AS PREBIOTICS

Galacto-oligosaccharides (GOSs) are nondigestible oligosaccharides (NDOs) composed of a variable number of galactosyl units (from two to nine) and one terminal glucose unit linked by different glycosidic bonds (see Fig. 4.1) that vary according to the enzyme and reaction conditions used for their synthesis (Gänzle, 2012), being mostly β1-4 and β1-6 (Coulier et al., 2009). GOSs are of current interest for human and animal health because of their well-reported beneficial effects.

There is considerable amount of information sustaining the prebiotic condition of GOS according to its accepted definition (see Section 2.4) (Gibson, 2004; Sangwan et al., 2011). Beneficial effects of GOS have been consistently proven in in vitro (Gibson and Fuller, 2000) and in vivo trials with model animals (Djouzi and Andrieux, 1997; Tzortzis et al., 2005a), human adults (Bouhnik et al., 2004; Depeint et al., 2008; Drakoularakou et al., 2010), infants (Moro et al., 2003; Knol et al., 2005; Fanaro et al., 2009), and elderly people (Walton et al., 2012). Prebiotic effect is dependent on the chain length, and better effects have been observed with tri- (GOS-3) and tetrasaccharides (GOS-4), which usually represent more than 90% of the oligosaccharides synthesized (Gopal et al., 2001; Rastall, 2006). In a comparative study with other dietary oligosaccharides using a prebiotic index combining stimulating effect on beneficial colonic bacteria (*Bifidobacteria* and *Lactobacilli*) and depressing effect on harmful ones (*Clostridia*), GOS was ranked fourth after lactulose, soybean oligosaccharides (SOSs) and isomalto-oligosaccharides (IMOSs) and well over xylo-oligosaccharides (XOSs), fructo-oligosaccharides (FOSs) and inulin (Palframan et al., 2003). Using a more refined prebiotic index, the measure of prebiotic effect (MPE), which is a lumped parameter considering substrate assimilation, changes in bacterial populations, and production of short-chain fatty acids (SCFA), GOS-FOS mixtures (9:1) scored first and GOS ranked second, well over FOS, SOS and IMOS; partially hydrolyzed guar gum, sunfiber, guar gum, and the control (sucrose) had negative MPE values (Vulevic et al., 2004). Significant evidence

Lactose-Derived Prebiotics. http://dx.doi.org/10.1016/B978-0-12-802724-0.00004-4

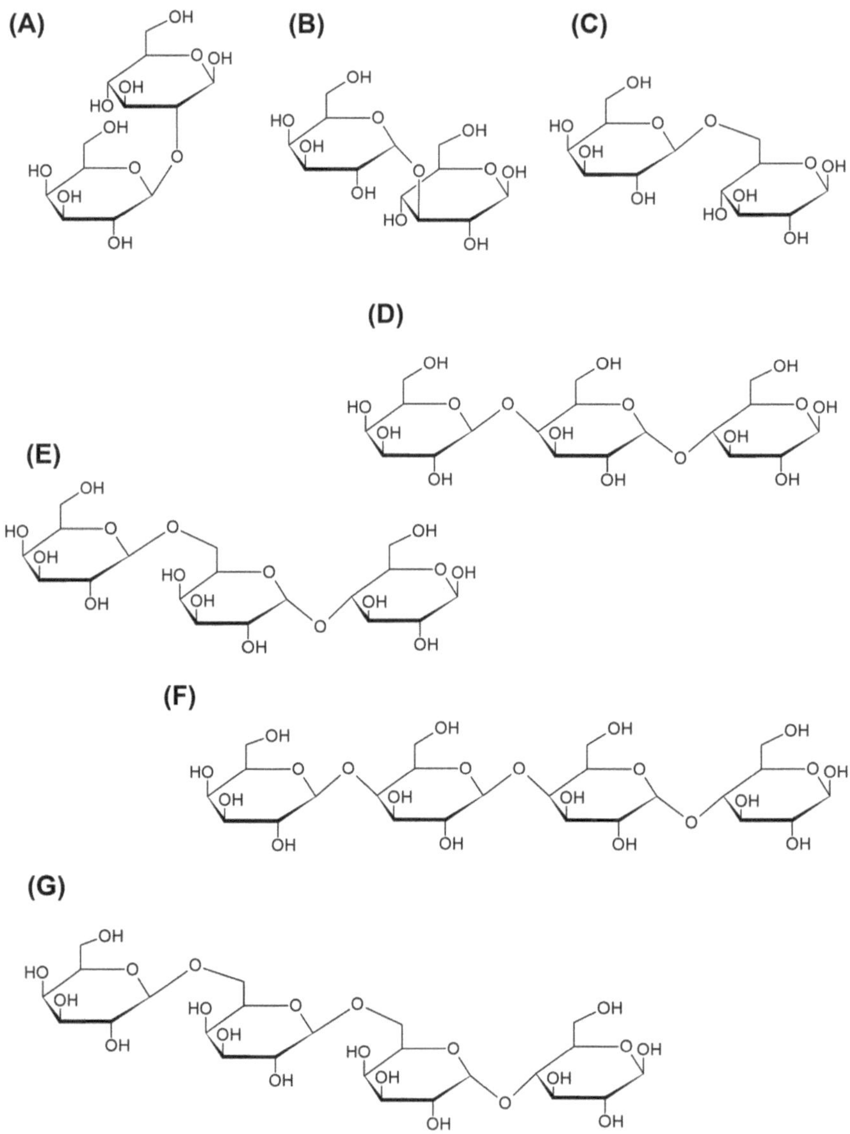

FIGURE 4.1 Chemical structure of most common galacto-oligosaccharides. (A): Gal (β1-2)Glu; (B): Gal(β1-3)Glu; (C): Gal(β1-6)Glu; (D): Gal(β1-4)Gal(β1-4)Glu; (E): Gal(β1-6)Gal(β1-4)Glu; (F): Gal(β1-4)Gal(β1-4)Gal(β1-4)Glu; (G): Gal(β1-6)Gal(β1-6)Gal(β1-4). (A) to (C): disaccharides (GOS-2); (D) and (E): trisaccharides (GOS-3); (F) and (G): tetrasaccharides (GOS-4).

about the prebiotic condition of GOS has come from dynamic in vitro colon model studies where the potential symbiotic action of GOS with probiotic strains, like *Bifidobacterium catenulatum*, *Bifidobacterium bifidum*, and *Lactobacillus gasseri*, was assessed (Maathuis et al., 2012).

Many health benefits have been claimed for GOS that are referred both to gut health and to immune modulation. Gut health promotion refers to the selective stimulation of beneficial bacteria (mainly *Bifidobacteria* and *Lactobacilli*) and inhibition of pathogens, representing the essence of its prebiotic condition; immune modulation refers not only to the indirect effect of GOS caused on gut microbiota but also to the direct interaction with immune cells. The confirmation of health claims for GOS-containing foods has been a powerful driving force for GOS consumption (Martínez-Villaluenga et al., 2008).

Stimulation of *Bifidobacteria* and to a lesser extent of *Lactobacilli* by GOS is well documented both in adults (Davis et al., 2010) and infants, and its synergistic effect with added probiotics (Piirainen et al., 2008) and in combination with other oligosaccharides (Scholtens et al., 2006; Scalabrin et al., 2012) has also been established. Inflammatory bowel disease resulting from an imbalance in the gut microbiota, which can lead to ulcerative colitis and Crohn's disease, can potentially be prevented by GOS, although studies with respect to this are scarce and not yet conclusive (Sangwan et al., 2011). Inhibition of proliferation of pathogenic bacteria by GOS has been attributed to SCFA formation that reduces the luminal pH of the colon below the value at which pathogens will proliferate (Bakker-Zierikzee et al., 2005). However, much of its effect has been associated with their antiadhesive properties that directly inhibit pathogen infection by mimicking their binding sites on the intestinal epithelial cells; in such a way, GOS molecules act as molecular receptor decoys (Kunz et al., 2000). Inhibition of *Escherichia coli* adherence to epithelial cells was higher in GOS than in other NDOs (Shoaf et al., 2006), and strong inhibition of *Salmonella typhimurium* cells was also reported for GOS (Tzortzis et al., 2005a; Searle et al., 2009). GOS consumption in laboratory animals produced an increase in mucins associated to the intestinal mucosa (Leforestier et al., 2009), which might be beneficial because these glycoproteins protect the intestinal epithelial cells from deleterious luminal compounds (Corfield et al., 2001).

Although not used as a laxative, GOSs have demonstrated effectiveness in the alleviation of constipation in infants (Mihatsch et al., 2006; Bongers et al., 2007), adults (Niittynen et al., 2007), and elderly people (Hamilton-Miller, 2004; Sairanen et al., 2007). Several mechanisms have been proposed to explain the effect of GOS on bowel motility improvement; most likely, this is due to the SCFA produced as a consequence of its fermentation by indigenous colonic bacteria (Nauta et al., 2010).

Reduction of the activity of several genotoxic enzymes, like β-glucosidase, β-glucuronidase, and arylsulfatase, involved in the formation of toxic and potentially carcinogenic compounds, has been demonstrated for GOS (McBain and Macfarlane, 2001); reduction of other harmful compounds, like ammonia, *p*-cresol and indole has also been reported after GOS intake (Mussatto and Mancilha, 2007). A recent review on this subject concluded that there is sound evidence suggesting a role of GOS in colorectal cancer prevention (Bruno-Barcena and Azcarate-Peril, 2015).

GOSs have also been claimed as stimulants of mineral absorption. Several studies with laboratory animals have shown the beneficial effect of GOS intake on calcium metabolism and bone health by promoting an increase in calcium absorption (Pérez-Conesa et al., 2007; Weaver et al., 2011). Studies with humans are scarce and not yet conclusive (van den Heuvel et al., 2000; Whisner et al., 2013). It has also been claimed its effectiveness in the prevention of allergic diseases (Kukkonen et al., 2007) and in reducing travelers' diarrhea (Drakoularakou et al., 2010).

A major part of the immune system resides in the gut and it has been proven that immune modulation of the intestinal tract is feasible by proper administration of prebiotics (Macfarlane et al., 2008). Fermentation of GOS produces SCFA from which butyrate, propionate, and acetate exert several immune responses (Sangwan et al., 2011); besides this indirect effect, direct interactions with immune cells have also been observed (Eiwegger et al., 2004, 2010). GOS administration to healthy elderly people showed that beyond its stimulatory effect on beneficial bacteria and detrimental effect on less beneficial bacteria, a significant positive effect on immune response was observed as evidenced by the increase in natural killer (NK) lymphocyte activity and phagocytosis, increase in anti-inflammatory cytokines secretion, and decrease in proinflammatory cytokines (Vulevic et al., 2008). A positive effect has been also shown for GOS in improving lactose digestion and alleviating symptoms in persons with lactose intolerance (Savaiano et al., 2013).

Beyond its good scoring as prebiotic, GOS has a unique property among NDOs that is related to human milk composition (Venema, 2012). Human milk oligosaccharides (HMOSs) are a complex mixture of oligosaccharides present in mother's milk that may amount to up to 1% of milk on a wet weight basis; they are exclusively synthesized in the mammary gland during lactation, and their amount and composition in human milk is quite different than in other animal species. More than 100 such oligosaccharides have been identified in mother's milk and their multiple beneficial physiological effects assessed (Bode, 2006, 2009), being a key determinant of the development of the primordial human intestinal microbiota. In fact, HMOSs have been considered as the first prebiotic in humans (Coppa et al., 2004) and its neutral fraction has proven to be the most relevant factor for the development of a healthy intestinal microbiota (Harmsen et al., 2000). Breast-fed infants' intestinal microbiota has a prevalence of *Bifidobacteria*, while infants fed with cow's milk formula exhibit a more mixed population with lower counts of *Bifidobacteria* and higher counts of *Clostridia* and *Enterococci*, which resembles more the profile of an adult intestinal microbiota (Playne and Crittenden, 2009). HMOS content in human milk is 10–100 times higher than the oligosaccharide content in the milk of the more relevant domestic animals; besides, such oligosaccharides are less complex and different in structure from HMOS (Boehm and Stahl, 2007). Therefore, infants fed with formulated cow's milk may have a serious deficiency in oligosaccharide ingestion that will have a negative effect on intestinal health and immune system buildup. Supplementation of milk formulas with HMOS is quite relevant but an impossible task due to its complex composition. Separation of lactose from oligosaccharides in farm animal milk by simulated moving bed chromatography has been suggested as a way of increasing their content of potentially beneficial oligosaccharides (Geisser et al., 2005). It was reported that goat's milk oligosaccharides resemble HMOS to a higher extent, especially with respect to the content of neutral and sialylated oligosaccharides, and was claimed as a better option than cow's milk for infant feeding (Martinez-Ferez et al., 2006). However, it has been shown that the effect of HMOS can be well mimicked by supplementing milk formulas with GOS (Ben et al., 2008). With such purpose, an oligosaccharide mixture composed of short-chain GOS (90%) and long-chain FOS (10%) was designed to closely match the molecular size distribution of HMOS (Moro et al., 2005). Clinical trials have demonstrated that many physiological parameters of infants fed with the GOS-enriched cow's milk formula are very much the same as in breast-fed infants (Fanaro et al., 2005; Knol et al., 2005) and no adverse effect has been observed (Moro et al., 2002). Stimulation of *Bifidobacteria* and *Lactobacilli* and depression of *Clostridia* and *Bacteroides* was demonstrated in infants fed with the GOS/FOS mixture and also in a GOS/FOS/pectin oligosaccharides mixture, minimizing the alteration of fecal microbiota after cessation of breast-feeding (Magne et al., 2008). Other beneficial health effects have been claimed for GOS/FOS mixtures in infant feeding, like reduction of intestinal and respiratory infections in healthy infants during the first year of age (Arslanoglu et al., 2007; Bruzzese et al., 2009), reduction of atopic dermatitis in term infants fed with the enriched formula during the first two weeks of life (Moro et al., 2006), and positive modulation of immunoglobulin profile in infants at risk of allergy (Arslanoglu et al., 2008; van Hoffen et al., 2009; Schouten et al., 2011), sustaining the immune-modulating capacity of prebiotics. Physiological parameters like fecal pH and SCFA profile and stool consistency and frequency were similar in enriched formula and breast-fed infants and quite distinct in infants fed with nonsupplemented cow's milk formula, without any adverse effect being detected in the former case (Playne and Crittenden, 2009). Altogether, it is concluded that GOS/FOS (9:1) very much resembles HMOS in its physiological effects on infants. GOS/FOS (9:1) mixture is considered safe and was approved in the European Union (EU) for infant formula supplementation at levels not exceeding 8 g/L (Braegger et al., 2011). GOS/FOS mixtures also produced a significant bifidogenic effect on pregnant women intestinal microbiota that was not transmitted to the neonates, not affecting fetal immunity (Shadid et al., 2007).

Mechanisms explaining the various health-promoting effects described herein have been sketched by Playne and Crittenden (2009).

With respect to safety issues and public acceptance, GOS has benefited by the inclusion of NDOs in the category of "dietary fiber" (Meyer, 2004), a concept that is easily linked to health benefit. In Japan, a pioneer in the concept of functional health foods, NDOs are benefited by being included in the category of food for specified health use (FOSHU) (Taniguchi, 2004). Safety of GOS intake has been unambiguously established for infants (Fanaro et al., 2009) and adults (Davis et al., 2010). Safety of GOS is sustained by its presence in human milk and its production from lactose in the gut by intestinal bacteria. Recommended effective dose of GOS in adults is 8–15 g/day, but significantly lower doses have produced beneficial effects in youngsters (Whisner et al., 2013) and infants (Ben et al., 2008), while in Japan doses as low as 2.5 g/day have proven effective in healthy adults and elderly persons (Sako et al., 1999). Intake in excess of 30 g/day may produce intestinal discomfort and diarrhea (Macfarlane et al., 2008). At levels of 0.3–0.4 g GOS/Kg body weight no adverse effects have been noticed, and adverse effects like diarrhea or flatulence are lower than with inulin or FOS at equivalent doses (Sako et al., 1999; Propst et al., 2003). It was recently reported that certain allergic reactions were observed in persons consuming GOS produced from *Bacillus circulans* and identified the tetrasaccharides present as potential allergens; they claimed having produced a hypoallergenic GOS produced with β-galactosidases from *Sporobolomyces singularis* and *Kluyveromyces lactis* (Kaneko et al., 2014).

GOS has a GRAS status in the United States, nonnovel food status in the EU, and FOSHU status in Japan (Tzortzis and Vulevic, 2009); infant formulas and follow-on formulas containing GOS can be categorized as "foods for particular nutritional uses" (PARNUTS) in the EU (Coppens et al., 2006).

In summary, GOSs are scientifically sustained prebiotics in compliance with the corresponding definition and guidelines (Pineiro et al., 2008), with high-scoring and excellent functional properties to be safely included in different food matrices; a distinctive feature of GOS is its role as HMOS substitute in formulas for infants.

4.2 ENZYMATIC SYNTHESIS OF GALACTO-OLIGOSACCHARIDES: WHOLE CELLS, FREE AND IMMOBILIZED ENZYMES

As pointed out in Section 3.1, chemical synthesis of oligosaccharides is a formidable task since each monosaccharide has one primary and four secondary hydroxyl groups so that the possible chemical structures are exceedingly high even for a short-chain oligosaccharide. Bonding between monosaccharide units can be established by each of these groups with either α or β configurations (Hanson et al., 2004). This is undesirable because only β configurations will be useful for conferring prebiotic condition, and in the case of GOS, whose structure is formed by galactose units and one terminal glucose unit, the prebiotic effect is dependent on the hydroxyl group involved in the glycosidic linkage, being β1-6 and β1-4 the more relevant (Casci and Rastall, 2006). A very complex mixture of GOS with different linkages with α and β configurations was formed by a reverse reaction during acid hydrolysis of lactose (Huh et al., 1991); therefore, despite the advances in the chemical synthesis of oligosaccharides (Sears and Wong, 2001; Seeberger and Werz, 2005), the poor specificity of the synthesis of GOS by chemical glycosylation makes it technologically unrealistic (Barreteau et al., 2006; Otieno, 2010; Konar et al., 2011). In fact, the synthesis of GOS is exclusively performed by biocatalysis due to its high selectivity (Crout and Vic, 1998), but also because of mild reaction conditions and process simplicity (Illanes et al., 2012). β-Galactosidases will catalyze only the synthesis of oligosaccharides with β-type linkages, which are the ones recalcitrant to degradation and hydrolysis in its passage through the upper gastrointestinal tract, which is on the basis of its prebiotic condition; this kind of linkage is also the one prevailing in HMOS (Boehm and Stahl, 2007).

GOS formation during the enzymatic hydrolysis of lactose was early detected (Roberts and McFarren, 1953; Asp et al., 1980) but at first considered a nuisance precisely because their indigestibility was nutritionally objectionable (Zárate and López-Leiva, 1990). Not much attention was paid then, since at the conditions favoring lactose hydrolysis the amount of GOS formed was actually quite low; however, characterization of these side products formed was done in reactions catalyzed both by bacterial (Toba et al., 1981; Mozaffar et al., 1985) and fungal (Burvall et al., 1979; Toba et al., 1985) β-galactosidases. It was soon realized that the amount of GOS can be increased significantly by increasing lactose concentration, and a technological potential was then envisaged associated to its health-promoting properties (López Leiva and Guzmán, 1995).

Essentially two types of bioprocesses have been developed for GOS synthesis: cell-free enzymes and whole cells. The former can be subdivided into those using soluble and those using immobilized enzymes; the latter can be subdivided according to the physiological condition of the cells: living cells, resting cells, permeabilized cells, and also immobilized cells (Park and Oh, 2010).

4.2.1 Synthesis of Galacto-Oligosaccharides With Whole Cell Systems

In principle, the use of whole cells seems advantageous for GOS synthesis when the β-galactosidase is intracellular or cell-bound, because it avoids the isolation of the enzyme from the producing cell that includes extraction and, eventually, purification, which are costly operations at productive scale. In the case of metabolically active cells, an additional benefit is that galactose and glucose can be consumed during cell growth, reducing their content in the final product and alleviating β-galactosidase inhibition on GOS formation, but with the disadvantage of the concomitant production of metabolites during fermentation, like glycerol and ethanol, and the presence of residual nutrients in the spent medium (Torres et al., 2010). An interesting scheme was early proposed in which GOS synthesis was conducted with toluene-treated cells of the yeast *Sterigmatomyces elviae* and then the crude product was fermented with the same strain to selectively remove the remaining monosaccharides (Onishi et al., 1995); later on the same yeast was used for GOS production in a cell recycle system allowing several cycles of production by adding only some minerals and thiamine, but with a significant reduction in the supply of nitrogenous compounds and yeast extract in the production medium (Onishi and Tanaka, 1998). Other yeasts have also been used for the whole cell synthesis of GOS, and thermotolerant yeasts were also screened for that purpose (Petrova and Kujumdzieva, 2010). Recently, GOS production by fermentation with whole cells of the yeasts *Pseudozyma tsukubaensis* and *Pichia fragilis* was optimized, obtaining GOS yields similar to those obtained in cell-free enzymatic synthesis; however, not many studies have been conducted for GOS production by fermentation with actively growing cells (Fai et al., 2015), and most of them have been conducted with resting or permeabilized cells.

GOS production with probiotic microbial cells is an appealing strategy (Splechtna et al., 2006; Gosling et al., 2010) and several studies have been conducted with strains like *Bifidobacterium infantis* RW-8120 (Roy et al., 2002) and *B. bifidum* NCIMB 41171. The latter is considered a promising GOS-producing strain allowing the synthesis of a GOS mixture with a stronger bifidogenic effect than commercially available GOS (Tzortzis et al., 2005b) and produced at comparable yields than obtained with isolated enzymes (Goulas et al., 2007). Later on, mathematical modeling and optimization of GOS synthesis with whole cells of *B. bifidum* NCIMB 41171 was reported (Osman et al., 2010).

In the case of intracellular yeast β-galactosidases, cell permeabilization with organic solvents has been used to increase lactose transport rate across the cell membrane, which is a major constraint due to its poor permeability to lactose (Numanoğlu and Sungur, 2004), so that organic solvents and mixtures thereof have been used for maximizing the expression of β-galactosidase activity (Kumari et al., 2013). An increased GOS yield is then expected for permeabilized cells, which has been proven with the yeast strains *Sterigmatomyces elviae* (Onishi et al., 1995), *Rhodotorula minuta* (Onishi and Yokozeki, 1996) and *Sirobasidium magnum* (Onishi et al., 1996). Later on, GOS production with isopropanol-permeabilized cells of the well-known β-galactosidase-producing yeast *Kluyveromyces marxianus*, taxonomically indistinct from *K. lactis* (Fonseca et al., 2008), was optimized in terms of lactose concentration, enzyme-substrate ratio, temperature, and pH, obtaining a rather low yield (16.5%) when compared with previous reports by Onishi, but with a significantly higher productivity of 27.6 g/L/h (Manera et al., 2010). GOS production with permeabilized cells of *K. lactis* was then compared to the production with two commercial *K. lactis* free β-galactosidases (Lactozym 3000 L HPG and Maxilact LGX 5000); at comparable conditions a higher maximum GOS yield was obtained with permeabilized cells, but yield declined sharply after that maximum, which did not occur with the soluble commercial enzymes, meaning that permeabilized cells had a higher tendency to synthesize GOS in the initial stage of reaction, but when GOS accumulates a higher tendency to hydrolyze the GOS formed occurred; this behavior was explained by the higher stability of the enzyme in the permeabilized cells than in the free stage (Rodriguez-Colinas et al., 2011). In order to depress the competing hydrolytic reactions, the synthesis of GOS with permeabilized cells of *K. marxianus* was conducted in supercritical CO_2 medium where the competition between synthesis and hydrolysis was less pronounced than in conventional aqueous medium, being an interesting option for developing a technology for GOS production (Manera et al., 2012). Synthesis of GOS was recently conducted with free and alginate-immobilized permeabilized cells of *K. marxianus*, immobilized cells expressing 85% of contacted β-galactosidase activity. Synthesis of GOS with both catalysts was done in repeated batch operation during 10 cycles and in the case of the immobilized cell catalyst synthesis was also done continuously for 10 days in a packed-bed reactor with a GOS productivity of 21.2 g/L/h and a GOS yield of 32% (Srivastava et al., 2016).

4.2.2 Synthesis of Galacto-Oligosaccharides With Isolated Enzymes

Whole-cell systems are obviously an alternative only with intracellular or cell-bound β-galactosidases; this is the case of yeast enzymes, while most β-galactosidases from molds and bacteria are extracellular in nature or excretable. Even in the former case, the potential advantages of whole-cell systems have to be counterbalanced by their drawbacks resulting from their low specificity and complex downstream operations, which become critical issues at productive scale (Woodley, 2006), so that isolated enzymes are in general a preferred option when no coenzymes are involved. Therefore, most technologies for GOS production are based on isolated enzymes that catalyze the reaction of synthesis with high regio- and stereospecificity (Crout and Vic, 1998).

As pointed out in Section 3.1, enzymes performing the synthesis of oligosaccharides within cell metabolism are glycosyltransferases, but they are not readily available, have stoichiometric coenzyme requirements, and their performance in vitro is poor, precluding their use as process biocatalysts (Contesini et al., 2013). β-Galactosidases (EC 3.1.2.23), on the other hand, are robust, readily available hydrolases not requiring coenzymes that, under proper conditions, can act in reverse catalyzing the formation of glycosidic bonds instead of their hydrolysis (Illanes, 2011). As mentioned in Section 1.3.1, β-galactosidases have been used for a long time in the hydrolysis of lactose in milk (Neuhaus et al., 2006) and dairy products (Perotti et al., 2012) so they are readily available, safe, and rather cheap enzymes. These enzymes can also catalyze the transgalactosylation of lactose leading to GOS synthesis, which represents a considerable bonus; however, reaction conditions are completely different than required for lactose hydrolysis since in this case high-lactose concentrations, elevated temperatures, and low-water activity in the reaction medium are required (Grosová et al., 2008).

β-galactosidase (lactase) is a rather ubiquitous enzyme, whose activity has been detected in animal, plant, and microbial cells (Husain, 2010). Microbial β-galactosidases are by far the most abundant and the only ones of technological significance. A variety of microorganisms, including mesophilic, thermophilic, hyperthermophilic, and psychrophilic bacteria and archaea, and mesophilic and thermophilic fungi and yeasts are β-galactosidase producers (Panesar et al., 2006; Park and Oh, 2010; Sheik Asraf and Gunasekaran, 2010). Most of them have been isolated and screened for their hydrolytic activity on lactose. Those whose β-galactosidases have been used for GOS synthesis are few; the most relevant among them are listed in Table 4.1, including the most cited bacteria, archaea, yeasts, and fungi covering mesophilic, thermophilic, and psychrophilic organisms. Thermostable enzymes are particularly desirable for GOS synthesis because of the dramatic increase of lactose solubility with temperature.

As shown, most cited microorganisms whose β-galactosidases have been used for GOS synthesis are *B. circulans*, *K. lactis*, and *Aspergillus oryzae*. Interestingly, enzymes from many probiotic strains have also been tested for GOS synthesis based on the fact that the GOS profile they synthesize is better for exerting their prebiotic effect. Many of such strains are not easily cultured, so that in many cases their genes have been cloned into suitable hosts for β-galactosidase production.

There is considerable interest in thermotolerant and psychroactive β-galactosidases, so that β-galactosidase genes from thermophilic and psychrophilic organisms have been cloned and expressed in suitable hosts (Ansari and Satar, 2012). Most cited recombinant organisms used for GOS synthesis are listed in Table 4.2. As can be seen, in many cases the use of recombinant strains has had the purpose of enzyme production from extremophiles that are poorly cultivated under laboratory conditions; most of them have used *E. coli* as host, which poses a problem for safe use in food applications (Oliveira et al., 2011). β-Galactosidases from nonculturable microorganisms have been obtained from metagenomic libraries using *Pichia pastoris* as host (Wang et al., 2012). In some cases, recombination has been used to produce an excretable β-galactosidase from a naturally intracellular enzyme, as occurred when the gene from a cytoplasmic *Geobacillus stearothermophilus* β-galactosidase was cloned and expressed in *Bacillus subtilis* (Xia et al., 2010). Despite the technological advantages of using recombinant strains in terms of productivity and ease of purification, it is premature to assess the impact of these recombinant β-galactosidases in the industrial production of GOS.

The following recent review articles covering different aspects of β-galactosidase GOS-producing strains amenable for GOS synthesis are recommended: Gosling et al. (2010), Park and Oh (2010), Sheik Asraf and Gunasekaran (2010), Torres et al. (2010), Panesar et al. (2011, 2013), Sangwan

TABLE 4.1 Main Microorganisms Producing β-Galactosidases for Galacto-Oligosaccharide Synthesis

Organism	Strain	References
Bacteria		
	Bacillus circulans	Mozaffar et al. (1988), Yanahira et al. (1995), Gosling et al. (2009, 2011), Palai and Bhattacharya (2013), Urrutia et al. (2013a,b), and Warmerdam et al. (2013a)
	Bacillus stearothermophilus	Chen et al. (2009)
	Bacillus sp.	Chakraborti et al. (2000)
	Bifidobacterium adolescentis	Rabiu et al. (2001)
	Bifidobacterium angulatum	Rabiu et al. (2001)
	Bifidobacterium bifidum	Depeint et al. (2008) and Osman et al. (2012)
	Bifidobacterium infantis	Roy et al. (2002)
	Bifidobacterium longum	Hsu et al. (2007)
	Enterobacter agglomerans	Lu et al. (2007)
	Enterobacter cloacae	Lu et al. (2009)
	Enterococcus faecium	Badarinath and Halami (2011)
	Escherichia coli	Reuter et al. (1999)
	Geobacillus stearothermophilus	Placier et al. (2009)
	Halothermothrix orenii	Hassan et al. (2015)
	Lactobacillus acidophilus	Nguyen et al. (2007)
	Lactobacillus fermentum	Sriphannam et al. (2012)
	Lactobacillus pentosus	Maischberger et al. (2010)
	Lactobacillus plantarum	Iqbal et al. (2010)
	Lactobacillus reuteri	Splechtna et al. (2006)
	Lactobacillus sakei	Iqbal et al. (2011) and Zhang et al. (2013)
	Lactobacillus sp.	Splechtna et al. (2007b)
	Streptococcus thermophilus	Crittenden and Playne (1996)
	Thermus sp.	Ladero et al. (2006)

Continued

TABLE 4.1 Main Microorganisms Producing β-Galactosidases for Galacto-Oligosaccharide Synthesis—cont'd

Organism	Strain	References
Archaea		
	Pyrococcus furiosus	Hansson and Adlercreutz (2001), Petzelbauer et al. (2002), and Bruins et al. (2003)
	Sulfolobus solfataricus	Reuter et al. (1999), Petzelbauer et al. (2002), Park et al. (2008), and Wu et al. (2013)
	Thermus sp.	Ladero et al. (2006)
Yeasts		
	Bullera singularis	Cho et al. (2003)
	Candida bombicola	Petrova and Kujumdzieva (2010)
	Cryptococcus laurentii	Ohtsuka et al. (1990)
	Guehomyces pullulans	Nakagawa et al. (2006)
	Kluyveromyces fragilis	Liu et al. (2012)
	Kluyveromyces lactis (*Kluyveromyces marxianus*)	Martínez-Villaluenga et al. (2008), Petrova and Kujumdzieva (2010), Kumari et al. (2011), Rodriguez-Colinas et al. (2011), Padilla et al. (2012), Braga et al. (2013), Klein et al. (2013), and Srivastava et al. (2015)
	Rhodotorula minuta	Onishi and Tanaka (1996) and Onishi and Yokozeki (1996)
	Sirobasidium magnum	Onishi et al. (1996) and Onishi and Tanaka (1997)
	Sporobolomyces singularis	Sakai et al. (2008)
	Sterigmatomyces elviae	Onishi et al. (1995) and Onishi and Tanaka (1998)
Molds		
	Aspergillus aculeatus	Rodriguez-Fernandez et al. (2011) and Frenzel et al. (2015)
	Aspergillus oryzae	Irazoqui et al. (2009), Vera et al. (2011), Urrutia et al. (2013a,b), and Fischer and Kleinschmidt (2015)
	Aureobasidium pullulans	Chi et al. (2009)

TABLE 4.1 Main Microorganisms Producing β-Galactosidases for Galacto-Oligosaccharide Synthesis—cont'd

Organism	Strain	References
	Penicillium expansum	Li et al. (2008)
	Penicillium simplicissimum	Cruz et al. (1999)
	Sporobolomyces singularis	Ishikawa et al. (2005)
	Sporotrichum thermophile	Petrova et al. (2009)
	Scopulariopsis sp.	dos Santos et al. (2009)
	Talaromyces thermophilus	Nakkharat and Haltrich (2006)

et al. (2011), and Nath et al. (2015). Properties of some of those enzymes as catalysts for GOS synthesis have been gathered by Benjamins (2014).

Protein-engineering tools have been used to improve β-galactosidase characteristics of importance for GOS synthesis. They have been mostly oriented to increase the transgalactosylation–hydrolysis activity ratio. Jørgensen et al. (2001) reported an increase in such ratio to 9:1 by truncating a β-galactosidase from *B. bifidum* from the C-terminal end and expressing the truncated enzyme variants in *E. coli*. A twofold increase in GOS yield at moderately low lactose concentration (10%) was obtained with a double mutant β-glucosidase of a hyperthermophilic strain of *Pyrococcus furiosus* using site-directed mutagenesis (Hansson et al., 2001). A significant increase in GOS yield to values higher than 50% was obtained by site-directed mutagenesis of a *Sulfolobus solfataricus* β-galactosidase when the mutant enzyme was expressed in *E. coli* (Wu et al., 2013). Site-directed mutagenesis has also been used with the purpose of increasing thermostability that, as previously said, is a key issue in GOS synthesis because of lactose solubility; substitution of Ile42 by polar amino acids significantly enhanced the thermostability of *G. stearothermophilus* β-galactosidase, but at the expense of reduced activity (Dong et al., 2015).

Directed evolution also has been applied to increase the transglycosylation activity of a β-galactosidase from *G. stearothermophilus* (Placier et al., 2009). A major concern in the enzymatic synthesis of GOS is the inhibition by the hydrolysis product galactose, which occurs at different levels in all producing strains (Chockchaisawasdee et al., 2004; Neri et al., 2009); in some cases, glucose is also an inhibitor and many strategies have been envisaged to avoid or reduce these effects. A fourfold increase in the inhibition constant by galactose was obtained by immobilization of a *Thermus* sp. β-galactosidase on Sepabeads (Pessela et al., 2003), and more than fiftyfold increase in glucose inhibition constant was obtained by cross-linking a *K. lactis* β-galactosidase (Mateo et al., 2004). Mutant enzymes from *Thermus thermophilus* obtained by site-directed mutagenesis oriented to the galactose binding site resulted in a reduction of

TABLE 4.2 Recombinant β-Galactosidases Used in Galacto-Oligosaccharide Synthesis

Gene Donor	Host	References
Alkalilactibacillus ikkense	*Escherichia coli*	Schmidt and Stougaard (2010)
Arthrobacter sp.	*E. coli*	Pawlak-Szukalska et al. (2014)
Aspergillus candidus	*Pichia pastoris*	Zheng et al. (2006)
Bacillus megaterium	*E. coli*	Juajun et al. (2011)
Bacillus stearothermophilus	*Bacillus subtilis*	Chen et al. (2008)
Bifidobacterium bifidum	*E. coli*	Goulas et al. (2009) and Osman et al. (2014)
Bifidobacterium breve	*E. coli*	Arreola et al. (2014)
Bifidobacterium infantis	*E. coli*	Hung and Lee (2002)
Caldicellulosiruptor sp.	*E. coli*	Liu et al. (2015)
Enterobacter cloacae	*E. coli*	Lu et al. (2009)
Lactobacillus acidophilus *Lactobacillus reuteri*	*Lactobacillus plantarum* *Lactobacillus sakei*	Halbmayr et al. (2008)
Lactobacillus plantarum	*E. coli*	Benavente et al. (2015)
Kluyveromyces lactis	*E. coli*	Kim et al. (2004a,b)
Paecilomyces aerugineus	*Pichia pastoris*	Katrolia et al. (2011)
Pyrococcus woesei	*E. coli*	Wanarska et al. (2005)
Sulfolobus solfataricus	*Pichia pastoris*	Sun et al. (2014)
Thermotoga maritima	*E. coli* *E. coli*	Ji et al. (2005) Kim et al. (2004a,b)
Thermus sp.	*E. coli*	Ladero et al. (2002)
Thermus thermophilus	*E. coli*	Nam et al. (2004)
Metagenomic library	*Pichia pastoris*	Wang et al. (2012)

galactose inhibition (Park and Oh, 2010). Site-directed mutagenesis of an *Aspergillus niger* gene and transformation of a *P. pastoris* host resulted in the production of a mutant β-galactosidase with a galactose inhibition constant of 6.5 mM, almost 10 times higher than in the parent enzyme (Hu et al., 2010). Another strategy to reduce inhibition of GOS synthesis by monosaccharides is based on in situ removal of inhibitors, either using membrane reactors for continuous removal (Matella et al., 2006; Sen et al., 2012) or adsorption from the reaction medium (Boon, 2000a), by partition in biphasic systems (del-Val and Otero, 2003) or by chemical transformation, as in the case of glucose removal by oxidation to gluconate with glucose oxidase and catalase (Cheng et al., 2006a). However, glucose effect is very much dependent on the enzyme origin; in some cases glucose inhibition was insignificant, and even a mild activation by glucose was reported in the case of *A. oryzae* β-galactosidase (Vera et al., 2011).

As generally outlined in Section 3.1, the enzymatic synthesis of GOS is a kinetically controlled reaction in which lactose acts both as donor and acceptor of the galactosyl moiety; actually, lactose is a very effective galactosyl donor (Planas and Faijes, 2002). Reaction is not restricted by equilibrium and depends on the kinetic characteristics of the β-galactosidases, being a flexible and successful strategy. GOS yields rarely exceed 40% and are usually in the range from 20% to 40%, with the notable exception of some hyperthermophilic enzymes with which yields as high as 50% are obtained as a consequence of using the very high lactose concentrations attainable at high temperatures (Wu et al., 2013). Such moderate GOS yields are obtained at lactose conversions in the range from 50% to 60%, and higher lactose conversions inevitably lead to lower GOS yields because of the increasing hydrolysis of the substrate and the synthesized product (Gosling et al., 2010). This situation is represented in Fig. 4.2.

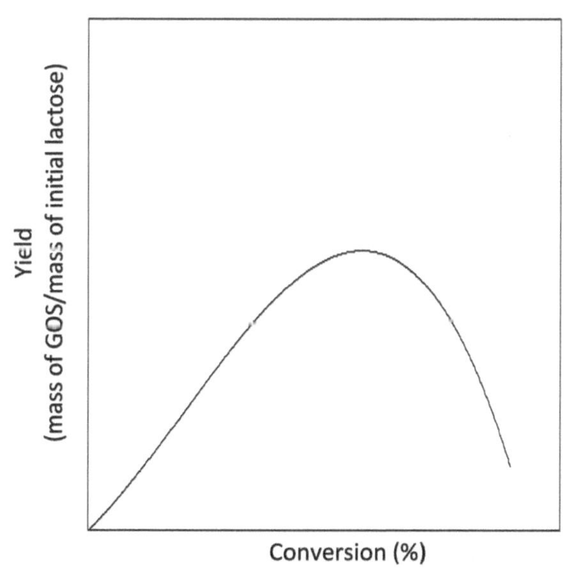

FIGURE 4.2 Relationship between GOS yield (mass of GOS produced per unit mass of initial lactose) and lactose conversion (percentage of initial lactose converted into products of transgalactosylation and hydrolysis).

The difficulty of obtaining higher yields implies that the synthesized product (raw GOS) contains a high proportion of residual lactose and monosaccharides that need to be removed to a variable extent according to the product specifications for use; this is highly demanding on downstream operations of purification, which is a major issue in GOS production, as will be analyzed in Section 4.4. There is agreement that the use of (very) high lactose concentrations is mandatory for GOS production, and concentrations in the range from 30% to 60% are used in most cases; concentrations of 10% or lower cannot be used because lactose hydrolysis will prevail. High substrate concentrations have a dual effect: an increase in the concentration of galactosyl precursor, which is desirable from kinetics consideration; and reduced water activity, which is also desirable by lowering the rate of lactose and GOS hydrolysis (Vera et al., 2012). Besides, a strong increase in enzyme stability occurs at increasing lactose concentrations (Warmerdam et al., 2013b). Decreasing water activity is a key issue as demonstrated by the positive effect on GOS yield when using organic solvents (Chen et al., 2001; Cruz-Guerrero et al., 2006; Srisimarat and Pongsawasdi, 2008), which, by the way, is hardly applicable in this case because of the limited solubility of lactose in such media.

Temperature is another key variable in GOS synthesis, having a dual effect: on the one hand, the ratio of synthesis to hydrolysis is dependent on temperature, and, in general, such ratio increases with temperature in a certain range; on the other hand, an indirect effect is that higher temperatures allow higher lactose concentrations, which is the most significant variable in GOS production (Boon et al., 2000b). In addition, viscosity of lactose solutions decreases sharply with temperature, which is another asset in terms of higher mass transfer rates, which is especially important in the case of using immobilized enzymes (Albayrak and Yang, 2002a). Optimal temperature is very much dependent on the enzyme origin, and temperature effects on activity and stability have to be optimally counterbalanced in each case, but the effects stated before suggest the convenience of using thermostable β-galactosidases (Park et al., 2008).

Microwave irradiation proved to be an interesting option for GOS synthesis since even though GOS yield did not increase significantly, product hydrolysis was significantly decreased. This behavior was observed both with *K. lactis* and *B. circulans* β-galactosidases and was explained by an early activation effect of microwaves on β-galactosidase followed by a rapid inactivation that prevented the hydrolysis of the GOS formed (Maugard et al., 2003; Kamerke et al., 2012).

Large-scale industrial producers of GOS are summarized in Table 4.3, but the catalysts that have been mostly reported for GOS production are the β-galactosidases from *A. oryzae*, *K. lactis*, and *B. circulans*. They have GRAS status and the former two have a long tradition of safe use both by the pharmaceutical and food industries in the hydrolysis of lactose (Booij, 1985; Gekas and López-Leiva, 1985; Ogawa and Shimizu, 2002). However, they exhibit very different properties and produce quite different GOS spectra when used as catalysts for lactose transgalactosylation (Boon and Janssen, 2000b). The enzyme from *A. oryzae*

TABLE 4.3 Main Industrial Producers of Galacto-Oligosaccharides

Brand Name	Producing Company	β-Galactosidase Producing Organism
Oligomate	Yakult Pharm Ind Co. (Japan)	*Sporobolomyces singularis, Kluyveromyces lactis*
Vivinal GOS	Friesland foods Domo (The Netherlands)	*Bacillus circulans*
Cup-Oligo	Nissin Sugar Mfg. Co. (Japan)	*Cryptococcus laurentii*
BiMuno	Clasado Ltd. (Malta)	*Bifidobacterium bifidum*
Bioligo GL 5700 IMF	Ingredion Incorporated (South Korea)	*B. circulans, Aspergillus oryzae*

produces GOS-3 as the main product with much smaller amounts of GOS-4 and negligible amounts of higher oligosaccharides, which is considered a good profile for prebiotic action; the enzyme is monomeric with an estimated molecular weight of 105,000 Da (Tanaka et al., 1975), although slightly higher molecular weights of 110,000 and 113,000 Da were later reported (Ogushi et al., 1980; Todorova-Balvay et al., 2006); the closely related β-galactosidase from *A. niger* has been better characterized, and molecular weights from 124,000 to 173,000 have been determined, differences being related to the different content of carbohydrates in the glycoprotein (Widmer and Leuba, 1979). The β-galactosidase from *K. lactis* produces significant amounts of GOS-2 and lesser amounts of higher oligosaccharides (Cheng et al., 2006a); it is a tetramer (dimer of dimers) whose 3D structure has been elucidated (Pereira-Rodríguez et al., 2012); it is a very potent lactose hydrolyzer, but when used for GOS synthesis a significant fraction is represented by disaccharides whose prebiotic potential is not fully disclosed (Martínez-Villaluenga et al., 2008; Rodríguez-Colinas et al., 2014). The β-galactosidase from *B. circulans* produces a wide distribution of oligosaccharide going from GOS-2 to GOS-5 and higher; it consists of four isoforms whose transgalactosylation–hydrolysis activity ratio varies from one to another (Warmerdam et al., 2013a). The β-galactosidases from *A. oryzae* and *B. circulans* exhibit high transgalactosylation activity, while those from *K. lactis* exhibit high hydrolytic activity but lower transgalactosylation activity (Panesar et al., 2006; Guerrero et al., 2015). Higher GOS yield is obtained with the *B. circulans* β-galactosidase (around 40%) than with the enzyme from *A. oryzae* (around 30%), but the latter is more stable allowing a higher operational temperature, being a well-known commodity enzyme that has been used for decades in lactose hydrolysis in milk and dairy products (Woychik and Wondolowski, 1973; Harju et al., 2012). Large-scale industrial producers of GOS are summarized in Table 4.3.

Synthesis of GOS at productive scale has been mostly conducted using refined lactose or concentrated whey permeate as substrate and enzymes free in solution as catalyst (Lamsal, 2012). Free enzymes are used most probably because (1) β-galactosidases are commodity enzymes whose impact on production cost may be of lesser significance than other aspects of the process; (2) the high concentrations of lactose required for synthesis may discourage the use of heterogeneous catalysis because of mass-transfer limitations. However, production with soluble enzymes is in most cases conducted in batch mode of operation, implying a very low efficiency of catalyst use (the enzyme is used once for a short period of time and then discharged) and inactivation and removal of the spent catalyst from the product stream, which are rather cumbersome operations at large scale, especially in this case where the catalyst has to be removed from a high-solids medium of high viscosity.

An alternative mode of operation with soluble enzymes is the production in continuous-stirred tank reactors (CSTR) provided with an ultrafiltration membrane device where the enzyme is retained while products are continually removed (Otieno, 2010). GOS yield was optimized in a CSTR using tubular ceramic membranes obtaining an optimum GOS yield of 38% with a commercial *Kluyveromyces fragilis* β-galactosidase, higher than obtained in conventional batch operation because of reduced hydrolysis of the formed product (Ebrahimi et al., 2010). Working in a similar system, better performance for CSTR was invoked than in conventional batch by avoiding enzyme inactivation and removal from the product stream (Pocedičová et al., 2010). Conventional batch operation was compared with operation in a CSTR provided with an external cross-flow polyethersulfone ultrafiltration membrane device to retain the enzyme; steady-state operation was achieved allowing to obtain constant conditions for each dilution rate tested, which may be advantageous for producing a more-defined GOS profile with respect to chain length and type of glycosidic bond (Splechtna et al., 2007a). A possible scheme of industrial production of GOS in concentrated syrup and powder forms is presented in Fig. 4.3.

Immobilization may solve most of the constraints inherent to the use of soluble catalysts, which explains the overwhelming amount of information related to the use of immobilized enzymes for GOS synthesis. Continuous (Albayrak and Yang, 2002a), fed-batch (Vera et al., 2014), or sequential batch (Sakai et al., 2008; Huerta et al., 2011) operation with robust and stable immobilized enzymes or cells is the way to go for improving GOS production at industrial level. This aspect will be reviewed in the next section.

4.2.3 Synthesis of Galacto-Oligosaccharides With Immobilized Enzymes

First reports on the use of immobilized β-galactosidase for upgrading cheese whey date back to 1977 (Leuba and Widmer, 1977). In the last decades, many studies have reported on the use of immobilized β-galactosidase for the synthesis

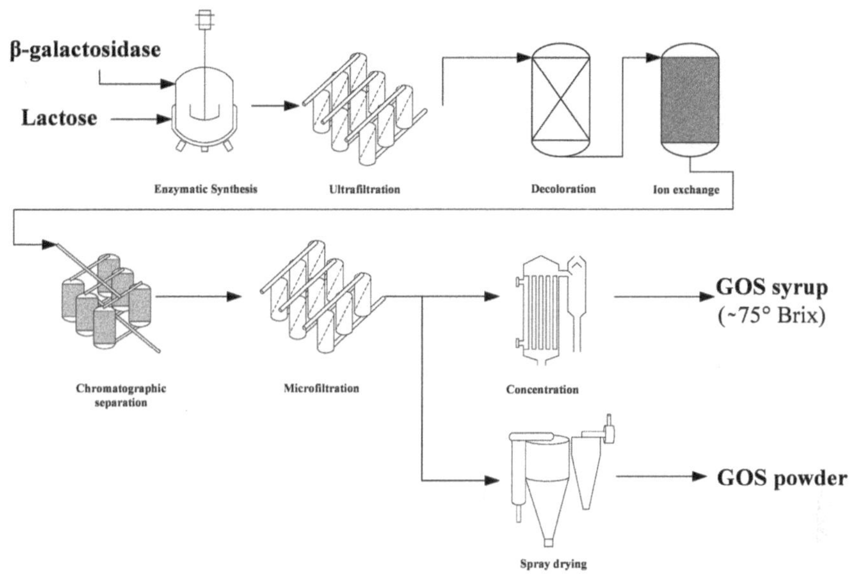

FIGURE 4.3 Process diagram of galacto-oligosaccharide production.

of GOS. Even though enzyme immobilization has been profusely developed and can be now considered a mature technology, most industrially applicable β-galactosidases are complex proteins consisting of more than one monomeric unit, which complicates its immobilization and stabilization. On the other hand, synthesis of GOS must be conducted at (very) high lactose concentration, which is a critical and nontrivial issue when selecting the immobilization system. Existing immobilization strategies, β-galactosidase molecular features according to its origin, and β-galactosidase immobilization strategies used for the synthesis of GOS will be reviewed.

4.2.3.1 Immobilization Strategies

First studies on enzyme immobilization as process catalysts date back to the late 1960s. In the following decades enzyme immobilization research bloomed and appeared as a most promising industrial biotechnology. Confronted with reality, industrial processes adopting this technology in the following decades were few; however, significant advances in immobilization strategies and support design have opened opportunities of using immobilized enzymes not only in conventional processes of hydrolysis or isomerization but also in organic synthesis. Early studies on enzyme immobilization were mostly focused on catalyst recovery and reuse and production of catalyst-free product streams. Recent work has been focused mostly on enzyme immobilization as a strategy for protein structure stabilization. Several studies are being reported that refer to support structure, protein surface configuration, enzyme–support interactions, catalyst recovery, and support regeneration after enzyme exhaustion with a view from a production process perspective. After more than four decades of development, enzyme immobilization is still facing many challenges with ample room for future opportunities especially in the area of organic synthesis processes and also in biomedical applications.

An ample spectrum of immobilization methodologies is now available (Mateo et al., 2007; Betancor and Luckarift, 2008) (see Fig. 4.4); however, the best methodology to use will be case dependent and many variables will be involved in optimizing the immobilization process.

Immobilization methods can be broadly classified as carrier-bound or carrier-free.

4.2.3.1.1 Carrier-Bound

This is the most used technique for enzyme immobilization and consists of the attachment or containment of the enzyme to (into) a solid support that allows an easy separation from the reaction medium producing in most cases an increase in enzyme stability. Many studies about enzyme immobilization by covalent and noncovalent bonding to solid supports and encapsulation have been developed (Gupta, 1991; Mateo et al., 2000). A powerful strategy of enzyme immobilization was developed based on the multipoint covalent attachment of amino acid residues of the enzyme protein to solid matrices, leading to very robust and stable catalysts (Guisán, 1988; Mateo et al., 2006). A more rational approach to the immobilization of enzymes to supports is now possible because of the advances in enzyme three-dimensional structure and in support microstructure determination that allow knowing the protein distribution within the support pores (Bolivar et al., 2016). Some of the most desirable properties of a support for enzyme immobilization are (Cantone et al., 2013; Datta et al., 2013; Zucca and Sanjust, 2014):

- high surface area
- wettability in the reaction medium

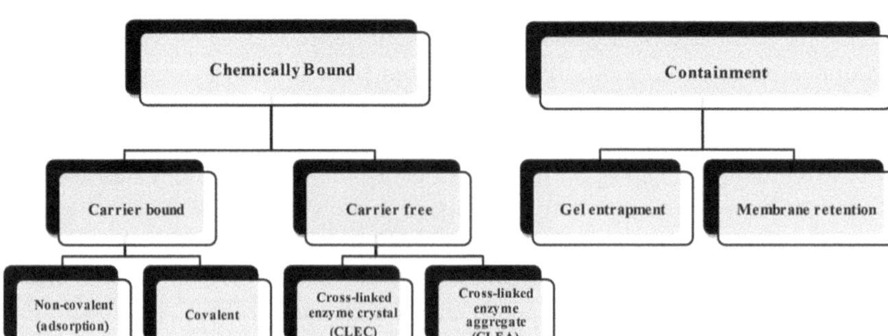

FIGURE 4.4 Strategies of enzyme immobilization.

- insolubility in the reaction medium
- chemical, thermal, and mechanical stability
- resistance to microbial attack
- compatibility with use (in the case of GOS, supports and all reagents used in their activation should be accepted for food use)
- availability at affordable price without restrictions of supply
- simplicity of fabrication
- regeneration after catalyst exhaustion

There is no support optimally complying with all these desirable characteristics, and a judicious choice of the support should balance pros and cons in each particular case.

It is important to note that in the case of multimeric enzymes (many β-galactosidases are in fact multimeric) all subunits must be stabilized since in most cases enzyme inactivation begins by subunit dissociation (Fig. 4.5). Traditional protocols of enzyme immobilization are not well suited for this kind of complex enzyme and additional cross-linking steps may be required for the stabilization of the whole enzyme structure (Balcao et al., 2001; Fernández-Lafuente et al., 2001, 2009; Betancor et al., 2003; Hidalgo et al., 2003; Pessela et al., 2004; Bolivar et al., 2010).

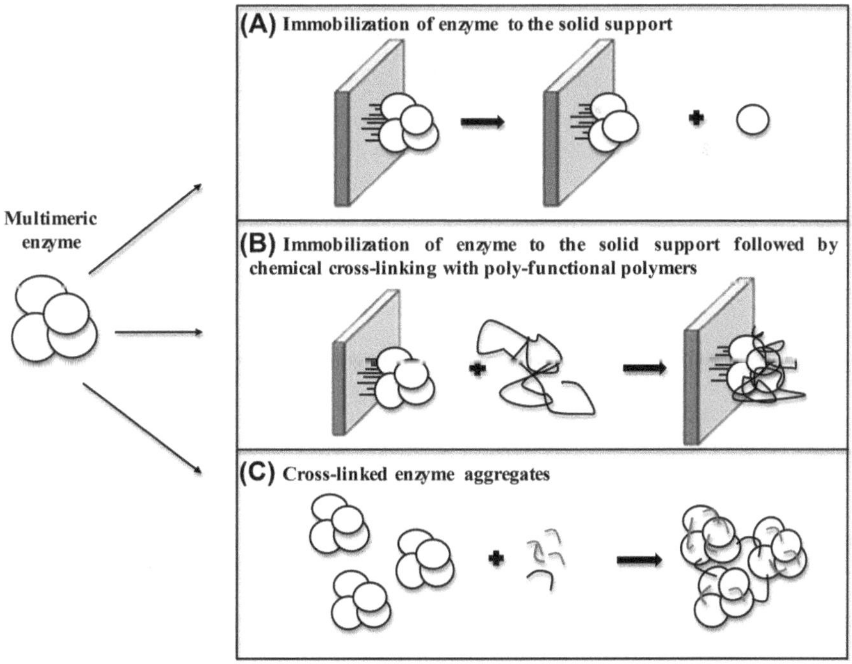

FIGURE 4.5 Strategies of immobilization of multimeric enzymes.

Options that have been proposed for the stabilization of multimeric enzymes are the immobilization in supports coated with ionic polymers at high-ionic strength allowing stronger interaction with the enzyme and deeper penetration of the enzyme in the polymeric matrix during immobilization (Pessela et al., 2005), and immobilization to the solid support followed by chemical cross-linking with polyfunctional polymers surrounding the enzyme (Fernández-Lafuente et al., 2009; Pessela et al., 2004).

Entrapment is the most reported system for cell immobilization. It consists of the retention of the enzyme within the inner cavities of a solid porous matrix usually formed by organic polymers, like polyacrylamide, collagen, alginate, resins, and polyvinyl alcohol (PVA). However, in the case of enzyme immobilization, leakage from the matrix is a problem due to the small molecular size of the proteins. Leakage can be avoided if the enzyme is first precipitated and cross-linked to form a protein aggregate of a size similar or higher than a microbial cell. This strategy has been successfully used in the case of penicillin acylase, as reported by Wilson et al. (2004a) and Sangeetha and Abraham (2008). Entrapment within semipermeable membranes can be considered in this category, even though the enzyme is not necessarily bound but could be simply contained by the membrane (Jochems et al., 2011).

4.2.3.1.2 Carrier-Free

In this case, the enzyme protein constitutes its own support so that high specific activities can be obtained with the advantage of not requiring an inert support that can be even more expensive than the enzyme itself (Cao et al., 2000). Cross-linked enzyme crystals (CLECs) are produced by cross-linking of enzyme crystals obtained from a highly purified enzyme (Vaghjiani et al., 2000), while cross-linked enzyme aggregates (CLEAs) are produced by cross-linking of enzyme protein aggregates produced by conventional nondenaturing protein precipitation techniques (Fig. 4.6). CLEAs combine the good properties of carrier-free systems (simplicity and high specific activity) with the advantage of much lower production cost since, as opposite to CLECs, no purified enzyme is required as starting material (Sheldon, 2011a; Cui and Jia, 2015)

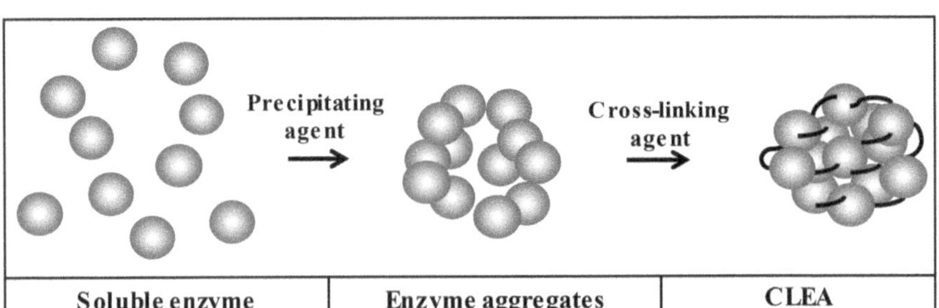

FIGURE 4.6 Scheme of preparation of cross-linked enzyme aggregates (CLEAs).

and several conventional techniques used for protein purification are applicable, obtaining high recovery yields of enzyme activity. CLEAs appear then as a promising technology for enzyme immobilization. A large number of technologically important enzymes have been produced in this format that should have an industrial impact in the forthcoming years, and special reactor configurations have been developed for their proper handling and recovery (Sheldon, 2011b). CLEAs are particularly well suited as an immobilization strategy for multimeric enzymes since cross-linking of the subunits is promoted (Wilson et al., 2004b).

4.2.3.2 Characteristics of β-Galactosidases Most Commonly Used in the Synthesis of GOS

Previous knowledge of some fundamental properties of the enzyme (distribution of surface amino acid residues, number of subunits if multimeric, molecular size, tendency to dimer formation, stability) is required for the proper adoption of an immobilization strategy.

As pointed out in Section 4.2.2, β-galactosidases having GRAS or equivalent status, which are therefore eligible for GOS production, are those from *A. oryzae*, *A. niger*, *K. lactis*, *K. fragilis*, and *B. circulans*. Some relevant characteristics of these enzymes are as follows:

- *A. oryzae* β-galactosidase: Its crystalline structure has been published recently; it is a monomeric multidomain enzyme with Glu200 as the acid/base catalyst and Glu298 as a nucleophile (Maksimainen et al., 2013). Its molecular mass is 113 kDa (Todorova-Balvay et al., 2006).
- *A. niger* β-galactosidase: It has been expressed in yeasts and purified by immobilized metal–ion affinity chromatography for crystallization experiments. The monomeric glycosylated enzyme has a molecular mass of 129 kDa (O'Connell and Walsh, 2010).
- *K. lactis* β-galactosidase: It has been purified and its crystal structure solved for the free enzyme and its galactosyl complex at 2.75 and 2.8 Å, respectively (Rojas et al., 2004). It is a tetramer, an assembly of dimers, with higher calculated dissociation energy for the dimers than for its assembly, which can explain that equilibrium may exist in solution between its dimeric and tetrameric forms. Two active centers are located at the interface within each dimer, in a narrow channel of 10 Å in width that makes the catalytic pockets accessible to the solvent (Pereira-Rodríguez et al., 2012). Molecular mass of each subunit is 118 kDa (Kim et al., 2003).
- *K. fragilis* β-galactosidase: It is a homodimer, each subunit having a molecular mass of 123 kDa (O'Connell and Walsh, 2007). Other authors have reported a tetrameric structure, but tetramer formation appears as nonessential for catalysis (Yoshida et al., 2010).
- *B. circulans* β-galactosidase: Several isoenzymes have been isolated in β-galactosidase crude extracts from *B. circulans* (Vetere and Paoletti, 1998;

Song et al., 2011; Warmerdam et al., 2013a). In the most recent study, four isoenzymes, designated as β-galA, β-galB, β-galC, and β-galD, were isolated, their molecular masses being 206, 181, 164, and 112 kDa, respectively (Warmerdam et al., 2013a). These isoenzymes differ in some properties, like specific activity and stability, but isolating them represents a high cost that may not be justified for technological purposes, so in practice the enzyme preparation containing the whole mixture of isoenzymes is used for GOS synthesis.

4.2.3.3 Immobilized β-Galactosidases Used in the Synthesis of GOS

Even though a large number of immobilized β-galactosidases have been developed as catalysts for GOS synthesis, it is believed that industrial production of GOS is produced mostly with soluble enzymes (Playne and Crittenden, 2009). This situation may reflect that at present the cost of the catalyst is not critical in terms of operating costs, but this situation will certainly change as competition gets stronger and prices of products are reduced. This is a challenge for developing more active and robust catalysts, improving its lifespan of use, and increasing the specific productivity of GOS synthesis. Most relevant strategies reported for the immobilization of β-galactosidases for GOS synthesis are described following.

4.2.3.3.1 Carrier-Bound Immobilized β-Galactosidase

This is the most reported strategy for developing catalysts for GOS synthesis. Table 4.4 summarizes the support materials and type of enzyme–support interaction used for the immobilization of β-galactosidases for GOS synthesis.

4.2.3.3.1.1 β-Galactosidase Immobilized on Silica Porous silica is an interesting support material for enzyme immobilization because of its high surface area that allows different pore geometries and pore size distributions; surface can be functionalized with different chemical groups to assist in the immobilization of the enzyme. The material has a high mechanical, thermal, and chemical stability being resistant to microbial degradation. This allows using it in several reactor configurations even at high shear rates without particle disintegration (Bernal et al., 2012; Magner, 2013). Mozaffar et al. (1988) studied the synthesis of GOS at pH 6, 40°C, and 46 g/L of lactose with a crude extract of *B. circulans* β-galactosidase immobilized by cross-linking with glutaraldehyde and immobilized by adsorption in porous silica followed by cross-linking. GOS yield with both catalysts was 40% obtained at lactose conversions of 60% and 55%, respectively.

4.2.3.3.1.2 β-Galactosidase Immobilized on Celite There is only one report on the use of this type of catalyst in the synthesis of GOS; the catalyst obtained was discarded for further studies because of its low protein load (Gaur et al., 2006).

TABLE 4.4 Carrier-Bound Immobilized β-Galactosidases Used for GOS Synthesis

Support	Enzyme-Support Interaction	Enzyme Source	References
Celite	Adsorption	*Aspergillus oryzae*	Gaur et al. (2006)
Duolite A-568, Ion Exchange Resin	Adsorption	*Kluyveromyces lactis*	Maugard et al. (2003)
EC-HFA1 Sepabeads	Covalent binding	*A. oryzae*	Huerta et al. (2011)
Silica	Covalent binding	*Bacillus circulans*	Mozaffar et al. (1988)
Eupergit C (oxirane-activated acrylic supports)	Covalent binding	*Talaromyces thermophilus* *B. circulans*	Nakkharat and Haltrich (2007) Warmerdam et al. (2014)
Polysiloxane–polyvinyl alcohol magnetic composite	Covalent binding	*K. lactis* *A. oryzae*	Neri et al. (2008) Neri et al. (2009)
Agarose (glyoxal groups)	Covalent binding	*A. oryzae*	Huerta et al. (2011)
Agarose (glyoxyl agarose and three heterofunctional supports: amino-, carboxy-, and chelate-glyoxyl agarose)	Covalent binding	*B. circulans*	Urrutia et al. (2013a,b)
Chitosan	Covalent binding	*A. oryzae*	Sheu et al. (1998)
Chitosan	Covalent binding	*B. circulans*	Urrutia et al. (2014)
Chitosan-coated magnetic Fe_3O_4 nanoparticles	Covalent binding	*A. oryzae*	Pan et al. (2009)
Chitosan	Covalent binding	*A. oryzae*	Gaur et al. (2006)
Chitosan activated with tris(hydroxymethyl) phosphine	Covalent binding	*B. circulans*	Cheng et al. (2006a)
Chitosan (Chitopearl BCW 3510)	Adsorption	*Bullera singularis*	Shin et al. (1998)
Chitosan-coated magnetic nanoparticles	Adsorption	Not reported	Chen and Duan (2015)

4.2.3.3.1.3 *β-Galactosidase Immobilized on Anion-Exchange Resins*

Maugard et al. (2003) immobilized β-galactosidase from *K. lactis* on DuoliteA-568 for using it as catalyst for the synthesis of GOS. Half-life under nonreactive conditions at 40°C was 15 h, representing a stabilization factor (ratio of half-life of the immobilized enzyme and the free-enzyme counterpart) of 7.5; under reactive conditions half-life was 40 h representing a stabilization factor of 2. Stability of the catalyst under microwave irradiation (residual activity after15 h at 12 W, 40°C) as thermal source was also determined. Microwave irradiation had no effect on the stability of the immobilized catalyst, but stability of the free enzyme was reduced at the same irradiation conditions. Better results were obtained in the synthesis of GOS under microwave irradiation than in conventional heating at comparable conditions, GOS yield being 38%. Selectivity of the enzyme for GOS synthesis was reported to increase 217-fold with respect to conventional synthesis (aqueous medium with free enzyme under conventional heating) by exposing the catalyst to microwave irradiation in the presence of organic solvents, such as hexanol. Recently Carević et al. (2016) used another anion-exchange resin (Purolite A109) that was chemically modified to immobilize a chemically modified β-galactosidase from *A. oryzae* and used the catalyst for the synthesis of GOS in 10 sequential batch and fluidized-bed operations at the end of which residual activity was 55% and 75% of initial activity, respectively. A product with 100 g/L of GOS was obtained with a yield of 25%, which is close to the usual values obtained with *A. oryzae* β-galactosidases.

4.2.3.3.1.4 β-Galactosidase Immobilized on Sepabeads Amino-epoxy Sepabeads is a readily available and highly reactive material with a defined particle and pore size, bearing excellent mechanical properties and producing highly stable biocatalysts (Mateo et al., 2007). Huerta et al. (2011) immobilized an *A. oryzae* β-galactosidase on EC-HFA1 Sepabeads with a protein immobilization yield of 87.1% and a yield of expressed activity of 42.3%, the catalyst having a specific activity of 1790 IU/g_{biocat}. Stabilization factor was 2.7 and 1.75 at 55°C and 60°C, respectively. Even though better results were obtained than with the free enzyme in the synthesis of GOS, performance was better with the enzyme immobilized by the same authors in other supports.

4.2.3.3.1.5 β-Galactosidase Immobilized on Eupergit Eupergit C 250 L is a well-known enzyme carrier consisting of porous acrylic beads with oxirane functionality. Covalent immobilization proceeds by the reaction of thiol and amino groups of the enzyme with the epoxide groups of the carrier (Boller et al., 2002; Sheldon, 2007). β-Galactosidase from *B. circulans* was immobilized on Eupergit C with 31% yield of expressed activity; even though this is a low yield, the catalyst obtained allowed the continuous production of GOS in a packed-bed reactor reflecting a good operational stability. Half-life of the catalyst at 38% (w/w) feed lactose concentration was 90 days at 50°C, and specific productivity of synthesis with the immobilized enzyme was six times higher than obtained with the free enzyme in batch operation (Warmerdam et al., 2014). In the case of β-galactosidase from *Talaromyces thermophilus* immobilized on Eupergit, part of the enzyme was merely adsorbed and was easily leached from the support; the other fraction was covalently bound and very stable throughout the operation (Nakkharat and Haltrich, 2007). The authors remarked that in the continuous conversion of lactose into GOS relatively more trisaccharides were produced than in discontinuous mode of operation with the same catalyst.

4.2.3.3.1.6 β-Galactosidase Immobilized on Polyvinyl Alcohol Even though PVA has been used mostly for enzyme encapsulation, as described later; it has also been used as support when the PVA particles formed are further functionalized with glutaraldehyde to covalently bind the enzyme. With such a catalyst, GOS was synthesized in 10 consecutive batches preserving 84% of the initial activity at the end of operation (Neri et al., 2009).

4.2.3.3.1.7 β-Galactosidase Immobilized on Agarose Agarose supports are remarkable for being highly porous, hydrophilic, devoid of functional groups producing nonspecific substrate or product binding, and commercially available (Brena and Batista-Viera, 2006). Agarose is a linear polysaccharide of repeating units of agarobiose, which is a dimer of D-galactose and 3-6-anhydro-L-galactopyranose. Agarose is a main component of agar, which is a polysaccharide extracted from seaweed (Laurienzo, 2010). Commercial agarose is now produced in cross-linked form with either epichlorohydrin or cyanogen bromide (Saudagar and Singhal, 2004). Enzyme immobilization in agarose gels is carried out after the functionalization of agarose with hydroxyl groups. The most common procedure of functionalization is the incorporation of linear aldehyde (glyoxyl) groups that bind the enzyme by multipoint formation of Schiff bases with the ε-amino groups of lysine residues in the enzyme surface (Guisán, 1998). When comparing different immobilized *A. oryzae* β-galactosidase for the synthesis of GOS, the glyoxyl-agarose immobilized enzyme was the best in terms of performance in GOS synthesis (GOS yield, lactose conversion, and product-to-enzyme mass ratio), even though expressed activity yield (42.7%) and stability under nonreactive conditions (half-life of 170 h at 50°C) were not the highest; operational stability after 10 consecutive batches at the optimized reaction conditions (half-life of 260 h) was significantly higher than under nonreactive conditions, which can be attributed to galactose protection (Huerta et al., 2011). In another study, agarose functionalized with different groups was used as support for *B. circulans* β-galactosidase. The evaluated supports were monofunctional glyoxyl agarose and three heterofunctional supports: amino-, carboxy-, and chelate-glyoxyl agarose. Best immobilization yield and catalyst stability were obtained for the glyoxyl-agarose immobilized enzyme whose immobilization protocol was optimized. Optimal conditions of immobilization were 28.9 mg/g and 36.4 h of contact, resulting in a biocatalyst with 595 IU/g and a stabilization factor of 89; GOS yield and productivity remained unchanged along 10 repeated batches (Urrutia et al., 2013a,b).

4.2.3.3.1.8 β-Galactosidase Immobilized on Chitosan Chitosan is a biopolymer of natural origin that has been widely used as support material for enzyme immobilization. It is an innocuous, hydrophilic, biocompatible, cheap, and readily available material that can be shaped in different forms (films, beads, powder). It bears functional hydroxyl and amino groups allowing its direct linkage to the enzyme, or else to be chemically modified for such purpose (Krajewska, 2004; Wang, 2012). Chitosan is made up of D-glucosamine and *N*-acetyl-D-glucosamine units linked by β-1–4 bonds; it is obtained by extraction and partial deacetylation of chitin from crustacean shells, which is an abundant residue from shellfish industries (Górecka and Jastrzębska, 2011). A product with more than 60% deacetylation is considered chitosan; the one with less is considered chitin (Teng, 2012). Shin et al. (1998) immobilized a β-galactosidase from *Bullera singularis* by adsorption on chitosan; binding occurred by ionic interaction between the tertiary amino groups of chitosan and the carboxylic groups of amino acid residues of the enzyme protein; hydrogen bonds and van der Waals forces were also involved. Maximum amount of protein bound was $50\,mg/g_{support}$. The catalyst was used in continuous reactor operation for 15 days at a space velocity of $0.08\,h^{-1}$; at space velocities higher than $0.2\,h^{-1}$ considerable desorption of the enzyme occurred. Sheu et al. (1998) compared the behavior of free and chitosan-immobilized *A. oryzae* β-galactosidase: the pH activity profiles were similar and the thermal stability was significantly improved in the immobilized system. However, GOS yield obtained with the immobilized enzyme was reduced by 10–20% with respect to the free enzyme, which was attributed to diffusional restrictions. Gaur et al. (2006) compared the performance of *A. oryzae* β-galactosidase immobilized on cross-linked chitosan with celite-immobilized and CLEAs of the same enzyme, with the chitosan-immobilized enzyme the one producing the highest yield in terms of GOS-3 synthesis. Stabilization factor of the catalyst at 60°C was 1.6, and it could be used in four sequential batches. In a recent study, chitosan was derivatized by two methodologies to design a robust *B. circulans* β-galactosidase catalyst for its further application in the synthesis of GOS (Urrutia et al., 2014). In the first one, chitosan was derivatized by cross-linking with glutaraldehyde and activation with epichlorohydrin; in the second one, cross-linking and activation were done with epichlorohydrin in a two-step process, favoring first the support cross-linking and then the support functionalization. The latter support allowed obtaining the catalyst with the best compromise between expressed activity and thermal stability. The immobilization conditions were optimized, obtaining a biocatalyst with 280 IU/g; immobilization yields in terms of expressed activity and protein were $17.3 \pm 0.4\%$ and $61.5 \pm 3.9\%$, respectively. Stabilization was strongly improved by immobilization: a stabilization factor of 449 was obtained at 60°C allowing the synthesis of GOS in repeated batch operation without affecting the GOS yield and product composition. Magnetic chitosan supports have been used for the immobilization of several enzymes (Xie and Wang, 2012; Pospiskova and Safarik, 2013); however, immobilization of β-galactosidase on chitosan-coated magnetic Fe_3O_4 has been reported only by Pan et al. (2009) and more recently by Chen and Duan (2015). In the latter case, the β-galactosidase from *A. oryzae* was immobilized in magnetic cross-linked chitosan microparticles and several variables were studied: concentration of cross-linking agent, concentration of enzyme, cross-linking time, pH of the enzyme solution, and pH of the glutaraldehyde solution. The catalyst was free of diffusional restrictions and retained 92% of its activity after 15 batches of GOS synthesis. Interestingly, the support was reused after enzyme activity exhaustion, showing that the recovered support maintained its properties after several cycles of protein desorption–resorption cycles. Cheng et al. (2006b) reported an immobilization yield of expressed activity of 62% for a *B. circulans* β-galactosidase immobilized in tris-(hydroxymethyl) phosphine (THP)–activated chitosan. After 13 days at 55°C, residual enzyme activity of the catalyst was 75%, while the residual activity of the free enzyme at the same conditions was only 25%. GOS yield obtained with the immobilized enzyme was 41%, close to the 43% obtained with the free enzyme.

4.2.3.3.1.9 β-Galactosidase Immobilized by Containment Several studies refer to the immobilization of β-galactosidase by entrapment for GOS synthesis, being it by encapsulation within gel particles or by containment in semipermeable membranes. A scheme of this immobilization strategy is illustrated in Fig. 4.7, and reported cases applied to the synthesis of GOS are listed in Table 4.5.

PVA is a most used material in several industrial applications because of its very good mechanical properties, chemical resistance, nontoxicity and low cost (Lozinsky and Plieva, 1998; Durieux et al., 2000; Czichocki et al., 2001;

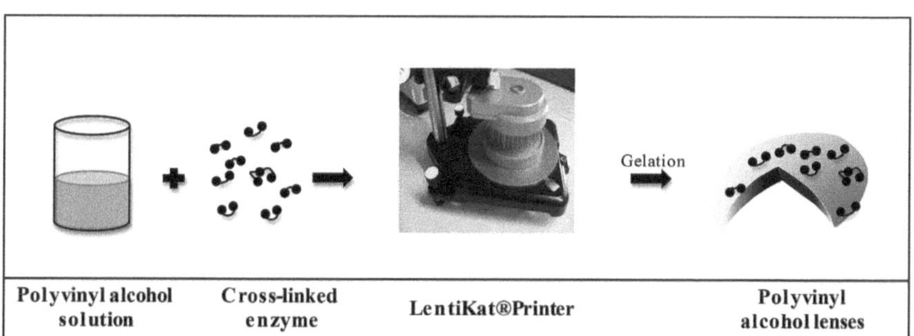

FIGURE 4.7 Scheme of β-galactosidase immobilization by containment in a polyvinyl alcohol gel matrix.

TABLE 4.5 Carrier-Bound Immobilized β-Galactosidase by Containment Applied to GOS Synthesis

Carrier	Enzyme-Support Interaction	Enzyme Source	References
Polyvinyl alcohol (PVA) lenses	Gel entrapment	*Aspergillus oryzae*	Jovanovic-Malinovska et al. (2012)
Sol–gel carriers	Gel entrapment	*A. oryzae*	Jovanovic-Malinovska et al. (2012)
Alginate	Gel entrapment	*Sporobolomyces singularis YIT*	Sakai et al. (2008)
	Gel entrapment	*Bacillus circulans*	Sen et al. (2014)
	Gel entrapment	*Aspergillus* sp.	Feng et al. (2010)
Polyvinylidene fluoride membrane	Cross-linking with glutaraldehyde	*B. circulans*	Palai et al. (2014)
Polymeric polyether sulfone, cellulose triacetate and thin film composite polyamide membrane	Covalent binding	*B. circulans*	Sen et al. (2011a)
Cellulose acetate membrane	Cross-linking	*A. oryzae*	Güleç et al. (2010)
	Cross-linking	*Kluyveromyces lactis*	Güleç (2013)
Cotton cloth membrane	Covalent binding to tosyl chloride activated support	*A. oryzae*	Albayrak and Yang (2002a)
	Polyethylenimine-coated support and glutaraldehyde cross-linking	*B. circulans*	Sanz-Valero (2009)
		A. oryzae	Matella et al. (2006) and Albayrak and Yang (2002b)
Membrane chromatography	Ionic binding	*K. lactis*	Engel et al. (2008)

Jahnz et al., 2001; Cattorini et al., 2009), which are highly desirable properties for an immobilization support. β-galactosidase was immobilized in PVA with a yield of recovered activity of 88.5% at a protein load of 5.32 $mg_{protein}/g_{support}$. PVA-immobilized β-galactosidase retained 95% of the initial activity after seven repeated batch operations of GOS synthesis and 51% of the initial activity after three months of storage. A maximum GOS yield of 30% was attained with a 40% (w/w) lactose solution fed at a rate of 10.8 mL/h, at pH 4.5 and 40°C. The maximum GOS productivity of 344 g/L/h was obtained at a flow rate of 28.7 mL/h. GOS-3 and GOS-4 were the major oligosaccharides formed. Using whey as raw material, GOS yield was 15% and product composition was 45% GOS-3, 40% GOS-4, and 15% GOS-5 (Jovanovic-Malinovska et al., 2012).

β-galactosidase from *A. oryzae* was immobilized in silica by sol–gel methodology with a yield of expressed activity of 73.8% and a specific activity of 200 $IU/g_{support}$ at a protein load of 5.32 $mg/g_{support}$. Stability was rather poor, 40% of initial activity being lost after seven repeated batches of GOS synthesis and 60% of activity lost after prolonged storage. Interestingly, optimum temperature for the immobilized enzyme was 60°C, 10°C higher than the optimum for the soluble enzyme (Jovanovic-Malinovska et al., 2012).

Sen et al. (2014) immobilized β-galactosidase in alginate with the purpose of maximizing the yield of expressed activity and the specific activity of the catalyst. Several variables were studied, and, at the best conditions, 25% of activity was lost during immobilization but the operational stability of the catalyst was not reported. On the other hand, Feng et al. (2010) studied some alternatives for improving the mechanical properties of the alginate-immobilized enzyme to make it suitable for the synthesis of GOS at process conditions. To that purpose, kaolin or diatomite were added prior to encapsulation to favor the subsequent cross-linking of the formed catalyst with glutaraldehyde or polyethylenimine; however, the mechanical properties of the catalyst were not improved, nor was the performance of the catalyst in the synthesis of GOS. Whole cells immobilized in alginate have also been used as catalysts for GOS synthesis. In this case, *S. singularis* cells were entrapped in the gel matrix and the enzyme was entrapped within the cell structure. In a single-batch reaction, GOS yield was 41% with free cells and 40% with immobilized cells containing β-galactosidase activity. Immobilized cells stably produced GOS for 20 batches (22 h/batch) at 55°C and pH 5.0 or 6.0 (Sakai et al., 2008).

Cotton fibers have been successfully used for β-galactosidase immobilization. Impressive results were reported for *A. oryzae* β-galactosidase immobilized on cotton cloth activated with tosyl chloride, obtaining a stabilization factor of 25 with a half-life higher than one year at 40°C. The catalyst was used in the synthesis of GOS in a continuous ultrafiltration membrane reactor fed with 200 g/L of lactose solution obtaining a GOS yield of 21% and a productivity of 80 g/L/h (Albayrak and Yang, 2002a). The same enzyme was immobilized on cotton cloth coated with polyethylenimine and cross-linked with glutaraldehyde; in this case, GOS yield obtained (20%) was only slightly lower than

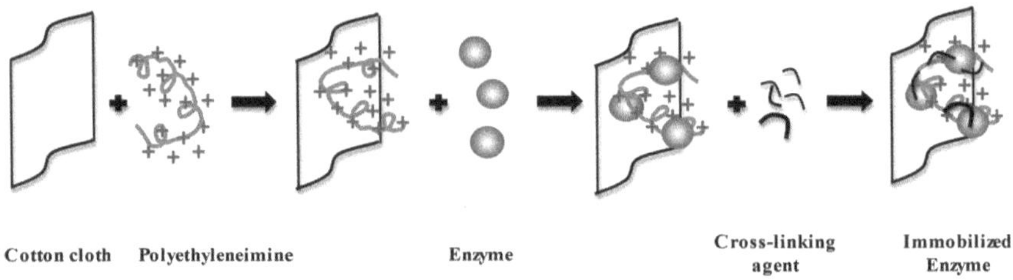

FIGURE 4.8 Scheme of immobilization of β-galactosidase on polyethylene-coated cotton cloth.

obtained with the free enzyme (22%) at the same reaction conditions; these values were obtained at 50% lactose conversion, which occurred at approximately 15 and 17 min of reaction, respectively; initial operation conditions were 270 g/L of initial lactose and 42,300 IU/L of enzyme (Matella et al., 2006). Albayrak and Yang (2002b) also immobilized *A. oryzae* β-galactosidase according to this strategy (see immobilization scheme in Fig. 4.8) and used the catalyst in the synthesis of GOS from a 400 g/L lactose solution at 40°C and pH 4.5, reaching a yield of 25% at 50% lactose conversion and a volumetric productivity of 6000 g/L/h. With the enzyme from *B. circulans* a high protein load was obtained (250 mg/$g_{support}$), but yield of expressed activity was only 35%. Product profile obtained with such catalyst in the synthesis of GOS at 400 g/L of lactose, pH 6, and 45°C was the same as obtained with the corresponding free enzyme; maximum GOS yield was 40%, obtained at 60% lactose conversion, and productivity was180 g/L/h (Sanz-Valero, 2009).

Sen et al. (2011a) immobilized β-galactosidase on polymeric membranes and proved that the functional groups in the membrane surface were modified altering the performance of the membrane with respect to carbohydrate separation and therefore affecting the GOS yield obtained in the membrane reactor. In this case, enzyme immobilization proceeded by protein adsorption followed by the cross-linking of the adsorbed protein. *B. circulans* β-galactosidase (Biolacta FN5) was immobilized in a TFC-SR2 NF membrane (polyamide thin-film composite with a microporous polysulfone supporting layer) obtaining a higher GOS yield (approximately 30%) than with the free enzyme (20%) and at the same time a lower monosaccharide concentration in the final product. β-galactosidase from *B. circulans* has been also immobilized in a microporous polyvinylidene fluoride membrane by covalent binding using glutaraldehyde. The immobilized enzyme retained 50% of its initial activity after 30 days of storage at 20°C, while the free enzyme was completely inactive after 21 days of storage at the same conditions. GOS synthesis was carried out in a laboratory-fabricated membrane reactor cell, whereby three immobilized enzyme membranes were housed in series and operated with partial recirculation loop. The reaction conditions were pH 6 and 40°C, at varying initial lactose concentrations and feed-flow rates. A maximum GOS yield of 30% was obtained at an initial lactose concentration of

50 g/L (Palai et al., 2014). Güleç et al. (2010) developed a three-step effective protocol for immobilizing *A. oryzae* β-galactosidase in cellulose acetate membrane: in the first step the membrane was contacted with polyethylenimine, in the second the enzyme was precipitated with polyethylenimine, and in the third the precipitated enzyme was cross-linked to the membrane with glutaraldehyde. Immobilization yield of expressed activity was 66% and the catalyst was quite stable under storage and operation conditions; half-life was close to one month at 30°C and 60 h at 60°C. Under operating conditions, the enzyme remained stable for 10 sequential batches of GOS synthesis at 60°C, inactivation progressively increasing in the following batches. Güleç (2013) compared the former strategy with adsorption and cross-linking. In the latter case, amino and thiol groups were chemically introduced in the membrane surface, then glutaraldehyde was added as a linker and the enzyme was contacted with the activated membrane. This method was based on the reaction of the protein thiol groups with disulfide oxide reactive structures on the cellulose acetate membrane surface created by plasma polymerization of 2-mercaptoethanol. Even though high protein loads were obtained on the membrane, enzyme activity was reduced, and also GOS yield was reduced when using this catalyst for GOS synthesis, which was attributed to the presence of active amino groups. Engel et al. (2008) immobilized *K. lactis* β-galactosidase on an anion-exchange membrane with a protein immobilization yield from 30% to 36% depending on the immobilization temperature. Expressed enzyme activity was higher when immobilized at 15°C. No data on operational stability of the catalyst was reported, but productivity of GOS production from 20% (w/w) lactose at 40°C and pH 7.0 in a continuous membrane chromatography reactor system was 98.7 g GOS/h/$cm^3_{membrane}$.

Enzymatic membrane reactors with free enzymes, where no enzyme–membrane interaction occurs, can also be considered as immobilization by containment. This strategy has also been used for GOS synthesis with *K. lactis* β-galactosidase (Foda and López-Leiva, 2000; Chockchaisawasdee et al., 2004; Czermak et al., 2004) where different operational variables have been studied for the optimization of GOS synthesis, but no information is reported on the operational stability of such catalysts.

4.2.4 Carrier-Free Immobilized β-Galactosidase

Immobilization of enzymes in carrier-free systems, especially CLEAs, has bloomed in recent years (Garcia-Galan et al., 2011; Sheldon and van Pelt, 2013; Cui and Jia, 2015). Despite this, there are only three publications dealing with the immobilization of β-galactosidase in carrier-free systems (see Table 4.6). In all of them, the stability of the enzyme was improved. However, Gaur et al. (2006) reported low GOS yields with CLEAs of *A. oryzae* β-galactosidase, concluding that the catalyst was better suited for lactose hydrolysis than for transgalactosylation. In the case of CLEAs of *K. lactis* β-galactosidase, an inert protein (albumin) was included as a protecting agent of the enzyme (Wang et al., 2014).

TABLE 4.6 Carrier-Free Immobilized β-Galactosidase

Enzyme-Support Interaction	Enzyme Source	References
Glutaraldehyde cross-linking	*Aspergillus oryzae*	Gaur et al. (2006)
Dialdehyde starch as cross-linking agent and bovine serum albumin as enzyme protective agent	*Kluyveromyces lactis*, *A. oryzae*	Wang et al. (2014)
Glutaraldehyde cross-linking	β-Galactosidase 1–3 from a marine metagenomic library	Li et al. (2015)

A very interesting recent publication on CLEAs of β-galactosidase obtained from a marine metagenomic library reported the synthesis of GOS with such catalyst showing an enhanced galactose tolerance after the immobilization, which could relieve the inhibitory effect of galactose benefiting GOS synthesis. The GOS content and product profile of the syrups produced with the free enzyme and CLEAs were similar. Furthermore, CLEAs maintained more than 80% of the activity after 10 sequential batches of GOS synthesis, GOS yield being higher than 50% at the end of the batch 10 (Li et al., 2015).

Advantages of using immobilized enzymes for GOS synthesis are beyond any doubt. Even though immobilization yields are in many cases only modest, this is compensated by the significant increase in operational stability that allows prolonged use of the catalyst both in continuous and sequential batch reactor operation. Higher stabilization factors are obtained by covalent immobilization of the enzyme to a solid support, as evidenced in the case of chitosan matrices. This is a quite appealing support for β-galactosidase immobilization being a cheap, biocompatible, readily available, and quite versatile matrix allowing the formation of pores of different dimensions and geometries and the activation by different functional groups. Cotton cloth has demonstrated excellent properties for β-galactosidase immobilization, and quite impressive stabilization factors have been reported allowing its use in prolonged continuous synthesis of GOS.

Immobilization of β-galactosidase has been profusely reported in the context of lactose hydrolysis, evaluating the effect of inhibitors and modulation factors on enzyme operational stability (Illanes et al., 1999, 2001); however, such studies are lacking with respect to GOS synthesis with immobilized β-galactosidase, despite their technological significance. One of the drawbacks of immobilized β-galactosidase is the rather severe diffusional restriction occurring in most cases when using it as catalyst for lactose hydrolysis; however, such restrictions should be less significant in the case of intrinsically slower reactions as the reactions of synthesis are when compared to hydrolysis. In theory, diffusional restrictions should be of lesser significance in the case of GOS synthesis, but this is yet to be demonstrated experimentally. Anyhow, catalyst design for the synthesis of GOS should not be biased by the guidelines suggested by immobilization of β-galactosidase from lactose hydrolysis.

These considerations highlight the fact that there is still ample room for process improvement in the synthesis of GOS. As stated before, the potential of immobilized catalysts may be not fully appreciated yet by the industry, which apparently uses the free enzyme with low efficiency, based on the low price of β-galactosidase; however, a stronger competition of producing companies can be forecasted, so that reducing the catalyst cost will probably become a competitive edge for GOS producers, opening up room for process improvement by using immobilized catalysts.

4.3 MECHANISMS AND OPTIMIZATION OF THE SYNTHESIS CATALYZED BY β-GALACTOSIDASES

As stated before, there are several strategies to synthesize GOS. However, the kinetically controlled synthesis of GOS by β-galactosidases is currently preferred at the commercial level. As a consequence, it is highly desirable to understand the molecular mechanism of the reaction catalyzed by β-galactosidase in order to realize how an enzyme traditionally employed in lactose hydrolysis can catalyze, under proper conditions, the reaction of galactose oligomerization. Furthermore, a molecular mechanism for the reaction of lactose transgalactosylation will allow understanding the fundamentals of the kinetic control in this reaction of synthesis and envisage the effect of environmental conditions on the reaction performance.

The catalytic mechanism of β-galactosidase has been mostly studied for the *E. coli* enzyme (Sinnott, 1990; Huber et al., 1994), being first described by Wallenfels and Malhotra in 1960 (Prenosil et al., 1987). Generally speaking, the mechanism of the reaction catalyzed by β-galactosidase implies the transferring of the terminal galactosyl moiety from a nonreducing β-galactoside to an acceptor containing a hydroxyl group. In the reaction of hydrolysis the acceptor is water, while in transgalactosylation a sugar, an alcohol, or a vitamin can be the acceptor (Sinnott, 1990; Nakayama and Amachi, 1999). The reaction mechanism is a double-displacement-type mechanism. Firstly, the substrate binds to the enzyme active site, then Glu 461 residue acts as catalyst protonating the oxygen in the glycosidic bond, producing the bond cleavage, the formation of a galactosyl-enzyme intermediate, and the release of the aglycone (glucose in the case of using lactose as substrate). Secondly, a hydroxyl-containing compound reacts with the galactosyl-enzyme complex yielding a galactose or a galactosyl-adduct, as corresponding, and frees the enzyme active site. The deglycosylation stage is also an acid-base catalytic mechanism assisted by a residue in the active site, probably a Glu 537 residue (Sinnott, 1990; Huber et al., 1994; Nakayama and Amachi, 1999). Particularly, in the synthesis of GOS lactose acts both as donor and acceptor of the galactosyl residue to yield trisaccharides, which can

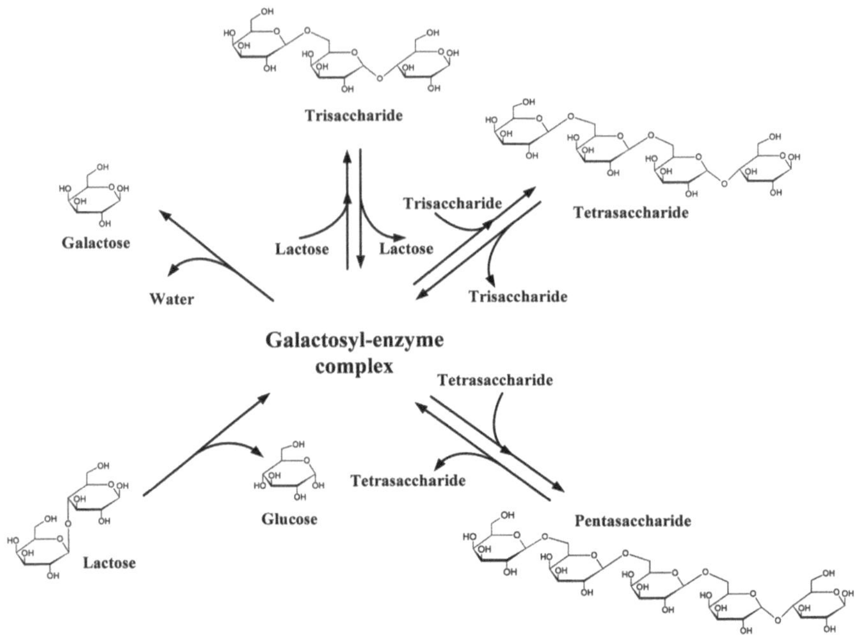

FIGURE 4.9 Reaction mechanism of galacto-oligosaccharide synthesis.

act as acceptor to yield tetrasaccharides, and so on. The reaction mechanism for synthesis of GOS is illustrated in Fig. 4.9.

A noticeable feature of the reaction of GOS synthesis is its kinetic control, which is a consequence of the β-galactosidase catalytic mechanism. During GOS synthesis the reactions of transgalactosylation and hydrolysis occur simultaneously. However, at the beginning of the reaction lactose concentration is high, therefore the transgalactosylation rate surpasses the hydrolysis rate causing an increase in GOS concentration until reaching a maximum value. The maximum GOS concentration is reached when transgalactosylation and hydrolysis rates are equal. At subsequent lactose conversions, hydrolysis rate surpasses transgalactosylation rate because of the lower lactose concentration, so GOS concentration decreases.

The β-galactosidase catalytic mechanism also sheds light about the effect of the reaction condition and reactor type in the synthesis of GOS. In this sense, the most important environmental variable is lactose initial concentration. Considering the mass action law, it is obvious that at higher lactose concentrations transgalactosylation reaction will be predominant so that GOS yield obtained will be higher (Iwasaki et al., 1996; Albayrak and Yang, 2002a; Vera et al., 2012). Other environmental variables, such as temperature and pH, have much lower effect on GOS yield but a significant effect on productivity and enzyme stability (Martínez-Villaluenga et al., 2008; Vera et al., 2011; Osman et al., 2014). Based on the mechanism of GOS synthesis, another important variable is the water activity of the reaction medium. It has been conclusively proven that GOS yield can be increased at low-water activity by depressing the hydrolytic reactions (Chen et al., 2001; Cruz-Guerrero et al., 2006; Srisimarat and Pongsawasdi, 2008). Water activity is usually tuned by using organic solvents, but application is constrained in this case because of the low solubility of lactose in organic media. Remarkably, Wang et al. (2012) working at lactose initial concentrations of 30% w/w obtained a 35% increase in GOS yield when using a cyclohexane/buffer 95/5 (v/v) mixture with respect to the one obtained in aqueous buffer medium. Reactor type will also influence the outcome of GOS synthesis. At equal initial or inlet lactose concentration (according to the case), batch reactor and continuous plug-flow reactor offer a better option than continuous-stirred tank reactor for GOS synthesis (Splechtna et al., 2007a) because the latter operates at the output lactose concentration, so the yield obtained is lower than in the former reactor configurations. Furthermore, β-galactosidase is subjected to strong monosaccharide inhibition so that its negative effect is stronger in continuous-stirred reactor tank than in batch reactor or plug-flow reactor (Splechtna et al., 2007a).

Additionally, the knowledge of the reaction mechanism allows the development of deterministic kinetic models for GOS synthesis, with the aim of optimizing the reaction by a wise choice of the reactor configuration and reaction condition. These deterministic models are usually derived from simplified reaction mechanisms that are, however, far from simple and formed by a set of nonlinear differential equations. The major issue with these models is the estimation of their parameters because of the difficulty of performing the customary experiments for their determination through the quantification of initial reaction rates with the different chemical species involved. Therefore, the estimation of such parameters by nonlinear regression is the most employed procedure, despite not being an easy task (Boon and Janssen, 2000b; Kim et al., 2004a;Vera et al., 2012; Palai and Bhattacharya, 2013). The kinetic models for GOS synthesis are in most cases badly conditioned, causing a large standard error in the estimated parameters and the least squares estimator to be unstable (Boon and Janssen, 2000b; Vera et al., 2012). On the other hand, the kinetic models are the result of a balance among a poor fit of the model to experimental data due to oversimplification and overfitting by considering a large number of parameters. Models including more parameters generally fit better to experimental data, but the greater number of degrees of freedom produces a large standard error on the estimated parameter (Gosling et al., 2010). Due to the complexity of mechanistic kinetic models, some researchers prefer stochastic models, mainly based on the response surface methodology, to describe and predict the effect of operational variables on the synthesis of GOS (Chen et al., 2002; Roy et al., 2002; Osman et al., 2010; Lisboa et al., 2012). These models are typically formed by polynomial functions of critical operation variables, eg, temperature, enzyme load, initial lactose concentration, and time. The main drawback of such models is the lack of any chemical background. Table 4.7 summarizes some of the models proposed for the synthesis of GOS and their main characteristics.

TABLE 4.7 Models Proposed for the Synthesis of Galacto-Oligosaccharides

Model Type	References	Enzyme Source	Species Considered	Number of Kinetic Constant
Deterministic	Iwasaki et al. (1996)	*Aspergillus oryzae*	Mo, Lac Tri, Tet, E, EDi, EDiDi, ETri, ETriTri, EDiTri,	14
	Boon and Janssen (2000b)	*A. oryzae, Bacillus circulans, Kluyveromyces lactis and Kluyveromyces fragilis*	Glu, Gal, Lac, Tri, E, EGlu, EGal, EGal*	6
	Kim et al. (2004a,b)	*K. lactis*	Glu, Gal, Lac, Di, Tri, E, EGal*	9
	Vera et al. (2011)	*A. oryzae*	Glu, Gal, Di, Tri, Tet, Pen, E, EDi, EGal, EGal*	8
	Palai and Bhattacharya (2013)	*B. circulans*	Glu, Gal, Lac, Tri, E, ELac, EGal*, EGlu.	11
	Martins and Lisbôa (2015)	*K. lactis*	Glu, Gal, Lac, Di, Tri, EGal*, EGalGlu	15
Model type	**Author**	**Enzyme source**	**Responses**	**Parameters**
Stochastic	Chen et al. (2002)	*A. oryzae*	Di, Tri, Tetra and total GOS	Lactose concentration, temperature, enzyme to substrate ratio and time
	Roy et al. (2002)	*Bifidobacterium infantis RW-8120*	Yield, GOS production	Lactose and cell concentration, temperature, time
	Osman et al. (2010)	*Bifidobacterium bifidum NCIMB 41171*	GOS yield, process efficiency, consumption yield of lactose, profile index of GOS, selectivity.	Enzyme amount, lactose concentration, time and temperature.
	Lisboa et al. (2012)	*K. lactis*	Yield, GOS concentration	Lactose and enzyme concentration, and temperature

Di, Disaccharide; *E*, free enzyme; *EGlu*, glucose-enzyme inhibition complex; *EGal*, galactose-enzyme inhibition complex; *EGal**, active galactosyl-enzyme complex; *Ei*, specie i-enzyme complex; *Eij*, specie i-specie j-enzyme complex; *Gal*, galactose; *Glu*, glucose; *Lac*, lactose; *Mo*, monosaccharides; *Tri*, trisaccharides; *Tet*, tetrasaccharides; *Pen*, pentasaccharides.

4.4 DOWNSTREAM PROCESSING FOR GALACTO-OLIGOSACCHARIDE PURIFICATION

A bottleneck in the enzymatic synthesis of GOS is the rather low yields attainable, rarely exceeding 40% (Gosling et al., 2010). Despite the efforts in pursuit of higher yields, results are modest and only in a few cases values higher than 50% have been obtained when using recombinant thermophilic β-galactosidases at temperatures over 60°C, taking advantage of the higher solubility of lactose that allows working at very high lactose concentrations (Park and Oh, 2010); however, the genes encoding the enzymes from such thermophilic organisms have to be cloned into suitable hosts, like *E. coli*, that are not allowed for food use, so that this is barely an option. Low-GOS yields make downstream processing of raw GOS a most critical issue in terms of processing cost, so that efforts to improve product purity and reduce purification costs are central to the economic feasibility of the process. GOS yields below 40% imply that a substantial fraction of the product (raw GOS) is residual lactose, glucose, and galactose. The presence of these compounds may be a restriction according to the intended use for the product, with the commercial partially purified products ranging from 50% to 70% total GOS (Torres et al., 2010). Purity of GOS will be dictated by economic considerations, but mostly by the restrictions imposed by the food that will carry it (Lamsal, 2012). Partially purified GOS is adequate as an ingredient in products like yogurt and cereal bars (see Section 4.5); however, GOSs are increasingly being used in special milks for newborns, elderly and lactose-intolerant people (Venema, 2012), where higher purity is certainly an asset. Purity of GOS will become a key issue in an increasingly competitive market but, despite its importance in the economy of GOS production, not much attention has been paid to downstream operations. Preliminary economic evaluation of a process for the production of prebiotic GOS showed that downstream operations represent a significant share of the total investment and total operation costs, being higher than the costs of upstream and reaction operations together (Tymczyszyn et al., 2014).

Several strategies for oligosaccharide fractionation have been applied to the purification of arabinoxylo-oligosaccharides (Park et al., 2001; Swennen et al., 2005), malto-oligosaccharides (MOS) and IMOS (Crittenden and Playne, 2002), XOS (Moure et al., 2006) and FOS (Kuhn and Maugeri Filho, 2010; Kuhn et al., 2011).

In the case of GOS, chromatographic separation, membrane fractionation, selective fermentation or bioconversion, precipitation, and adsorption have been mostly used and will be reviewed in the next sections. GOS purification by extraction with supercritical CO_2 was also proposed (Montañés et al., 2009) and claimed to be an economically viable alternative for GOS purification (Montañés et al., 2012). Also, removal of lactose from a raw GOS produced by transgalactosylation of lactose with *S. solfataricus* β-galactosidase was done by oxidizing it to lactobionic with a cellobiose dehydrogenase using

2-6-dichloro-indophenol as redox mediator. A GOS product with 97% purity was obtained, but recovery of GOS was low and the system is complex and expensive to be scaled up (Splechtna et al., 2001). Selective extraction of sugars with hydrophobic ionic liquids has been used for the separation of glucose, fructose, sucrose, and lactose (Rosatella et al., 2009), but its application to GOS purification, although promising, has not been reported yet.

4.4.1 Purification of Galacto-Oligosaccharides by Chromatographic Techniques

Liquid chromatography is a very powerful technique for oligosaccharide purification. However, it has been mostly used as an analytical tool rather than as a preparative method, mostly because its scale-up at production level is costly and cumbersome (Kovács et al., 2014). Cationic ion exchangers have been mostly used as chromatographic matrices for oligosaccharide purification, having a stronger affinity for monosaccharides than for oligosaccharides (Dendene et al., 1995; Nobre et al., 2009). This type of exchanger was used for the purification of raw GOS produced with immobilized *A. oryzae* β-galactosidase in a packed-bed reactor (Sanz-Valero, 2009); a strong cationic exchanger in the hydrogen ionic form was selected for better performance in the separation of GOS (GOS-3, GOS-4, and GOS-5) from lactose (Wiśniewsky et al., 2014). Size-exclusion chromatography was also used for GOS purification and proved to be the best choice for clear-cut separation of GOS from different sizes, but the operation is costly, productivity is low, and considerable dilution of the product occurs so that it is not recommended for productive-scale purification of GOS (Hernández et al., 2009). An interesting approach was to combine chromatography with a previous step of lactose removal: a raw GOS produced by lactose transgalactosylation with *Lactobacillus reuteri* β-galactosidase was enzymatically treated with cellobiose dehydrogenase to oxidize the remaining lactose to lactobionic acid, and then monosaccharides and lactobionic acid were removed by ion exchange and size-exclusion chromatography, obtaining a highly purified GOS with less than 1% contaminant monosaccharides, lactose, and lactobionic acid, with 60% recovery of GOS in the purification step (Maischberger et al., 2008).

Chromatographic separation can be done in batch mode or in continuous operation. The latter was used for the purification of FOS using a rotating annular chromatographic column (Takahashi and Goto, 1994). Simulated moving bed (SMB) chromatography is considered the state of the art technology for oligosaccharide fractionation, being a better choice than batch chromatography in terms of eluent consumption, productivity, product dilution, and flexibility of operation (Kovács et al., 2014). SMB is a continuous chromatographic process consisting of multiple fixed-bed chromatographic columns where the inlet and outlet ports to the unit are switched periodically in the direction of the fluid flow simulating, in a discontinuous manner, the continuous countercurrent movement of the solid phase occurring in a true moving bed chromatographic process (Geisser et al., 2005). SMB, originally developed for the petrochemical industry, is now widely used, being a powerful tool for complex separations in organic chemistry, like the resolution of the enantiomeric forms of a chiral compound (Rajendran et al., 2009). SMB chromatography was evaluated for FOS purification from a crude mixture containing 40% digestible sugars, and a simulation was done for a plant producing 24,000 tons of purified FOS (with less than 5% of mono- and disaccharides) using SMB chromatography as the main downstream operation (Vaňková et al., 2008). SMB chromatography has been considered a most promising technology for the large-scale purification of FOS (Nobre et al., 2015).

SMB chromatography was applied to the purification of a raw GOS produced by galactose condensation with a thermophilic β-glycosidase from *P. furiosus*. A highly pure GOS was produced, in part due to the process of synthesis (in this case no glucose is produced) and in part due to the high resolution of SMB chromatography; however, significant dilution of the product is produced (Bruins, 2003). A commercial GOS product containing 59% GOS was purified by SMB chromatography using a unit equipped with eight columns packed with a cation exchanger (ionic sulfonated cross-linked styrenedivinylbenzene copolymer) and a 64-port multifunctional rotary valve. A raffinate containing more than 99% GOS was obtained while the extract contained more than 99% of mono- and disaccharides (Wiśniewski et al., 2013). SMB chromatography is a high-cost operation representing a significant share of operating costs (Sanz-Valero, 2009) but may be the choice when GOS of high purity is required (Vanneste et al., 2011) and presumably is the ongoing operation at industrial level.

4.4.2 Purification of Galacto-Oligosaccharides by Membrane Nanofiltration

Membrane separation processes are classified according to the particle size of the compounds to be separated in: microfiltration, ultrafiltration, nanofiltration, and reverse osmosis. According to the molecular size of the oligosaccharides to be fractionated, the operation corresponds to nanofiltration. Membrane separation processes offer some advantages over other options in terms of energy consumption, low temperature operation, no requirements of additives, and facile integration to other downstream operations (Pinelo et al., 2009). The main constraint of membrane separation processes is the concentration of solids to be handled because the operation is cumbersome at high solids concentrations due to membrane fouling and concentration polarization effect (Van der Bruggen et al., 2008; Vegas et al., 2008). However, advances in materials science have allowed constructing robust membranes that withstand high pressures and harsh chemicals (Meng et al., 2009), and advances in nanotechnology have allowed constructing nanoparticle-based polymeric and ceramic membranes that are particularly resistant to fouling (Kim and Van der Bruggen, 2010).

Different cellulose acetate membranes were tested for the fractionation of model saccharide mixtures, obtaining in the best case a twofold increase in the monosaccharide-to-trisaccharide ratio in the permeate; total saccharide concentrations were however low (less than 5 g/L) and the effect of increasing concentration was not evaluated (Catarino et al., 2008). Fractionation of GOS by nanofiltration has been a matter of research only recently. Early attempts of commercial GOS fractionation in flat sheet asymmetric cellulose acetate membranes allowed obtaining average retentions of 90%, 75%, and 15% for GOS-3, disaccharides, and monosaccharides, respectively, the selectivity of the membrane being severely affected by the total concentration of sugars (Goulas et al., 2002). Fractionation of commercial GOS preparations by nanofiltration in spiral wound membrane modules operated in full recycle mode was done at rather low transmembrane pressures (below 8 bar) evaluating the effect of temperature, pressure, and feed concentration on GOS fractionation (Feng et al., 2009). Removal of 90.5% of monosaccharides and 52.5% of lactose was obtained at optimum conditions (50°C and 6 bar), and, at such conditions, the purified product contained 54.5% of GOS, which represents a 1.5-fold purification; however, feed concentration was below 60 g/L of carbohydrates, which is almost one order of magnitude below the concentration obtained after lactose transgalactosylation, so results, being good, are of little practical meaning until the effect of higher concentrations is assessed. Botelho-Cunha et al. (2010) assumed this task by working at 150 and 300 g/L of total carbohydrates (closer to actual concentrations in raw GOS) with a cellulose acetate membrane module at transmembrane pressures from 2 to 14 bar. At 150 g/L they observed complete retention of GOS-3 (probably also for higher GOS, but no data on those were reported) within the whole range of pressure, while retention of disaccharides and monosaccharides increased smoothly with transmembrane pressure within such range; at 2 bar retention of GOS-2 was close to 90%, retention of lactose was 80%, and retention of monosaccharides was somewhat higher than 60%. Interestingly, they observed different retentions for lactose (glucose-galactose β1-4) and GOS-2 (mostly glucose-galactose β1-6). At 300 g/L of total carbohydrates, retention for all of them was reduced: at 2 bar retentions were close to 90%, close to 80%, close to 70%, and close to 40% for GOS-3, GOS-2, lactose, and monosaccharides, respectively. As can be appreciated, results in terms of GOS purification are rather modest (a purification factor of 1.33 at 300 g/L and 2 bar can be calculated from the reported data). Nanofiltration with polyethersulfone membranes of a 100 g/L model solution with 17% GOS allowed a 1.8-fold purification and a GOS recovery of 61% (Michelon et al., 2014), which are results needing substantial improvement. These results altogether clearly exemplify the limits of this strategy for GOS purification and the significant reduction in selectivity with the concentration of carbohydrates. Actually, in a comparative study with other strategies for GOS purification, membrane separation was ruled out (Hernández et al., 2009).

Beyond the classical problems of membrane fractionation related to fouling and cleaning (Astudillo et al., 2010), a major drawback of GOS purification by nanofiltration is the significant reduction of membrane selectivity as the concentration of carbohydrates is increased. In practical terms, raw GOS (usually containing between 40% and 50% w/w solids concentrations) has to be diluted at least twice for obtaining an acceptable separation of GOS from monosaccharides, which implies significant water consumption and energy to remove it. Compromises between GOS recovery and sugars removal, and between selectivity and throughput, are inevitable and set up the limits for this technology.

Strategies for improving membrane selectivity may be working at varying transmembrane pressure according to the volume reduction factor (feed-to-concentrate volume ratio) and using multistage nanofiltration devices combining membranes of different properties. Prehydrolysis of raw GOS to remove most of the lactose content will help by allowing a neater separation of GOS from contaminant sugars (mostly monosaccharides). All these strategies are now being undertaken. Other strategies also have been proposed, like vibrating membrane modules, electrofiltration, and pulse feeding (Pinelo et al., 2009). Cascading of membrane modules has been proposed as a strategy to solve the compromise between purification and recovery of GOS. A yield of GOS plus disaccharide recovery of 88% was obtained with only 19% of monosaccharide when four consecutive nanofiltration steps were performed using polyethersulfone membranes (Goulas et al., 2003); as seen, separation of lactose from GOS was not possible with this system so that a prehydrolysis step should be necessary for making the system effective for GOS purification. In a simulation study a five-stage cascade system was determined to be optimal in terms of purification and recovery (Patil et al., 2014). Vanneste et al. (2011) proposed that membrane cascade operation is a valuable alternative to SMB chromatography for obtaining oligosaccharides of high purity. However, increase in product quality should be counterbalanced by the high investment and operating costs (Kovács et al., 2014). A theoretical basis for membrane fractionation is still under development (Iritani, 2013; Astudillo-Castro, 2015) and it has been pointed out as a limiting factor for GOS purification (Pinelo et al., 2009).

Continuous synthesis of GOS with simultaneous removal of monosaccharides from the reaction medium by nanofiltration has been tried (Pruksasri, 2007) but it is still to be developed as a mature technology.

The problem of microbial contamination of nanofiltration membranes during operation has been recently addressed by conducting the fractionation of raw GOS (200 g/L of carbohydrates) obtained with *Lactobacillus delbrueckii* β-galactosidase at low (5°C) and high (60°C) temperatures; operating at 5°C was considered advantageous with respect to GOS yield, but increased viscosity of the feed stream was a problem limiting permeate flux (Pruksasri et al., 2015).

Membrane technology has been used also for GOS synthesis in continuous ultrafiltration reactors, but in this case the purpose was to retain the enzyme rather than using the separation unit for GOS fractionation (Ebrahimi et al.,

2010; Nath et al., 2013). Quite recently, a continuous ultrafiltration membrane reactor (UFMR) coupled to a nanofiltration unit was used for GOS production with *K. lactis* β-galactosidase and purification; a raw GOS containing 33% GOS was obtained in the UFMR and a partially purified product with 57% GOS was obtained after nanofiltration, which corresponds to a 1.7-fold purification and a product of quality compared to some commercial GOS products (Ren et al., 2015).

4.4.3 Purification of Galacto-Oligosaccharides by Selective Fermentation

Among the technological options for raw GOS purification, selective fermentation (bioconversion) appears very attractive. The idea underlying it is that monosaccharides and some disaccharides are readily metabolized by most microorganisms while higher oligosaccharides are not. One of the earliest reports refer to the purification of several oligosaccharides, namely FOS, MOS, IMOS, and gentio-oligosaccharides by fermentation with immobilized *Zymomonas mobilis* cells where glucose, fructose, and sucrose were significantly removed while ethanol was produced along with a small amount of sorbitol (Crittenden and Playne, 2002). This strategy was not applied for GOS purification due to the narrow carbon substrate range of this bacterium. Selective fermentation with the yeast *P. pastoris* was used to remove monosaccharides from a raw FOS produced from sucrose with a β-fructofuranosidase form *Aspergillus japonicus* (Yang et al., 2008).

In the case of GOS purification the idea is simple: residual lactose and monosaccharides are readily fermentable sugars by most microorganisms, while GOS are not, so that selective fermentation is a feasible strategy for purification. Glucose and galactose, which are the most important side products of lactose transgalactosylation, are efficiently metabolized, the former by the glycolytic route or some alternative catabolic route and the latter mostly by the Leloir pathway, which is more energy demanding and in which four enzymes are involved: galactose mutarotase, galactokinase, galactose-1-phosphate uridyltransferase, and UDP-galactose-4-epimerase. Residual lactose, which represents a significant fraction of raw GOS because of the modest GOS yields attainable, can be fermented by those microorganisms bearing a β-galactosidase gene. Yeast strains are the choice because they have GRAS status and have been used safely in the food sector throughout history.

Saccharomyces strains (bakers' and brewers' yeasts) are unable to metabolize lactose by lacking the β-galactosidase operon, but can efficiently remove monosaccharides (Velagapudi et al., 2007), while higher oligosaccharides are poorly consumed or not consumed at all (Yoon et al., 2003). More than 90% of glucose was removed from a raw GOS produced with permeabilized cells of *B. bifidum* by fermentation with *S. cerevisiae* cells; however, galactose was barely consumed and of course no lactose consumption occurred so that purification was modest, but the merit is that fermentation was conducted on an undiluted raw GOS with 450–500 g/L of carbohydrates without affecting yeast metabolism. GOS produced by lactose transgalactosylation with *Penicillium expansum* immobilized β-galactosidase was purified by bioconversion with immobilized *S. cerevisiae* cells; less than 2% of GOS was degraded but the final product still had a purity below 40% because no lactose was removed (Li et al., 2008). Interestingly, the immobilized cells could be reused in 19 repeated batches. Despite the inability of lactose removal, the use of *S. cerevisiae* is attractive because strains are readily available and even commercial bakers' yeast or spent brewers' yeast can be used. This is a key aspect since one of the limitations of this technology is the rather high yeast-to-carbohydrate mass ratio required, so that the cost of cells has a strong impact on operational costs. A commercial GOS preparation was purified by selective fermentation with commercial bakers' yeast, monosaccharides being completely removed while residual lactose and GOS remained unaltered; however, a slight increase in disaccharides was observed after 24 h of treatment as the consequence of trehalose formation during fermentation (Hernández et al., 2009); the authors compared this strategy to others used for GOS purification and proposed it as a sound alternative for the removal of monosaccharide from actual commercial GOS products. It has to be kept in mind though that the maximum achievable GOS purity by *S. cerevisiae* fermentation is about 40%; a typical composition of a raw GOS is 28% GOS, 40% lactose, and 32% monosaccharides (Li et al., 2008), so assuming a complete removal of monosaccharides lactose will represent about 60% of the total carbohydrates in the purified product and maximum purification will be lower than 1.5-fold.

Yeasts from the genus *Kluyveromyces* appear as more appropriate for raw GOS purification because they are endowed with an inducible lactose assimilating system (Dickson et al., 1979), so that they can in principle remove most of the contaminant sugars in raw GOS. Despite the potential for scale-up and removal of contaminant sugars, this strategy has not received much attention as a technological option for raw GOS purification, and, in fact, reality is far away from theoretical expectations. Li et al. (2008) used immobilized cells of *K. lactis* to purify the raw GOS produced by *P. expansum* and indeed a GOS purity higher than 95% was obtained with significant removal of lactose and monosaccharides, but more than 20% of GOS was also degraded. Immobilized *K. lactis* cells were reused in repeated batch operation: in the first batch GOS purity was 97.5% but after four batches it dropped to 75% and after six batches purity was 25%, meaning that the cells were completely inactive, as opposite to the behavior of *S. cerevisiae* cells under similar conditions. This is not surprising since *Kluyveromyces* are sensitive to osmotic stress, which is the case at high-sugar concentrations (Ozmihci and Kargi, 2007). Besides, rates of unwanted sugar removal are much lower than with *S. cerevisiae* so that the productivity of purification is lower. Carbohydrate concentration used was 20%, meaning that raw GOS was diluted prior to fermentation, which is a limiting factor for this strategy; we have proved that efficient sugars removal by *Kluyveromyces* should be done at carbohydrate concentrations not exceeding 20%, which implies more than a twofold dilution of the raw GOS. The purification of a crude GOS (20% GOS) produced by lactose transgalactosylation with a *Bacillus* sp. β-galactosidase and also by the mixed action of the β-galactosidase with a commercial glucose oxidase-catalase preparation was done by fermentation with *K. marxianus* cells (Cheng et al., 2006a). In both cases, most of the monosaccharides and residual lactose were removed, obtaining a GOS of high purity; the addition of glucose oxidase-catalase increased GOS purity and GOS recovery only slightly from 97% to 98% and from 31% to 32%, respectively; again the raw GOS was diluted to 20% carbohydrates and supplementary nutrients were added in the fermentation step.

Purification by selective fermentation has been considered a technology not yet mature, most limitations arising from the following facts: (1) in most cases, raw GOS has to be diluted to allow for efficient consumption of contaminating sugars; (2) additional nutrients are required for fermentation, which increases costs and contaminates the product; (3) purified product contains metabolites of fermentation that should be removed as required for the food product in which the prebiotic will be incorporated (Kovács et al., 2014). The first two aspects have been addressed by considering the bioconversion of the raw GOS without any nutrient supplementation and without dilution; working with an undiluted raw GOS produced by transgalactosylation of a 50% w/w lactose solution with a commercial *A. oryzae* β-galactosidase, a product with 95% GOS, was obtained by bioconversion with *K. marxianus* cells (Guerrero et al., 2014). A critical variable is the cell-to-carbohydrate mass ratio and even though very high purification yields could be obtained in the mass ratio range from 0.1 to 0.7, the time of bioconversion increased significantly at the lower ratios, so that the productivity of purification was severely reduced. This is a question to be addressed before making it a technologically sound operation for GOS purification. Cell reuse could solve this problem if a suitable immobilization system is developed in which cells remain active in prolonged operation when using high-carbohydrate concentrations; however, in the case of *K. marxianus* cells this is yet to be proven. Some work done with immobilized cells of a thermophilic strain of *K. marxianus* in the alcoholic fermentation of whey is encouraging because of the acceptable stability of the yeast under harsh operating conditions (Kourkoutas et al., 2002). The use of *S. cerevisiae* is probably a better option from a technological perspective, but purification attainable is modest because lactose in not metabolized, so that a complementary operation for its degradation is required.

GOS purification has been also attempted using probiotic bacteria. A crude GOS preparation obtained with a β-galactosidase from a native strain of *Streptococcus thermophilus* was purified by fermentation with the probiotic bacterium *Lactobacillus helveticus* (Messaoudi et al., 2011), obtaining a product with more than 90% GOS, with a considerable reduction in lactose and galactose

(Sangwan et al., 2014). Being done with the purpose of evaluating prebiotic potential, not much information was given with respect to the concentrations of carbohydrates in the fermentation step or the cells-to-carbohydrates mass ratio used, so that its technological significance cannot be appreciated. Even so, the use of probiotic strains for raw GOS purification is worthwhile exploring.

One-pot reaction and purification is an appealing strategy that has been applied to other systems, like the simultaneous saccharification and fermentation processes for the production of second-generation ethanol (Koppram and Olsson, 2014; Tran et al., 2015). This strategy is in principle applicable to the simultaneous transgalactosylation of lactose with β-galactosidase and purification of the product with yeast strains.

4.4.4 Purification of Galacto-Oligosaccharides by Selective Precipitation and Adsorption

Differences in solubility of sugars and oligosaccharides in various solvents have been profusely reported and can be applicable as a strategy for GOS purification (Brito and Giulietti, 2007; Montañés et al., 2007). In principle, fractionation based on solubility has the potential advantage over other strategies of simplicity and cost-effectiveness during scale-up. Even though purification of lactose by precipitation has been used for removing it from its products of hydrolysis (Bourne et al., 1983) and also for the production of highly pure lactose (Gänzle et al., 2008), purification of GOS by selective precipitation has been seldom reported.

Extraction of galactose-containing oligosaccharides (mostly stachyose and raffinose) from soybeans was performed with absolute ethanol in a 1:10 mass ratio. The precipitate formed was dissolved in distilled water and further purified by filtration with diatomaceous earth and charcoal. After precipitation, recovery of soybean oligosaccharides was 42% and a 50-fold purification was obtained (Espinosa-Martos and Rupérez, 2006). In a related field, fractionation of arabinoxylo-oligosaccharides was performed by ethanol precipitation and proved to be more selective than membrane separation, the degree of polymerization of the oligosaccharides precipitated depending on the ethanol concentration used (Swennen et al., 2005).

GOS purification by fractional precipitation with ethanol was recently claimed to be the first report on such strategy. Two sequential precipitations at 90% ethanol and 28 g/L total sugars allowed increasing GOS percentage from 15% in raw GOS to 75%, while reducing monosaccharides from 48% in raw GOS to 4%; however, despite these good results in terms of purification, recovery of GOS was below 10%. Recovery could be increased to more than 95% when concentration was increased to 81 g/L, but purification was severely reduced, setting up a compromise between purification and recovery even at the rather low carbohydrate concentrations used, which implies a significant dilution of raw GOS (Sen et al., 2011b). It is to be assumed that at higher carbohydrate concentrations precipitation will become nonselective and the system will be inapplicable. In our own experience, carbohydrate concentrations should be below 100 g/L to obtain a significant purification factor, so that, despite its advantages in terms of simplicity, low cost, and scalability, fractional precipitation can only be considered in connection with other more powerful fractionation systems. A comprehensive study on GOS purification by fractionation with different precipitating agents is still lacking.

Separation of sugars by adsorption into ion-exchange resins and activated carbon has been used with success. Several ion-exchange resins were effective as matrices for the separation by selective adsorption of glucose and fructose from agroindustrial wastes (Luz et al., 2008) and have also been used for the separation of lactose, lactulose, and galactose, as mentioned in Section 4.4.1. They developed a two-stage adsorption process: in the first step, a column with the resin in its K^+ form was used at 55°C to selectively remove galactose; and in a second stage a column with the same resin was used at 15°C to separate lactose from lactulose.

Cheaper matrices for adsorption are recommended, such as zeolite and activated carbon. Zeolites have been used in the separation and purification of isomaltose (Berensmeier and Buchholz, 2004; Holtkamp and Scholl, 2011) and FOS (Kuhn and Maugeri Filho, 2010), having a high exchange capacity, selectivity, and good mechanical properties. Activated carbon is a readily available cheap material with a high surface area and pore volume that is commonly used in the food industry (Roy, 1995) and in particular in sugar and oligosaccharide fractionation (Redmond and Packer, 1999). Most of the surface in activated carbon is nonpolar so that the more hydrophobic compounds interact more strongly. The hydrophobicity of the carbohydrate is related to the number of CH groups in its structure; therefore, oligosaccharides will be more strongly adsorbed than disaccharides and monosaccharides. Selective removal of monosaccharides from FOS has been accomplished in packed-bed columns of activated carbon (Nobre et al., 2012); fractions enriched in FOS were obtained by desorption with ethanol from 10% to 40% v/v, delivering 93% pure FOS with 75% recovery. Adsorption in activated carbon has also been used for the purification of XOS (Montané et al., 2006; Wang and Lu, 2013). Oligosaccharides from honey have been fractionated by selective adsorption in activated carbon and elution in water–ethanol mixtures; at 10% v/v ethanol monosaccharides were completely removed, and at 50% v/v ethanol oligosaccharides of polymerization degree from 3 to 14 were recovered, even though most of them were tri- and tetrasaccharides (Morales et al., 2006).

With respect to GOS purification, Hernández et al. (2009) claimed that there was no previous report on the purification of GOS by adsorption in activated carbon and reported that selective desorption with 1% ethanol allowed the complete separation of monosaccharides from di-, tri-, and tetrasaccharides. They studied the selective elution with different concentrations of ethanol and showed a compromise between GOS purity and recovery: at 8% v/v ethanol 90% of GOS was recovered but also 20% of disaccharides were recovered alongside, while at 10% v/v ethanol disaccharides were completely removed but GOS recovery dropped to 53%. The latter allowed a GOS product of high purity at the expense of significant GOS loss. This compromise between purity and recovery in GOS purification by carbon adsorption has not been optimized yet; type of eluent, eluent concentration, total carbohydrate concentration, and temperature are key variables.

In situ product removal by selective adsorption of GOS in activated carbon during the transgalactosylation of lactose with *B. circulans* β-galactosidase allowed an increase in GOS yield from 33% to 43% when GOS-3 was removed along with synthesis (Boon, 2000a). To our knowledge, in situ product removal during GOS synthesis by selective adsorption has not been explored any further.

As can be appreciated, there is no systematic work on GOS purification by selective precipitation or adsorption; information is scarce and scattered so a comprehensive study of this operation as applied for GOS purification is still to be done to fully appreciate its technological potential. However, they are simple, scalable, and low-cost operations that may well be combined with other strategies, like selective fermentation or membrane separation, to attain high levels of GOS purification.

In summary, several technologies are challenging the ongoing chromatographic operation for GOS purification. Its implementation at productive level will be effective if clear advantages are obtained on a cost-basis analysis. In our appraisal, selective bioconversion with suspended or immobilized yeast cells and nanofiltration with a lactose prehydrolysis step are the most promising.

4.5 APPLICATIONS IN FOOD AND FEED

GOS are well-established prebiotics so their use is mostly framed within the functional food segment. GOSs, as well as other NDOs, have been considered as prebiotic ingredients in a wide variety of food products, namely, yogurt (Sairanen et al., 2007), cereal bars (Buttriss and Stokes, 2008), baked goods (Sangwan et al., 2011), soups and sauces (Macfarlane et al., 2008), baby foods (Rivero-Urgell and Santamaria-Orleans, 2001; Angus et al., 2005), confectionery (Patel and Goyal, 2011), buttermilk (Čurda et al., 2006), and beverages (Shioiri et al., 2006); it has also been considered a valuable component in clinical nutrition (Mussatto and Mancilha, 2007).

GOSs have excellent properties for their incorporation into different food matrices: they are extremely stable during long-term storage at room temperature, heat and acid pH stable, highly soluble in water, moderately sweet, noncariogenic, and bland tasting (Mlichová and Rosenberg, 2006; Charalampopoulos and Rastall, 2012). Stability under conditions of food processing is a distinctive advantage of GOS over FOS (Huebner et al., 2008). Applications of GOS rely not only in its prebiotic condition but also in the functional properties it can confer to the foods to which it is added (Angus et al., 2005). In fact, in many cases it is precisely these functional properties that have opened opportunities for the introduction of GOS into food products. For example, in bakery products GOS provides functional properties like moisture retention and low-caloric content, contributing to texture and taste; in acid fruit and soft drinks it enhances flavor without altering the taste and clarity of the products (Franck, 2008). GOS, as any other NDO, is considered as dietary fiber so it has been used for its laxative properties; a yogurt was formulated with GOS, prunes, and linseed and used as a functional food to alleviate constipation in elderly people (Sairanen et al., 2007).

GOS is also attractive as a functional sweetener because it is noncariogenic, has excellent taste, is acid resistant, and moderately sweet (Voragen, 1998; Sako et al., 1999; Splechtna et al., 2006), improving the texture and mouthfeel of foods providing bulk properties similar to sucrose (Torres et al., 2010). GOS, even at high concentrations, does not crystallize, as different from lactose and other sugars, so that this property can be used advantageously in those food products where crystal formation is undesirable (Torres et al., 2011). Another interesting feature of GOS is its cryoprotective effect in the preservation of prebiotic *Lactobacillus* strains (Tymczyszyn et al., 2011).

The dairy industry is a privileged sector for the application of GOS, which is itself a product derived from it. Yogurt has been considered an excellent food matrix for GOS supplementation (Varzakas and Arvanitoyannis, 2010), where it can be added before or after fermentation. High stability of GOS at low pH ensures that it will not be degraded during yogurt elaboration and storage and will be fully available at the point of consumption; in addition, GOS-containing yogurt was considered smoother and creamier than plain yogurt (Sangwan et al., 2011). Adding a preincubation stage with GOS-producing *Bifidobacteria*, a yogurt containing 0.7% GOS was obtained when using *B. infantis*; survival of the probiotic and stability of the prebiotic during storage were high (Lamoureux et al., 2002). A yogurt containing GOS was designed for consumption by elderly people, obtaining a stimulation of the *Bifidobacteria* population and improvement in gut functioning (Surakka et al., 2009). In another study, however, consumption of GOS-containing yogurt by people of different ages had no significant effect on the diversity and temporal stability of *Lactobacilli* and *Bifidobacteria* (Maukonen et al., 2008). Recently, β-galactosidase from *K. lactis* and the probiotic strain *Lactobacillus acidophilus* were added to allow for GOS formation during manufacture and storage, obtaining a product enriched in GOS (starter cultures and probiotic added did not degrade it) and depleted in lactose so as to obtain a synbiotic product for lactose intolerants (Vénica et al., 2015). Fluid milk has also been considered as a food matrix for GOS enrichment. A process was developed for producing GOS-enriched milk in which milk proteins were separated by ultrafiltration and the permeate was treated with *A. oryzae* β-galactosidase for the transgalactosylation of lactose, milk being then reconstituted by adding back the milk proteins; this process performed better than the direct transgalactosylation of lactose in milk where the activity of the enzyme was much lower than in permeate and secondary reactions with milk proteins affected the quality of the enriched milk (Chen et al., 2002). Recently, several commercial β-galactosidase preparations were tested for the transgalactosylation of skim milk to produce GOS-enriched milk. The best performance was obtained with a commercial *K. lactis* preparation with which 95% of the lactose was depleted producing an enriched milk with 7 g/L of GOS, which is within the range of HMOS in human milk (Rodriguez-Colinas et al., 2014). Low-lactose concentration in fluid milk (50–60 g/L) is a major drawback for GOS synthesis in milk so that enzymes with a very high transgalactosylation-to-hydrolytic activity ratio will be needed for this purpose.

The most significant application of GOS is in the production of special milks for infants. As previously stated, GOSs are similar in structure and in action to HMOS, so that in this case no other NDOs can be used for such purpose. Supplementing infant formulas with GOS allows obtaining a oligosaccharide profile resembling HMOS so that an intestinal microbiota in formula-fed infants can be developed similar to that of breast-fed infants during the early days of life, promoting a beneficial effect on the development of the immune system (Ben et al., 2004; Veereman-Wauters, 2005). Production of enriched milk formulas with a 9:1 combination of GOS and FOS is now well established at the industrial level (see for instance: www.heinznurture.com.au); this product is of social significance by allowing an adequate nutrition in this critical period to infants that cannot be breast-fed by their mothers (Vandenplas et al., 2008; Donovan et al., 2009). GOS is also being used in follow-up milks for children of 1 year of age or older (Chatchatee et al., 2014; Li et al., 2014). It is clear that GOS will play a key role in achieving gold-standard (same structural and functional benefits as HMOS) milk formulas and follow-up milks (Riva et al., 2005). Supplementation of institutional milk formulas with GOS may have a significant impact on third-world countries where governmental programs of milk distribution for infants are based on cow's milk with the consequent problems of intestinal microbiota implantation, immune system development, and episodes of diarrhea that render such programs scarcely effective.

The animal feed market is huge and still expanding. Functional feed is a major issue in animal nutrition nowadays as applied both to farm animals and pets and also in aquaculture (Van Loo and Vancraeynest, 2008). Prebiotics (and probiotics) are gaining a place in animal feeding, mostly pushed by the ban of antibiotics as food promoters (Gaggìa et al., 2010) that has prompted the development of alternatives, among which the use of prebiotics outstands (Huyghebaert et al., 2011). FOS and inulin have been mostly used as prebiotics in animal feeding (Flickinger and Fahey, 2002; Hajati and Rezaei, 2010). However, GOS is being increasingly used in the feeding of poultry (Patterson and Burkholder, 2003; Biggs et al., 2007; Jung et al., 2008), swine (Houdijk et al., 1998; Lind Mikkelsen et al., 2003; Tzortzis et al., 2005a; Modesto et al., 2009), foal (Vendrig et al., 2014), sheep (Mwenya et al., 2004; Santoso et al., 2004), and steer (Mwenya et al., 2005). Despite these advances, it is unclear if GOS will find a competitive edge in the animal feed sector.

Beyond the food and feed sector, GOS has been proposed as an ingredient in cosmetics and pharmaceuticals (Qiang et al., 2009); topical applications have been developed based on the stimulation of healthy skin bacteria (Bockmühl et al., 2007; Krutmann, 2009; Ouwehand et al., 2010).

In summary, GOSs are being increasingly used in a wide variety of food and also in some nonfood products. Inulin and FOS are GOS competitors in many of their fields of applications, except in the case of special milk for infants where GOS shares striking and distinctive similarities with HMOS; in this case, formulas containing GOS:FOS mixtures, in which GOS is 90% of the oligosaccharide mixture, already represent a significant share of the market of formulated milks for infants and is, at least for the moment, the most significant field for GOS application.

REFERENCES

Albayrak, N., Yang, S.T., 2002a. Production of galacto-oligosaccharides from lactose by *Aspergillus oryzae* β-galactosidase immobilized on cotton cloth. Biotechnology and Bioengineering 77, 8–19.

Albayrak, N., Yang, S.T., 2002b. Immobilization of β-galactosidase on fibrous matrix by polyethyleneimine for production of galacto-oligosaccharides from lactose. Biotechnology Progress 18, 240–251.

Angus, F., Smart, S., Shortt, C., 2005. Prebiotic ingredients with emphasis on galacto-oligosaccharides and fructooligosaccharides. In: Tamine, A. (Ed.), Probiotic Dairy Products. Blackwell Publishing Ltd, Oxford, pp. 120–137.

Ansari, S.A., Satar, R., 2012. Recombinant β-galactosidases – past, present and future: a mini review. Journal of Molecular Catalysis B: Enzymatic 81, 1–6.

Arslanoglu, S., Moro, G.E., Boehm, G., 2007. Early supplementation of prebiotic oligosaccharides protects formula-fed infants against infections during the first 6 months of life. The Journal of Nutrition 137, 2420–2424.

Arslanoglu, S., Moro, G.E., Schmitt, J., Tandoi, L., Rizzardi, S., Boehm, G., 2008. Early dietary intervention with a mixture of prebiotic oligosaccharides reduces the incidence of allergic manifestations and infections during the first two years of life. The Journal of Nutrition 138, 1091–1095.

Asp, N.G., Burval, A., Dahlqist, A., Hallgren, P., Lundblad, A., 1980. Oligosaccharide formation during hydrolysis of lactose with *Saccharomyces lactis* lactase (Maxilact ®). II. Oligosaccharide structures. Food Chemistry 5, 147–153.

Astudillo, C., Parra, J., González, S., Cancino, B., 2010. A new parameter for membrane cleaning evaluation. Separation and Purification Technology 73, 286–293.

Astudillo-Castro, C.L., 2015. Limiting flux and critical transmembrane pressure determination using an exponential model: the effect of concentration factor, temperature, and cross-flow velocity during casein micelle concentration by microfiltration. Industrial and Engineering Chemistry Research 54, 414–425.

Arreola, S.L., Intanon, M., Suljic, J., Kittl, R., Pham, N.H., Kosma, P., Haltrich, D., Nguyen, T.H., 2014. Two β-galactosidases from the human isolate *Bifidobacterium breve* DSM 20213: molecular cloning and expression, biochemical characterization and synthesis of galacto-oligosaccharides. PLoS One 9 (8), e104056. http://dx.doi.org/10.1371/journal.pone.0104056.

Badarinath, V., Halami, P., 2011. Purification of new β-galactosidase from *Enterococcus faecium* MTCC 5153 with transgalactosylation activity. Food Biotechnology 25, 225–239.

Bakker-Zierikzee, A.M., Alles, M.S., Knol, J., Kok, F.J., Tolboom, J.J.M., Bindels, J.G., 2005. Effects of infant formula containing a mixture of galacto- and fructo-oligosaccharides or viable *Bifidobacterium animalis* on the intestinal microflora during the first 4 months of life. British Journal of Nutrition 94, 783–790.

Balcao, V.M., Mateo, C., Fernández-Lafuente, R., Malcata, F.X., Guisán, J.M., 2001. Structural and functional stabilization of L-asparaginase via multisubunit immobilization onto highly activated supports. Biotechnology Progress 17, 537–542.

Barreteau, H., Delattre, C., Michaud, P., 2006. Production of oligosaccharides as promising new food additive generation. Food Technology and Biotechnology 44, 323–333.

Ben, X.M., Zhou, X.Y., Zhao, W.H., Yu, W.L., Pan, W., Zhang, W.L., Wu, S.M., Van Beusekom, C.M., Schaafsma, A., 2004. Supplementation of milk formula with galacto-oligosaccharides improves intestinal micro-flora and fermentation in term infants. Chinese Medical Journal (England) 117, 927–931.

Ben, X.M., Li, J., Feng, Z.T., Shi, S.Y., Lu, Y.D., Chen, R., Zhou, X.Y., 2008. Low level of galacto-oligosaccharide in infant formula stimulates growth of intestinal *Bifidobacteria* and *Lactobacilli*. World Journal of Gastroenterology 14, 6564–6568.

Benavente, R., Pessela, B.C., Curiel, J.A., de las Rivas, B., Muñoz, R., Guisán, J.M., Mancheño, J.M., Cardelle-Cobas, A., Ruiz-Matute, A.I., Corzo, N., 2015. Improving properties of a novel β-galactosidase from *Lactobacillus plantarum* by covalent immobilization. Molecules 20, 7874–7889.

Benjamins, F., 2014. Galacto-Oligosaccharide Synthesis Using Immobilized β-Galactosidase (Doctoral thesis). Rijksuniversiteit Groningen.

Berensmeier, S., Buchholz, K., 2004. Separation of isomaltose from high sugar concentrated enzyme reaction mixture by dealuminated β-zeolite. Separation and Purification Technology 38, 129–138.

Bernal, C., Sierra, L., Mesa, M., 2012. Improvement of thermal stability of β-galactosidase from *Bacillus circulans* by multipoint covalent immobilization in hierarchical macro-mesoporous silica. Journal of Molecular Catalysis B: Enzymatic 84, 166–172.

Betancor, L., Hidalgo, A., Fernández-Lorente, G., Mateo, C., Fernández-Lafuente, R., Guisán, J.M., 2003. Preparation of a stable biocatalyst of bovine liver catalase using immobilization techniques. Biotechnology Progress 19, 763–767.

Betancor, L., Luckarift, H.R., 2008. Bioinspired enzyme encapsulation for biocatalysis. Trends in Biotechnology 26, 566–572.

Biggs, P., Parsons, C.M., Fahey, G.C., 2007. The effects of several oligosaccharides on growth performance, nutrient digestibilities, and cecal microbial populations in young chicks. Poultry Science 86, 2327–2336.

Bockmühl, D., Jassoy, C., Nieveler, S., Scholtyssek, R., Wadle, A., Waldmann-Laue, M., 2007. Prebiotic cosmetics: an alternative to antibacterial products. International Journal of Cosmetic Science 29, 63–64.

Bode, L., 2006. Recent advances on structure, metabolism and function of human milk oligosaccharides. The Journal of Nutrition 136, 2127–2130.

Bode, L., 2009. Human milk oligosaccharides: prebiotics and beyond. Nutrition Reviews 67, S183–S191.

Boehm, G., Stahl, B., 2007. Oligosaccharides from milk. The Journal of Nutrition 137, 847S–849S.

Bolivar, J.M., Mateo, C., Grazu, V., Carrascosa, A.V., Pessela, B.C., Guisan, J.M., 2010. Heterofunctional supports for the one-step purification, immobilization and stabilization of large multimeric enzymes: amino-glyoxyl versus amino-epoxy supports. Process Biochemistry 45, 1692–1698.

Bolivar, J.M., Eisl, I., Nidetzky, B., 2016. Advanced characterization of immobilized enzymes as heterogeneous biocatalysts. Catalysis Today 259, 66–80.

Boller, T., Meier, C., Menzler, S., 2002. Eupergit oxirane acrylic beads: how to make enzymes fit for biocatalysis. Organic Process Research and Development 6, 509–519.

Bongers, M.E.J., de Lorijn, F., Reitsma, J.B., Groeneweg, M., Taminiau, J.A.J.M., Benninga, M.A., 2007. The clinical effect of a new infant formula in term infants with constipation: a double-blind, randomized cross-over trial. Nutrition Journal 6, 8. http://dx.doi.org/10.1186/1475-2891-6-8.

Booij, C.J., 1985. Use of lactose in the pharmaceutical and chemical industry. International Journal of Dairy Technology 38, 105–109.

Boon, M.A., van't Riet, K., Janssen, A.E.M., 2000a. Enzymatic synthesis of oligosaccharides: product removal during a kinetically controlled reaction. Biotechnology and Bioengineering 70, 411–420.

Boon, M.A., Janssen, A.E.M., van't Riet, K., 2000b. Effect of temperature and enzyme origin on the enzymatic synthesis of oligosaccharides. Enzyme and Microbial Technology 26, 271–281.

Botelho-Cunha, V.A., Mateus, M., Petrus, J.C.C., de Pinho, M.N., 2010. Tailoring the enzymatic synthesis and nanofiltration fractionation of galacto-oligosaccharides. Biochemical Engineering Journal 50, 29–36.

Bouhnik, Y., Raskine, L., Simoneau, G., Vicaut, E., Neut, C., Flourié, B., Brouns, F., Bornet, F.R., 2004. The capacity of nondigestible carbohydrates to stimulate fecal bifidobacteria in healthy humans: a double-blind, randomized, placebo-controlled, parallel-group, dose-response relation study. American Journal of Clinical Nutrition 80, 1658–1664.

Bourne, J.R., Hegglin, M., Prenosil, J.E., 1983. Solubility and selective crystallization of lactose from solutions of its hydrolysis products glucose and galactose. Biotechnology and Bioengineering 25, 1625–1639.

Braegger, C., Chmielewska, A., Decsi, T., Kolacek, S., Mihatsch, W., Moreno, L., Pieścik, M., Puntis, J., Shamir, R., Szajewska, H., Turck, D., van Goudoever, J., 2011. Supplementation of infant formula with probiotics and/or prebiotics: a systematic review and comment by the ESPGHAN committee on nutrition. Journal of Pediatric Gastroenterology and Nutrition 52, 238–250.

Braga, A.R.C., Manera, A.P., Ores, J.D.C., Francisco, L.S., Kalil, S.J., 2013. Kinetics and thermal properties of crude and purified β-galactosidase with potential for the production of galactooligosaccharides. Food Technology and Biotechnology 51, 45–52.

Brena, B., Batista-Viera, F., 2006. Immobilization of enzymes: a literature survey. In: Guisan, J.M. (Ed.), Immobilization of Enzymes and Cells, second ed. Humana Press Inc., New Jersey, pp. 15–30.

Brito, A.B., Giulietti, M., 2007. Study of lactose crystallization in water-acetone solutions. Crystal Research and Technology 42, 583–588.

Bruins, M.E., 2003. Oligosaccharide Production With Thermophilic Enzymes (Doctoral thesis). Wageningen Universiteit.

Bruins, M.E., Strubel, M., van Lieshout, J.F.T., Janssen, A.E.M., Boom, R.M., 2003. Oligosaccharide synthesis by the hyperthermostable β-glucosidase from *Pyrococcus furiosus*: kinetics and modelling. Enzyme and Microbial Technology 33, 3–11.

Bruno-Barcena, J.M., Azcarate-Peril, M.A., 2015. Galacto-oligosaccharides and colorectal cancer: feeding our intestinal probiome. Journal of Functional Foods 12, 92–108.

Bruzzese, E., Volpicelli, M., Squeglia, V., Bruzzese, D., Salvini, F., Bisceglia, M., Lionetti, P., Cinquetti, M., Iacono, G., Amarri, S., Guarino, A., 2009. A formula containing galacto- and fructo-oligosaccharides prevents intestinal and extra-intestinal infections: an observational study. Clinical Nutrition 28, 156–161.

Burvall, A., Asp, N.G., Dahlqvist, A., 1979. Oligosaccharide formation during hydrolysis of lactose with *Saccharomyces lactis* lactase (MaxilactR): Part 1 – quantitative aspects. Food Chemistry 4, 243–250.

Buttriss, J.L., Stokes, C.S., 2008. Dietary fibre and health: an overview. Nutrition Bulletin 33, 186–200.

Cantone, S., Ferrario, V., Corici, L., Ebert, C., Fattor, D., Spizzo, P., Gardossi, L., 2013. Efficient immobilisation of industrial biocatalysts: criteria and constraints for the selection of organic polymeric carriers and immobilisation methods. Chemical Society Reviews 42, 6262–6276.

Cao, L., Van Rantwijk, F., Sheldon, R.A., 2000. Cross-linked enzyme aggregates: a simple and effective method for the immobilization of penicillin acylase. Organic Letters 2, 1361–1364.

Carević, M., Corović, M., Mihailović, M., Banjanac, K., Milisavljević, A., Veličković, D., Bezbradica, D., 2016. Galacto-oligosaccharide synthesis using chemically modified β-galactosidase from *Aspergillus oryzae* immobilised onto macroporous amino resin. International Dairy Journal 54, 50–57.

Casci, T., Rastall, R.A., 2006. Manufacture of prebiotic oligosaccharides. In: Gibson, G.R., Rastall, R.A. (Eds.), Prebiotics: Development and Applications. John Wiley, New York, pp. 29–55.

Catarino, I., Minhalma, M., Beal, L.L., Mateus, M., de Pinho, M.N., 2008. Assessment of saccharide fractionation by ultrafiltration and nanofiltration. Journal of Membrane Science 312, 34–40.

Cattorini, S., Marques, M.P.C., Carvalho, F., Chheub, V., Cabral, J.M.S., Fernandes, P., 2009. Lentikat®-based biocatalysts: effective tools for inulin hydrolysis. Chemical and Biochemical Engineering Quarterly 23, 429–434.

Chakraborti, S., Sani, R.K., Banerjee, U.C., Sobti, R.C., 2000. Purification and characterization of a novel β-galactosidase from *Bacillus* sp MTCC 3088. Journal of Industrial Microbiology and Biotechnology 24, 58–63.

Charalampopoulos, D., Rastall, R.A., 2012. Prebiotics in foods. Current Opinion in Biotechnology 23, 187–191.

Chatchatee, P., Lee, W.S., Carrilho, E., Kosuwon, P., Simakachorn, N., Yavuz, Y., Schouten, B., Graaff, P.L., Szajewska, H., 2014. Effects of growing-up milk supplemented with prebiotics and LCPUFAs on infections in young children. Journal of Pediatric Gastroenterology and Nutrition 58, 428–437.

Chen, S.C., Duan, K.J., 2015. Production of galactooligosaccharides using β-galactosidase immobilized on chitosan-coated magnetic nanoparticles with Tris (hydroxymethyl) phosphine as an optional coupling agent. International Journal of Molecular Sciences 16, 12499–12512.

Chen, S.X., Wei, D.Z., Hu, Z.H., 2001. Synthesis of galacto-oligosaccharides in AOT/isooctane reverse micelles by β-galactosidase. Journal of Molecular Catalysis B: Enzymatic 16, 109–114.

Chen, C.S., Hsu, C.K., Chiang, B.H., 2002. Optimization of the enzymic process for manufacturing low-lactose milk containing oligosaccharides. Process Biochemistry 38, 801–808.

Chen, W., Chen, H., Xia, Y., Zhao, J., Tian, F., Zhang, H., 2008. Production, purification, and characterization of a potential thermostable galactosidase for milk lactose hydrolysis from *Bacillus stearothermophilus*. Journal of Dairy Science 91, 1751–1758.

Chen, W., Chen, H.Y., Yang, J., Zhao, J., Tian, F., Zhang, H.P., Zhang, H., 2009. Immobilization of recombinant thermostable β-galactosidase from *Bacillus stearothermophilus* for lactose hydrolysis in milk. Journal of Dairy Science 92, 491–498.

Cheng, C.C., Yu, M.C., Cheng, T.C., Sheu, D.C., Duan, K.J., Tai, W.L., 2006a. Production of high-content galacto-oligosaccharide by enzyme catalysis and fermentation with *Kluyveromyces marxianus*. Biotechnology Letters 28, 793–797.

Cheng, T.C., Duan, K.J., Sheu, D.C., 2006b. Application of tris (hydroxymethyl) phosphine as a coupling agent for β-galactosidase immobilized on chitosan to produce galactooligosaccharides. Journal of Chemical Technology and Biotechnology 81, 233–236.

Chi, Z., Wang, F., Chi, Z., Yue, L., Liu, G., Zhang, T., 2009. Bioproducts from *Aureobasidium pullulans*, a biotechnologically important yeast. Applied Microbiology and Biotechnology 82, 793–804.

Cho, Y.J., Shin, H.J., Bucke, C., 2003. Purification and properties of a galactooligosaccharide producing β-galactosidase from *Bullera singularis*. Biotechnology Letters 25, 2107–2111.

Chockchaisawasdee, S., Athanosopoulos, V.I., Niranjan, K., Rastall, R.A., 2004. Synthesis of galacto-oligosaccharide from lactose using β-galactosidase from *Kluyveromyces lactis*: studies on batch and continuous UF membrane-fitted bioreactors. Biotechnology and Bioengineering 89, 434–443.

Contesini, F.J., de Alencar Figueira, J., Kawaguti, H.K., de Barros Fernandes, P.C., de Oliveira Carvalho, P., Nascimento, M.G., Sato, H.H., 2013. Potential applications of carbohydrases immobilization in the food industry. International Journal of Molecular Science 14, 1335–1369.

Coppa, G.V., Bruni, S., Morelli, L., Soldi, S., Gabrielli, O., 2004. The first prebiotics in humans: human milk oligosaccharides. Journal of Clinical Gastroenterology 38, S80–S83.

Coppens, P., Da Silva, M.F., Pettman, S., 2006. European regulations on nutraceuticals, dietary supplements and functional foods: a framework based on safety. Toxicology 221, 59–74.

Corfield, A.P., Carroll, D., Myerscough, N., Probert, C.S.J., 2001. Mucins in the gastrointestinal tract in health and disease. Frontiers in Bioscience 6, 1321–1357.

Coulier, L., Timmermans, J., Bas, R., Van Den Dool, R., Haaksman, I., Klarenbeek, B., Slaghek, T., Van Dongen, W., 2009. In-depth characterization of prebiotic galacto-oligosaccharides by a combination of analytical techniques. Journal of Agricultural and Food Chemistry 23, 8488–8495.

Crittenden, R.G., Playne, M.J., 1996. Production, properties and applications of food-grade oligosaccharides. Trends in Food Science and Technology 7, 353–361.

Crittenden, R.G., Playne, M.J., 2002. Purification of food-grade oligosaccharides using immobilized cells of *Zymomonas mobilis*. Applied Microbiology and Biotechnology 58, 297–302.

Crout, D.H., Vic, G., 1998. Glycosidases and glycosyl transferases in glycoside and oligosaccharide synthesis. Current Opinion in Chemical Biology 2, 98–111.

Cruz, R., D'Arcadia Cruz, V., Belote, J.G., De Oliveira Khenayfes, M., Dorta, C., Dos Santos Oliveira, L.H., Ardiles, E., Galli, A., 1999. Production of transgalactosylated oligosaccharides (TOS) by galactosyltransferase activity from *Penicillium simplicissimum*. Bioresource Technology 70, 165–171.

Cruz-Guerrero, A.E., Gómez-Ruiz, L., Viniegra-González, G., Bárzana, E., García-Garibay, M., 2006. Influence of water activity in the synthesis of galactooligosaccharides produced by a hyperthermophilic β-glycosidase in an organic medium. Biotechnology and Bioengineering 93, 1123–1129.

Cui, J.D., Jia, S.R., 2015. Optimization protocols and improved strategies of cross-linked enzyme aggregates technology: current development and future challenges. Critical Reviews in Biotechnology 35, 15–28.

Čurda, L., Rudolfová, J., Štětina, J., Dryák, B., 2006. Dried buttermilk containing galactooligosaccharides – process layout and its verification. Journal of Food Engineering 77, 468–471.

Czermak, P., Ebrahimi, M., Grau, K., Netz, S., Sawatzki, G., Pfromm, P.H., 2004. Membrane-assisted enzymatic production of galactosyl-oligosaccharides from lactose in a continuous process. Journal of Membrane Science 232, 85–91.

Czichocki, G., Dautzenberg, H., Capan, E., Vorlop, K.D., 2001. New and effective entrapment of polyelectrolyte-enzyme-complexes in LentiKats. Biotechnology Letters 23, 1303–1307.

Datta, S., Christena, L.R., Rajaram, Y.R.S., 2013. Enzyme immobilization: an overview on techniques and support materials. 3 Biotechnology 3, 1–9.

Davis, L.M.G., Martínez, I., Walter, J., Hutkins, R., 2010. A dose dependent impact of prebiotic galactooligosaccharides on the intestinal microbiota of healthy adults. International Journal of Food Microbiology 144, 285–292.

del-Val, M.I., Otero, C., 2003. Biphasic aqueous media containing polyethylene glycol for the enzymatic synthesis of oligosaccharides from lactose. Enzyme and Microbial Technology 33, 118–126.

Dendene, K., Guihard, L., Balannec, B., Bariou, B., 1995. Study of separation of lactose, lactulose and galactose by liquid chromatography using cationic ion-exchange resin columns. Chromatographia 41, 561–567.

Depeint, F., Tzortzis, G., Vulevic, J., I'Anson, K., Gibson, G.R., 2008. Prebiotic evaluation of a novel galactooligosaccharide mixture produced by the enzymatic activity of *Bifidobacterium bifidum* NCIMB 41171, in healthy humans: a randomized, double-blind, crossover, placebo-controlled intervention study. American Journal of Clinical Nutrition 87, 785–791.

van der Bruggen, B., Mänttäri, M., Nyström, M., 2008. Drawbacks of applying nanofiltration and how to avoid them: a review. Separation and Purification Technology 63, 251–263.

Dickson, R.C., Dickson, L.R., Markin, J.S., 1979. Purification and properties of an inducible β-galactosidase isolated from the yeast *Kluyveromyces lactis*. Journal of Bacteriology 137, 51–61.

Djouzi, Z., Andrieux, C., 1997. Compared effects of three oligosaccharides on metabolism of intestinal microflora in rats inoculated with a human fecal flora. British Journal of Nutrition 78, 313–324.

Dong, Y.N., Chen, H.Q., Sun, Y.H., Zhang, H., Chen, W., 2015. A differentially conserved residue (Ile42) of GH42 β-galactosidase from *Geobacillus stearothermophilus* BgaB is involved in both catalysis and thermostability. Journal of Dairy Science 98, 2268–2276.

Donovan, S., Gibson, G., Newburg, D., 2009. Prebiotics in Infant Nutrition. Mead Johnson Nutrition, Glenview.

dos Santos, R., Simiqueli, A.P.R., Pastore, G.M., 2009. Production of galactooligosaccharide by *Scopulariopis* sp. Ciência e Tecnologia de Alimentos 29, 682–689.

Drakoularakou, A., Tzortzis, G., Rastall, R.A., Gibson, G.R., 2010. A double-blind placebo-controlled randomized human study assessing the capacity of a novel galacto-oligosaccharide mixture in reducing travellers' diarrhea. European Journal of Clinical Nutrition 64, 146–152.

Durieux, A., Nicolay, X., Simon, J.P., 2000. Continuous malolactic fermentation by *Oenococcus oeni* entrapped in LentiKats. Biotechnology Letters 22, 1679–1684.

Ebrahimi, M., Placido, L., Engel, L., Shams Ashaghi, K., Czermak, P., 2010. A novel ceramic membrane reactor system for the continuous synthesis of oligosaccharides. Desalination 250, 1105–1108.

Eiwegger, T., Stahl, B., Schmitt, J., Boehm, G., Gerstmayr, M., Pichler, J., Dehlink, E., Loibichler, C., Urbanek, R., Szépfalusi, Z., 2004. Human milk-derived oligosaccharides and plant-derived oligosaccharides stimulate cytokine production of cord blood T-cells *in vitro*. Pediatric Research 56, 536–540.

Eiwegger, T., Stahl, B., Schmitt, J., Boehm, G., Dehlink, E., Urbanek, R., Szépfalusi, Z., 2010. Prebiotic oligosaccharides: *in vitro* evidence for gastrointestinal epithelial transfer and immunomodulatory properties. Pediatric Allergy and Immunology 21, 1179–1188.

Engel, L., Ebrahimi, M., Czermak, P., 2008. Membrane chromatography reactor system for the continuous synthesis of galactosyl-oligosaccharides. Desalination 224, 46–51.

Espinosa-Martos, I., Rupérez, P., 2006. Soybean oligosaccharides. Potential as new ingredients in functional foods. Nutrición Hospitalaria 21, 92–96.

Fai, A.E.C., Simiqueli, A.P.R., Ghiselli, G., Pastore, G.M., 2015. Sequential optimization approach for prebiotic galactooligosaccharides synthesis by *Pseudozyma tsukubaensis* and *Pichia kluyveri*. LWT – Food Science and Technology 63, 1214–1219.

Fanaro, S., Boehm, G., Garssen, J., Knol, J., Mosca, F., Stahl, B., Vigi, V., 2005. Galacto-oligosaccharides and long-chain fructo-oligosaccharides as prebiotics in infant formulas: a review. Acta Pædiatrica 94, 22–26.

Fanaro, S., Marten, B., Bagna, R., Vigi, V., Fabris, C., Peña-Quintana, L., Argüelles, F., Scholz-Ahrens, K.E., Sawatzki, G., Zelenka, R., Schrezenmeir, J., de Vrese, M., Bertino, E., 2009. Galacto-oligosaccharides are bifidogenic and safe at weaning: a double-blind randomized multicenter study. Journal of Pediatric Gastroenterology and Nutrition 48, 82–88.

Feng, Y.M., Chang, X.L., Wang, W.H., Ma, R.Y., 2009. Separation of galacto-oligosaccharides mixture by nanofiltration. Journal of the Taiwan Institute of Chemical Engineers 40, 326–332.

Feng, Y., Chang, X., Wang, W., Ma, R., 2010. Artificial cells. Stabilities of immobilized β-galactosidase of *Aspergillus* sp. AF for the optimal production of galactooligosaccharides from lactose. Blood Substitutes and Biotechnology 38, 43–51.

Fernández-Lafuente, R., Hernández-Ruiz, O., Mateo, C., Terreni, M., Alonso, J., García-López, J., Moreno, M.A., Guisán, J.M., 2001. Stabilization of a tetrameric enzyme (α-amino acid ester hydrolase from *Acetobacter turbidans*, enables a very improved performance of ampicillin synthesis. Journal of Molecular Catalysis B: Enzymatic 11, 633–638.

Fernandez-Lafuente, R., 2009. Stabilization of multimeric enzymes: strategies to prevent subunit dissociation. Enzyme and Microbial Technology 45, 405–418.

Fischer, C., Kleinschmidt, T., 2015. Synthesis of galactooligosaccharides using sweet and acid whey as a substrate. International Dairy Journal 48, 15–22.

Flickinger, E.A., Fahey Jr., G.C., 2002. Pet food and feed applications of inulin, oligofructose and other oligosaccharides. British Journal of Nutrition 87, S297–S300.

Foda, M.I., Lopez-Leiva, M., 2000. Continuous production of oligosaccharides from whey using a membrane reactor. Process Biochemistry 35, 581–587.

Fonseca, G.G., Heinzle, E., Wittmann, C., Gombert, A.K., 2008. The yeast *Kluyveromyces marxianus* and its biotechnological potential. Applied Microbiology and Biotechnology 79, 339–354.

Franck, A., 2008. Food applications of prebiotics. In: Gibson, G.R., Roberfroid, M.B. (Eds.), Handbook of Prebiotics. Taylor & Francis, Boca Raton, pp. 437–448.

Frenzel, M., Zerge, K., Clawin-Rädecker, I., Lorenzen, P.C., 2015. Comparison of the galacto-oligosaccharide forming activity of different β-galactosidases. LWT – Food Science and Technology 60, 1068–1071.

Gaggìa, F., Mattarelli, P., Biavati, B., 2010. Probiotics and prebiotics in animal feeding for safe food production. International Journal of Food Microbiology 141, S15–S28.

Gänzle, M.G., 2012. Enzymatic synthesis of galacto-oligosaccharides and other lactose derivatives (hetero-oligosaccharides) from lactose. International Dairy Journal 22, 116–122.

Gänzle, M.G., Haase, G., Jelen, P., 2008. Lactose: crystallization, hydrolysis and value-added derivatives. International Dairy Journal 18, 685–694.

Garcia-Galan, C., Berenguer-Murcia, A., Fernandez-Lafuente, R., Rodrigues, R.C., 2011. Potential of different enzyme immobilization strategies to improve enzyme performance. Advanced Synthesis and Catalysis 353, 2885–2904.

Gaur, R., Pant, H., Jain, R., Khare, S.K., 2006. Galacto-oligosaccharide synthesis by immobilized *Aspergillus oryzae* β-galactosidase. Food Chemistry 97, 426–430.

Geisser, A., Hendrich, T., Boehm, G., Stahl, B., 2005. Separation of lactose from human milk oligosaccharides with simulated moving bed chromatography. Journal of Chromatography A 1092, 17–23.

Gekas, V., López-Leiva, M., 1985. Hydrolysis of lactose: a literature review. Process Biochemistry 20, 2–12.

Gibson, G.R., Fuller, R., 2000. Aspects of in vitro and in vivo research approaches directed toward identifying probiotics and prebiotics for human use. The Journal of Nutrition 130, 391S–395S.

Gibson, G.R., 2004. From probiotics to prebiotics and a healthy digestive system. Journal of Food Science 69, 141–143.

Gopal, P.K., Sullivan, P.A., Smart, J.B., 2001. Utilisation of galacto-oligosaccharides as selective substrates for growth by lactic acid bacteria including *Bifidobacteriium lactis* DR-10 and *Lactobacillus rhamnosus* DR20. International Dairy Journal 11, 19–25.

Górecka, E., Jastrzębska, M., 2011. Immobilization techniques and biopolymer carriers. Biotechnology and Food Science 75, 65–86.

Gosling, A., Alftrén, J., Stevens, G.W., Barber, A.R., Kentish, S.E., Gras, S.L., 2009. Facile pretreatment of *Bacillus circulans* β-galactosidase increases the yield of galactosyl oligosaccharides in milk and lactose reaction systems. Journal of Agricultural and Food Chemistry 57, 11570–11574.

Gosling, A., Stevens, G.W., Barber, A.R., Kentish, S.E., Gras, S.L., 2010. Recent advances refining galactooligosaccharide production from lactose. Food Chemistry 121, 307–318.

Gosling, A., Stevens, G.W., Barber, A.R., Kentish, S.E., Gras, S.L., 2011. Effect of the substrate concentration and water activity on the yield and rate of the transfer reaction of β-galactosidase from *Bacillus circulans*. Journal of Agricultural and Food Chemistry 59, 3366–3372.

Goulas, A.K., Kapasakalidis, P.G., Sinclair, H.R., Rastall, R.A., Grandison, A.S., 2002. Purification of oligosaccharides by nanofiltration. Journal of Membrane Science 209, 321–335.

Goulas, A.K., Grandison, A.S., Rastall, R.A., 2003. Fractionation of oligosaccharides by nanofiltration. Journal of the Science of Food and Agriculture 837, 675–680.

Goulas, A., Tzortzis, G., Gibson, G.R., 2007. Development of a process for the production and purification of α- and β-galactooligosaccharides from *Bifidobacterium bifidu*m NCIMB 41171. International Dairy Journal 17, 648–656.

Goulas, T., Goulas, A., Tzortzis, G., Gibson, G.R., 2009. Comparative analysis of four β-galactosidases from *Bifidobacterium bifidum* NCIMB41171: purification and biochemical characterization. Applied Microbiology and Biotechnology 82, 1079–1088.

Grosová, Z., Rosenberg, M., Rebroš, M., 2008. Perspectives and applications of immobilized β-galactosidase in food industry – a review. Czech Journal of Food Science 26, 1–14.

Guerrero, C., Vera, C., Novoa, C., Dumont, J., Acevedo, F., Illanes, A., 2014. Purification of highly concentrated galacto-oligosaccharide preparations by selective fermentation with yeasts. International Dairy Journal 39, 78–88.

Guerrero, C., Vera, C., Conejeros, R., Illanes, A., 2015. Transgalactosylation and hydrolytic activities of commercial preparations of β-galactosidase for the synthesis of prebiotic carbohydrates. Enzyme and Microbial Technology 70, 9–17.

Guisán, J.M., 1988. Aldehyde-agarose gels as activated supports for immobilization-stabilization of enzymes. Enzyme and Microbial Technology 10, 375–382.

Güleç, H.A., 2013. Immobilization of β-galactosidase from *Kluyveromyces lactis* onto polymeric membrane surfaces: effect of surface characteristics. Colloids and Surfaces B: Biointerfaces 104, 83–90.

Güleç, H.A., Gürdaş, S., Albayrak, N., Mutlu, M., 2010. Immobilization of *Aspergillus oryzae* β-galactosidase on low-pressure plasma-modified cellulose acetate membrane using polyethyleneimine for production of galactooligosaccharide. Biotechnology and Bioprocess Engineering 15, 1006–1015.

Gupta, M.N., 1991. Thermostabilization of proteins. Biotechnology and Applied Biochemistry 4, 1–11.

Hajati, H., Rezaei, M., 2010. The application of prebiotics in poultry production. International Journal of Poultry Science 9, 298–304.

Halbmayr, E., Mathiesen, G., Nguyen, T.H., Maischberger, T., Peterbauer, C.K., Eijsink, V.G.H., Haltrich, D., 2008. High-level expression of recombinant β-galactosidases in *Lactobacillus plantarum* and *Lactobacillus sakei* using a Sakacin P-based expression system. Journal of Agricultural and Food Chemistry 56, 4710–4719.

Hamilton-Miller, J.M.T., 2004. Probiotics and prebiotics in the elderly. Postgraduate Medical Journal 80, 447–451.

Hanson, S., Best, M., Bryan, M.C., Wong, C.H., 2004. Chemoenzymatic synthesis of oligosaccharides and glycoproteins. Trends in Biochemical Sciences 29, 656–663.

Hansson, T., Adlercreutz, P., 2001. Optimization of galactooligosaccharide production from lactose using beta-glycosidases from hyperthermophiles. Food Biotechnology 15, 79–97.

Hansson, T., Kaper, T., van der Oost, J., de Vos, W., Adlercreutz, P., 2001. Improved oligosaccharide synthesis by protein engineering of β-glucosidase CelB from hyperthermophilic *Pyrococcus furiosus*. Biotechnology and Bioengineering 73, 203–210.

Harju, M., Kallioinen, H., Tossavainen, O., 2012. Lactose hydrolysis and other conversions in dairy products: technological aspects. International Dairy Journal 22, 104–109.

Harmsen, H.J.M., Wildeboer–Veloo, A.C.M., Raangs, G.C., Wagendorp, A.A., Klijn, N., Bindels, J.G., Welling, G.W., 2000. Analysis of intestinal flora development in breast-fed and formula-fed infants by using molecular identification and detection methods. Journal of Pediatric Gastroenterology and Nutrition 30, 61–67.

Hassan, N., Nguyen, T.H., Intanon, M., Kori, L.D., Patel, B.K.C., Haltrich, D., Divne, C., Tan, T.C., 2015. Biochemical and structural characterization of a thermostable β-glucosidase from *Halothermothrix orenii* for galacto-oligosaccharide synthesis. Applied Microbiology and Biotechnology 99, 1731–1744.

Hernández, O., Ruiz Matute, A., Olano, A., Moreno, F.J., Sanz, L., 2009. Comparison of fractionation techniques to obtain prebiotic galactooligosaccharides. International Dairy Journal 19, 531–536.

Hidalgo, A., Betancor, L., López-Gallego, F., Moreno, R., Berenguer, J., Fernández-Lafuente, R., Guisán, J.M., 2003. Design of an immobilized preparation of catalase from *Thermus thermophilus* to be used in a wide range of conditions: structural stabilization of a multimeric enzyme. Enzyme and Microbial Technology 33, 278–285.

Holtkamp, M., Scholl, S., 2011. Adsorption properties of BEA zeolites and their aluminum phosphate extrudates for purification of isomaltose. Adsorption 17, 801–811.

Houdijk, J.G.M., Bosch, M.W., Verstegen, M.W.A., Berenpas, H.J., 1998. Effects of dietary oligosaccharides on the growth performance and faecal characteristics of young growing pigs. Animal Feed Science and Technology 71, 35–48.

Hsu, C.A., Lee, S.L., Chou, C.C., 2007. Enzymatic production of galacto-oligosaccharides by β-galactosidase from *Bifidobacterium longum* BCRC 15708. Journal of Agricultural and Food Chemistry 55, 2225–2230.

Hu, X., Robin, S., O'Connell, S., Walsh, G., Wall, J.G., 2010. Engineering of a fungal β-galactosidase to remove product inhibition by galactose. Applied Microbiology and Biotechnology 87, 1773–1782.

Huber, R.E., Gupta, M.N., Khare, S.K., 1994. The active site and mechanism of the β-galactosidase from *Escherichia coli*. International Journal of Biochemistry 26, 309–318.

Huebner, J., Wehling, R.L., Parkhurst, A., Hutkins, R.W., 2008. Effect of processing conditions on the prebiotic activity of commercial prebiotics. International Dairy Journal 18, 287–293.

Huerta, L.M., Vera, C., Guerrero, C., Wilson, L., Illanes, A., 2011. Synthesis of galacto-oligosaccharides at very high lactose concentrations with immobilized β-galactosidases from *Aspergillus oryzae*. Process Biochemistry 46, 245–252.

Huh, K.T., Toba, T., Adachi, S., 1991. Oligosaccharide structures formed during acid hydrolysis of lactose. Food Chemistry 39, 39–49.

Hung, M.N., Lee, B.H., 2002. Purification and characterization of a recombinant β-galactosidase with transgalactosylation activity from *Bifidobacterium infantis* HL96. Applied Microbiology and Biotechnology 58, 439–445.

Husain, Q., 2010. β-Galactosidase and their potential applications: a review. Critical Reviews in Biotechnology 30, 41–62.

Huyghebaert, G., Ducatelle, R., Van Immerseel, F., 2011. An update on alternatives to antimicrobial growth promoters for broilers. The Veterinary Journal 187, 182–188.

Illanes, A., Wilson, L., Raiman, L., 1999. Design of immobilized enzyme reactors for the continuous production of fructose syrup from whey permeate. Bioprocess Engineering 21, 509–515.

Illanes, A., Wilson, L., Tomasello, G., 2001. Effect of modulation of enzyme inactivation on temperature optimization for reactor operation with chitin-immobilized lactase. Journal of Molecular Catalysis B: Enzymatic 11, 531–540.

Illanes, A., 2011. Whey upgrading by enzyme biocatalysis. Electronic Journal of Biotechnology 14 (6). http://dx.doi.org/10.2225/vol14-issue6-fulltext-11.

Illanes, A., Cauerhff, A., Wilson, L., Castro, G.R., 2012. Recent trends in biocatalysis engineering. Bioresource Technology 115, 48–57.

Iqbal, S., Nguyen, T.H., Nguyen, T.T., Maischberger, T., Haltrich, D., 2010. β-Galactosidase from *Lactobacillus plantarum* WCFS1: biochemical characterization and formation of prebiotic galacto-oligosaccharides. Carbohydrate Research 345, 1408–1416.

Iqbal, S., Nguyen, T.H., Nguyen, H.A., Nguyen, T.T., Maischberger, T., Kittl, R., Haltrich, D., 2011. Characterization of a heterodimeric GH2 β-galactosidase from *Lactobacillus sakei* Lb790 and formation of prebiotic galacto-oligosaccharides. Journal of Agricultural and Food Chemistry 59, 3803–3811.

Irazoqui, G., Giacomini, C., Batista-Viera, F., Brena, B.M., Cardelle-Cobas, A., Corzo, N., Jimeno, M.L., 2009. Journal of Agricultural and Food Chemistry 57, 11302–11307.

Iritani, E., 2013. A review on modeling of pore-blocking behaviors of membranes during pressurized membrane filtration. Drying Technology 31, 146–162.

Ishikawa, E., Sakai, T., Ikemura, H., Matsumoto, K., Abe, H., 2005. Identification, cloning, and characterization of a *Sporobolomyces singularis* β-galactosidase-like enzyme involved in galacto-oligosaccharide production. Journal of Bioscience and Bioengineering 99, 331–339.

Iwasaki, K., Nakajima, M., Sakao, S., 1996. Galacto-oligosaccharide production from lactose by an enzymic batch reaction using β-galactosidase. Process Biochemistry 31 (1), 69–76.

Jahnz, U., Wittlich, P., Prüsse, U., Vorlop, K., 2001. New matrices and bioencapsulation processes. In: Hofman, M., Thonart, P. (Eds.), Engineering and Manufacturing for Biotechnology, Focus on Biotechnology. Academic Publishers, Dordrech, pp. 293–307.

Ji, E.S., Park, N.H., Oh, D.K., 2005. Galacto-oligosaccharide production by a thermostable recombinant β-galactosidase from *Thermotoga maritima*. World Journal of Microbiology and Biotechnology 21, 759–764.

Jochems, P., Satyawali, Y., Diels, L., Dejonghe, W., 2011. Enzyme immobilization on/in polymeric membranes: status, challenges and perspectives in biocatalytic membrane reactors (BMRs). Green Chemistry 13, 1609–1623.

Jørgensen, F., Hansen, O.C., Stougaard, P., 2001. High efficiency synthesis of oligosaccharides with a truncated β-galactosidase from *Bifidobacterium bifidum*. Applied Microbiology and Biotechnology 576, 647–652.

Jovanovic-Malinovska, R., Fernandes, P., Winkelhausen, E., Fonseca, L., 2012. Galacto-oligosaccharides synthesis from lactose and whey by β-galactosidase immobilized in PVA. Applied Biochemistry and Biotechnology 168, 1197–1211.

Juajun, O., Nguyen, T.H., Maischberger, T., Iqbal, S., Haltrich, D., Yamabhai, M., 2011. Cloning, purification, and characterization of β-galactosidase from *Bacillus licheniformis* DSM 13. Applied Microbiology and Biotechnology 89, 645–654.

Jung, S.J., Houde, R., Baurhoo, B., Zhao, X., Lee, B.H., 2008. Effects of galacto-oligosaccharides and a *Bifidobacteria lactis*-based probiotic strain on the growth performance and fecal microflora of broiler chickens. Poultry Science 87, 1694–1699.

Kamerke, C., Pattky, M., Huhn, C., Elling, L., 2012. Synthesis of UDP-activated oligosaccharides with commercial β-galactosidase from *Bacillus circulans* under microwave irradiation. Journal of Molecular Catalysis B: Enzymatic 79, 27–34.

Kaneko, K., Watanabe, Y., Kimura, K., Matsumoto, K., Mizobuchi, T., Onoue, M., 2014. Development of hypoallergenic galacto-oligosaccharides on the basis of allergen analysis. Bioscience, Biotechnology and Biochemistry 78, 100–108.

Katrolia, P., Yan, Q., Jia, H., Li, Y., Jiang, Z., Song, C., 2011. Molecular cloning and high-level expression of a β-galactosidase gene from *Paecilomyces aerugineus* in *Pichia pastoris*. Journal of Molecular Catalysis B: Enzymatic 69, 112–119.

Kim, C.S., Ji, E.S., Oh, D.K., 2003. Expression and characterization of *Kluyveromyces lactis* β-galactosidase in *Escherichia coli*. Biotechnology Letters 25, 1769–1774.

Kim, C.S., Ji, E.S., Oh, D.K., 2004a. A new kinetic model of recombinant β-galactosidase from *Kluyveromyces lactis* for both hydrolysis and transgalactosylation reactions. Biochemical and Biophysical Research Communications 316, 738–743.

Kim, C.S., Ji, E.S., Oh, D.K., 2004b. Characterization of a thermostable recombinant β-galactosidase from *Thermotoga maritima*. Journal of Applied Microbiology 97, 1006–1014.

Kim, J., Van der Bruggen, B., 2010. The use of nanoparticles in polymeric and ceramic membrane structures: review of manufacturing procedures and performance improvement for water treatment. Environmental Pollution 158, 2335–2349.

Klein, M.P., Fallavena, L.P., Schöffer, J.N., Ayub, M.A.Z., Rodrigues, R.C., Ninow, J.L., Hertz, P.F., 2013. High stability of immobilized β-D-galactosidase for lactose hydrolysis and galactooligosaccharides synthesis. Carbohydrate Polymers 95, 465–470.

Knol, J., Scholtens, P., Kafka, C., Steenbakkers, J., Groß, S., Helm, K., Klarczyk, M., Schöpfer, H., Böckler, H.M., Wells, J., 2005. Colon microflora in infants fed formula with galacto- and fructo-oligosaccharides: more like breast-fed infants. Journal of Pediatric Gastroenterology and Nutrition 40, 36–42.

Konar, E., Sarkar, S., Singhal, R.S., 2011. Galactooligosaccharides: chemistry, production, properties, market status and applications. Trends in Carbohydrate Research 3, 1–16.

Koppram, R., Olsson, L., 2014. Combined substrate, enzyme and yeast feed in simultaneous saccharification and fermentation allow bioethanol production from pretreated spruce biomass at high solids loadings. Biotechnology for Biofuels. 7, 54. http://www.biotechnologyforbiofuels.com/content/7/1/54.

Kourkoutas, Y., Dimitropoulou, S., Kanellaki, M., Marchant, R., Nigam, P., Banat, I.M., Koutinas, A.A., 2002. High-temperature alcoholic fermentation of whey using *Kluyveromyces marxianus* IMB3 yeast immobilized on delignified cellulosic material. Bioresource Technology 82, 177–181.

Kovács, Z., Benjamins, E., Grau, K., Rehman, A.U., Ebrahimi, M., Czermak, P., 2014. Recent developments in manufacturing oligosaccharides with prebiotic functions. Advances in Biochemical Engineering and Biotechnology 143, 257–295.

Krajewska, B., 2004. Application of chitin- and chitosan-based materials for enzyme immobilization: a review. Enzyme and Microbial Technology 35, 126–139.

Krutmann, J., 2009. Pre- and probiotics for human skin. Journal of Dermatological Science 54, 1–5.

Kuhn, R.C., Maugeri Filho, F., 2010. Separation of fructooligosaccharides using zeolite fixed bed columns. Journal of Chromatography B 878, 2023–2028.

Kuhn, R.C., Palacio, L., Prádanos, P., Hernández, A., Maugeri Filho, F., 2011. Selection of membranes for purification of fructooligosaccharides. Desalination and Water Treatment 27, 18–24.

Kukkonen, K., Savilahti, E., Haahtela, T., Juntunen-Backman, K., Korpela, R., Poussa, T., Tuure, T., Kuitunen, M., 2007. Probiotics and prebiotic galactooligosaccharides in the prevention of allergic diseases: a randomized, double-blind, placebo-controlled trial. Journal of Allergy and Clinical Immunology 119, 192–198.

Kumari, S., Panesar, P.S., Panesar, R., 2011. Production of β-galactosidase using novel yeast isolate from whey. International Journal of Dairy Science 6, 150–157.

Kumari, S., Panesar, P.S., Bera, M.B., Panesar, R., 2013. Permeabilization of a newly isolated *Kluyveromyces* sp. for the preparation of whole cell biocatalysts with β-galactosidase activity. International Journal of Food and Nutritional Sciences 2, 22–26.

Kunz, C., Rudloff, S., Baier, W., Klein, N., Strobel, S., 2000. Oligosaccharides in human milk: structural, functional, and metabolic aspects. Annual Review of Nutrition 20, 699–722.

Ladero, M., Santos, A., García, J.L., Carrascosa, A.V., Pessela, B.C.C., García-Ochoa, F., 2002. Studies on the activity and the stability of β-galactosidases from *Thermus* sp strain T2 and from *Kluyveromyces fragilis*. Enzyme and Microbial Technology 30, 392–405.

Ladero, M., Ruiz, G., Pessela, B.C.C., Vian, A., Santos, A., Garcia-Ochoa, F., 2006. Thermal and pH inactivation of an immobilized thermostable β-galactosidase from *Thermus* sp. strain T2: comparison to the free enzyme. Biochemical Engineering Journal 31, 14–24.

Lamoureux, L., Roy, D., Gauthier, S.F., 2002. Production of oligosaccharides in yogurt containing *Bifidobacteria* and yogurt cultures. Journal of Dairy Science 85, 1058–1069.

Lamsal, B.P., 2012. Production, health aspects and potential food uses of dairy prebiotic galactooligosaccharides. Journal of the Science of Food and Agriculture 92, 2020–2028.

Laurienzo, P., 2010. Marine polysaccharides in pharmaceutical applications: an overview. Marine Drugs 8, 2435–2465.

Leforestier, G., Blais, A., Blachier, F., Marsset-Baglieri, A., Davila-Gay, A.M., Perrin, E., Tome, D., 2009. Effects of galacto-oligosaccharide ingestion on the mucosa-associated mucins and sucrase activity in the small intestine of mice. European Journal of Nutrition 48, 457–464.

Leuba, J.L., Widmer, F., 1977. Immobilization of the β-galactosidase from *Aspergillus niger* on chitosan. Journal of Solid-Phase Biochemistry 2, 257–271.

Li, Z., Xiao, M., Lu, L., Li, Y., 2008. Production of non-monosaccharide and high-purity galactooligosaccharides by immobilized enzyme catalysis and fermentation with immobilized yeast cells. Process Biochemistry 43, 896–899.

Li, F., Jin, X., Liu, B., Zhuang, W., Scalabrin, D., 2014. Follow-up formula consumption in 3- to 4-year-olds and respiratory infections: an RCT. Pediatrics 133, e1533–e1540.

Li, L., Li, G., Cao, L.C., Ren, G.H., Kong, W., Wang, S.D., Guo, G.S., Liu, Y.H., 2015. Characterization of the cross-linked enzyme aggregates of a novel β-galactosidase, a potential catalyst for the synthesis of galacto-oligosaccharides. Journal of Agricultural and Food Chemistry 63, 894–901.

Lind Mikkelsen, L., Jakobsen, M., Borg Jensen, B., 2003. Effects of dietary oligosaccharides on microbial diversity and fructo-oligosaccharide degrading bacteria in faeces of piglets post-weaning. Animal Feed Science and Technology 109, 133–150.

Lisboa, C.R., de Simoni Martinez, L., Trindade, R.A., de Almeida Costa, F.A., Fernandes de Medeiros Burkert, J., Veiga Burkert, C.A., 2012. Response surface methodology applied to the enzymatic synthesis of galacto-oligosaccharides from cheese whey. Food Science and Biotechnology 21, 1519–1524.

Liu, H., Liu, J., Tan, B., Zhou, F., Qin, Y., Yang, R., 2012. Covalent immobilization of *Kluyveromyces fragilis* β-galactosidase on magnetic nanosized epoxy support for synthesis of galacto-oligosaccharide. Bioprocess and Biosystems Engineering 35, 1287–1295.

Liu, Z., Zhao, C., Deng, Y., Huang, Y., Liu, B., 2015. Characterization of a thermostable recombinant β-galactosidase from a thermophilic anaerobic bacterial consortium YTY-70. Biotechnology and Biotechnological Equipment 29, 547–554.

López Leiva, M.H., Guzman, M., 1995. Formation of oligosaccharides during enzymic hydrolysis of milk whey permeates. Process Biochemistry 30, 757–762.

Lozinsky, V.I., Plieva, F.M., 1998. Polyvinylalcohol cryogels employed as matrices for cell immobilization. Overview of recent research and developments. Enzyme and Microbial Technology 23, 224–242.

Lu, L., Xiao, M., Xu, X., Li, Z., Li, Y., 2007. A novel β-galactosidase capable of glycosyl transfer from *Enterobacter agglomerans* B1. Biochemical and Biophysical Research Communications 356, 78–84.

Lu, L., Xiao, M., Li, Z., Li, Y., Wang, F., 2009. A novel transglycosylating β-galactosidase from *Enterobacter cloacae* B5. Process Biochemistry 44, 232–236.

Luz, D.A., Rodrigues, A.K.O., Silva, F.R.C., Torres, A.E.B., Cavalcante, C.L., Brito, E.S., Azevedo, D.C.S., 2008. Adsorptive separation of fructose and glucose from an agroindustrial waste of cashew industry. Bioresource Technology 99, 2455–2465.

Maathuis, A.J.H., van den Heuvel, E.G., Schoterman, M.H.C., Venema, K., 2012. Galacto-oligosaccharides have prebiotic activity in a dynamic in vitro colon model using a ^{13}C-labeling technique. The Journal of Nutrition 142, 1205–1212.

Macfarlane, G.T., Steed, H., Macfarlane, S., 2008. Bacterial metabolism and health-related effects of galacto-oligosaccharides and other prebiotics. Journal of Applied Microbiology 104, 305–344.

Magne, F., Hachelaf, W., Suau, A., Boudra, G., Bouziane-Nedjadi, K., Rigottier-Gois, L., Touhami, M., Desjeux, J.F., Pochart, P., 2008. Effects on faecal microbiota of dietary and acidic oligosaccharides in children during partial formula feeding. Journal of Pediatric Gastroenterology and Nutrition 46, 580–588.

Magner, E., 2013. Immobilisation of enzymes on mesoporous silicate materials. Chemical Society Reviews 42, 6213–6222.

Maischberger, T., Nguyen, T.H., Sukyai, P., Kittl, R., Riva, S., Ludwig, R., Haltrich, D., 2008. Production of lactose-free galacto-oligosaccharide mixtures: comparison of two cellobiose dehydrogenases for the selective oxidation of lactose to lactobionic acid. Carbohydrate Research 343, 2140–2147.

Maischberger, T., Leitner, E., Nitisinprasert, S., Juajun, O., Yamabhai, M., Nguyen, T.H., Haltrich, D., 2010. β-Galactosidase from *Lactobacillus pentosus*: purification, characterization and formation of galacto-oligosaccharides. Biotechnology Journal 5, 838–847.

Maksimainen, M.M., Lampio, A., Mertanen, M., Turunen, O., Rouvinen, J., 2013. The crystal structure of acidic β-galactosidase from *Aspergillus oryzae*. International Journal of Biological Macromolecules 60, 109–115.

Manera, A.P., De Almeida Costa, F.A., Rodrigues, M.I., Kalil, S.J., Maugeri Filho, F., 2010. Galacto-oligosaccharides production using permeabilized cells of *Kluyveromyces marxianus*. International Journal of Food Engineering 6, 1–13.

Manera, A.P., Zabot, G.L., Oliveira, J.V., de Oliveira, D., Mazutti, M.A., Kalil, S.J., Treichel, H., Maugeri Filho, F., 2012. Enzymatic synthesis of galactooligosaccharides using pressurised fluids as reaction medium. Food Chemistry 133, 1408–1413.

Martinez-Ferez, A., Rudloff, S., Guadix, A., Henkel, C.A., Pohlentz, G., Boza, J.J., Guadix, E.M., Kunz, C., 2006. Goats' milk as a natural source of lactose-derived oligosaccharides: isolation by membrane technology. International Dairy Journal 16, 173–181.

Martínez-Villaluenga, C., Cardelle-Cobas, A., Corzo, N., Olano, A., Villamiel, M., 2008. Optimization of conditions for galactooligosaccharide synthesis during lactose hydrolysis by β-galactosidase from *Kluyveromyces lactis* (Lactozym 3000 L HPG). Food Chemistry 107, 258–264.

Martins, A.R., Lisbôa, C.R., 2015. Mathematical model for the conversion of lactose and synthesis of galacto-oligosaccharides (GOS) with simultaneous reversible inhibition by glucose and galactose. International Journal of Engineering Research and Technology 4, 299–305.

Matella, N.J., Dolan, K.D., Lee, Y.S., 2006. Comparison of galactoologisaccharide production in free-enzyme ultrafiltration and in immobilized-enzyme systems. Journal of Food Science 71, C363–C368.

Mateo, C., Abián, O., Fernández-Lafuente, R., Guisán, J.M., 2000. Increase in conformational stability of enzymes immobilized on epoxy-activated supports by favoring additional multipoint covalent attachment. Enzyme and Microbial Technology 26, 509–515.

Mateo, C., Monti, R., Pessela, B.C.C., Fuentes, M., Torres, R., Guisán, J.M., Fernández-Lafuente, R., 2004. Immobilization of lactase from *Kluyveromyces lactis* greatly reduces the inhibition promoted by glucose. Full hydrolysis of lactose in milk. Biotechnology Progress 20, 1259–1262.

Mateo, C., Palomo, J.M., Fuentes, M., Betancor, L., Grazu, V., López-Gallego, F., Pessela, B.C.C., Hidalgo, A., Fernández-Lorente, G., Fernández-Lafuente, R., Guisán, J.M., 2006. Glyoxyl agarose: a fully inert and hydrophilic support for immobilization and high stabilization of proteins. Enzyme and Microbial Technology 39, 274–280.

Mateo, C., Grazu, V., Palomo, J.M., Lopez-Gallego, F., Fernandez-Lafuente, R., Guisan, J.M., 2007. Immobilization of enzymes on heterofunctional epoxi supports. Nature Protocols 2, 1022–1032.

Maugard, T., Gaunt, D., Legoy, M.D., Besson, T., 2003. Microwave-assisted synthesis of galacto-oligosaccharides from lactose with immobilized β-galactosidase from *Kluyveromyces lactis*. Biotechnology Letters 25, 623–629.

Maukonen, J., Matto, J., Kajander, K., Mattila-Sandholm, T., Saarela, M., 2008. Diversity and temporal stability of fecal bacterial populations in elderly subjects consuming galacto-oligosaccharide containing probiotic yoghurt. International Dairy Journal 18, 386–395.

McBain, A.J., Macfarlane, G.T., 2001. Modulation of genotoxic enzyme activities by non-digestible oligosaccharide metabolism in in vitro human gut bacterial ecosystems. Journal of Medical Microbiology 50, 833–842.

Meng, F., Chae, S.R., Drews, A., Kraume, M., Shin, H.S., Yang, F., 2009. Recent advances in membrane bioreactors (MBRs): membrane fouling and membrane material. Water Research 43, 1489–1512.

Messaoudi, M., Violle, N., Bisson, J.F., Desor, D., Javelot, H., Rougeot, C., 2011. Beneficial psychological effects of a probiotic formulation (*Lactobacillus helveticus* R0052 and *Bifidobacterium longum* R0175) in healthy human volunteers. Gut Microbes 2, 256–261.

Meyer, P.D., 2004. Nondigestible oligosaccharides as dietary fiber. Journal of AOAC International 87, 718–726.

Michelon, M., Manera, A.P., Limoeiro Carvalho, A., Maugeri Filho, F., 2014. Concentration and purification of galacto-oligosaccharides using nanofiltration membranes. International Journal of Food Science and Technology 49, 1953–1961.

Mihatsch, W.A., Hoegel, J., Pohlandt, F., 2006. Prebiotic oligosaccharides reduce stool viscosity and accelerate gastrointestinal transport in preterm infants. Acta Pædiatrica 95, 843–848.

Mlichová, Z., Rosenberg, M., 2006. Current trends of β-galactosidase application in food technology. Journal of Food and Nutrition Research 45, 47–54.

Modesto, M., D'Aimmo, M.R., Stefanini, I., Trevisi, P., De Filippi, S., Casini, L., Mazzoni, M., Bosi, P., Biavati, B., 2009. A novel strategy to select *Bifidobacterium* strains and prebiotics as natural growth promoters in newly weaned pigs. Livestock Science 122, 248–258.

Montané, D., Nabarlatz, D., Martorell, A., Torné-Fernández, V., Fierro, V., 2006. Removal of lignin and associated impurities from xylo-oligosaccharides by activated carbon adsorption. Industrial and Engineering Chemistry Research 45, 2294–2302.

Montañés, F., Olano, A., Ibáñez, E., 2007. Modeling solubilities of sugars in alcohols based on original experimental data. AIChE Journals 23, 2411–2418.

Montañés, F., Olano, A., Reglero, G., Ibáñez, E., Fornari, T., 2009. Supercritical technology as an alternative to fractionate prebiotic galactooligosaccharides. Separation and Purification Technology 66, 383–389.

Montañés, F., Fornari, T., Olano, A., Ibáñez, E., 2012. Isolation of prebiotic carbohydrates by supercritical fluid extraction. Scaling-up and economical feasibility. Journal of Chromatography A 150, 92–98.

Morales, V., Sanz, M.L., Olano, A., Corzo, N., 2006. Rapid separation on activated charcoal of high oligosaccharides in honey. Chromatographia 3–4, 233–238.

Moro, G., Minoli, I., Mosca, M., Fanaro, S., Jelinek, J., Stahl, B., Boehm, G., 2002. Dosage-related bifidogenic effects of galacto- and fructooligosaccharides in formula-fed term infants. Journal of Pediatric Gastroenterology and Nutrition 34, 291–295.

Moro, G.E., Mosca, F., Miniello, V., Fanaro, S., Jelinek, J., Stahl, B., Boehm, G., 2003. Effects of a new mixture of prebiotics on faecal flora and stool in term infants. Acta Pediatrica 91, 77–79.

Moro, G.E., Stahl, B., Fanaro, S., Jelinek, J., Boehm, G., Coppa, G.V., 2005. Dietary prebiotic oligosaccharides are detectable in the faeces of formula-fed infants. Acta Pædiatrica 94, 27–30.

Moro, G., Arslanoglu, S., Stahl, B., Jelinek, J., Wahn, U., Boehm, G., 2006. A mixture of prebiotic oligosaccharides reduces the incidence of atopic dermatitis during the first six months of age. Archives of Disease in Childhood 91, 814–819.

Moure, A., Gullón, P., Domínguez, H., Parajó, J.C., 2006. Advances in the manufacture, purification and applications of xylo-oligosaccharides as food additives and nutraceuticals. Process Biochemistry 41, 1913–1923.

Mozaffar, Z., Nakanishi, K., Matsuno, R., 1985. Formation of oligosaccharides during hydrolysis of lactose in milk using β-galactosidase from *Bacillus circulans*. Journal of Food Science 50, 1602–1606.

Mozaffar, Z., Nakanishi, K., Matsuno, R., 1988. Production of oligosaccharides by glutaraldehyde treated and immobilized beta-galactosidase from *Bacillus circulans*. Biotechnology Letters 10, 805–808.

Mussatto, S.I., Mancilha, I.M., 2007. Non-digestible oligosaccharides: a review. Carbohydrate Polymers 68, 587–597.

Mwenya, B., Santoso, B., Sar, C., Gamo, Y., Kobayashi, T., Arai, I., Takahashi, J., 2004. Effects of including β1-4 galacto-oligosaccharides, lactic acid bacteria or yeast culture on methanogenesis as well as energy and nitrogen metabolism in sheep. Animal Feed Science and Technology 115, 313–326.

Mwenya, B., Sar, C., Santoso, B., Kobayashi, T., Morikawa, R., Takaura, K., Umetsu, K., Kogawa, S., Kimura, K., Mizukoshi, H., Takahashi, J., 2005. Comparing the effects of β1-4 galactooligosaccharides and L-cysteine to monensin on energy and nitrogen utilization in steers fed a very high concentrate diet. Animal Feed Science and Technology 118, 19–30.

Nakagawa, T., Ikehata, R., Uchino, M., Miyaj, T., Takano, K., Tomizuka, N., 2006. Cold-active acid β-galactosidase activity of isolated psychrophylic-basidiomycetous yeast *Guehomyces pullulans*. Microbiological Research 161, 75–79.

Nakayama, T., Amachi, T., 1999. β-galactosidase, enzymology. In: Flickinger, M.C., Drew, S.W. (Eds.), Encyclopedia of Bioprocess Technology Fermentation, Biocatalysis and Bioseparation. John Wiley & Sons Inc, New York, pp. 1291–1305.

Nakkharat, P., Haltrich, D., 2006. Purification and characterization of an intracellular enzyme with β-glucosidase and β-galactosidase activity from the thermophilic fungus *Talaromyces thermophilus* CBS 236.58. Journal of Biotechnology 123, 304–313.

Nakkharat, P., Haltrich, D., 2007. β-Galactosidase from *Talaromyces thermophilus* immobilized on to Eupergit C for production of galacto-oligosaccharides during lactose hydrolysis in batch and packed-bed reactor. World Journal of Microbiology and Biotechnology 23, 759–764.

Nam, E.S., Choi, J.W., Lim, J.H., Hwang, S.K., Jung, H.J., Kang, S.K., Cho, K.K., Choi, Y.J., Ahn, J.K., 2004. β-Galactosidase gene of *Thermus thermophilus* KNOUC112 isolated from hot springs of a volcanic area in New Zealand: identification of the bacteria, cloning and expression of the gene in *Escherichia coli*. Asian-Australian Journal of Animal Science 17, 1591–1598.

Nath, A., Bhattacharjee, C., Chowdhury, R., 2013. Synthesis and separation of galacto-oligosaccharides using membrane bioreactor. Desalination 316, 31–41.

Nath, A., Verasztó, B., Basak, S., Koris, A., Kovács, Z., Vatai, G., 2015. Synthesis of lactose-derived nutraceuticals from dairy waste whey – a review. Food and Bioprocess Technology. http://dx.doi.org/10.1007/s11947-015-1572-2 Published Online: 11 September.

Nauta, A., Bakker-Zierikzee, A.M., Schoterman, M.C.H., 2010. Galacto-oligosaccharides. In: Cho, S.S., Finocchiaro, E.T. (Eds.), Handbook of Prebiotics and Probiotics Ingredients. Health Benefits and Food Applications. CRC Press, Boca Raton, pp. 75–93.

Neri, D.F.M., Balcao, V.M., Carneiro-da-Cunha, M.G., Carvalho, L.G., Teixeira Jr., J.R., 2008. Immobilization of β-galactosidase from *Kluyveromyces lactis* onto a polysiloxane-polyvinyl-alcohol magnetic (mPOS-PVA) composite for lactose hydrolysis. Catalysis Communications 9, 2334–2339.

Neri, D.F.M., Balcão, V.M., Costa, R.S., Rocha, I.C.A.P., Ferreira, E.M.F.C., Torres, D.P.M., Rodrigues, L.R.M., Carvalho Jr., L.B., Teixeira, J.A., 2009. Galacto-oligosaccharides production during lactose hydrolysis by free *Aspergillus oryzae* β-galactosidase and immobilized on magnetic polysiloxane-polyvinyl alcohol. Food Chemistry 115, 92–99.

Neuhaus, W., Novalin, S., Klimacek, M., Splechtna, B., Petzelbauer, I., Szivak, A., Kulbe, K.D., 2006. Optimization of an innovative hollow-fiber process to produce lactose-reduced skim milk. Applied Biochemistry and Biotechnology 134, 1–14.

Nguyen, T.H., Splechtna, B., Krasteva, S., Kneifel, W., Kulbe, K.D., Divne, C., Haltrich, D., 2007. Characterization and molecular cloning of a heterodimeric β-galactosidase from the probiotic strain *Lactobacillus acidophilus* R22. FEMS Microbiology Letters 269, 136–144.

Niittynen, L., Kajander, K., Korpela, R., 2007. Galacto-oligosaccharides and bowel function. Scandinavian Journal of Food and Nutrition 51, 62–66.

Nobre, C., Santos, M.J., Dominguez, A., Torres, D., Rocha, O., Peres, A.M., Rocha, I., Ferreira, E.C., Teixeira, J.A., Rodrigues, R.L., 2009. Comparison of adsorption equilibrium of fructose, glucose and sucrose on potassium gel-type and macroporous sodium ion-exchange resins. Analytica Chimica Acta 654, 71–76.

Nobre, C., Teixeira, J.A., Rodrigues, L.R., 2012. Fructo-oligosaccharides purification from a fermentative broth using an activated charcoal column. New Biotechnology 29, 395–401.

Nobre, C., Teixeira, J.A., Rodrigues, R.L., 2015. New trends and technological challenges in the industrial production and purification of fructo-oligosaccharides. Critical Reviews in Food Science and Nutrition 55, 1444–1455.

Numanoğlu, Y., Sungur, S., 2004. β-Galactosidase from *Kluyveromyces lactis* cell disruption and enzyme immobilization using a cellulose–gelatin carrier system. Process Biochemistry 39, 703–709.

O'Connell, S., Walsh, G., 2007. Purification and properties of a beta-galactosidase with potential application as a digestive supplement. Applied Biochemistry and Biotechnology 141, 1–14.

O'Connell, S., Walsh, G., 2010. A novel acid-stable, acid-active β-galactosidase potentially suited to the alleviation of lactose intolerance. Applied Microbiology and Biotechnology 86, 517–524.

Ogawa, J., Shimizu, S., 2002. Industrial microbial enzymes: their discovery by screening and use in large-scale production of useful chemicals in Japan. Current Opinion in Biotechnology 13, 367–375.

Ogushi, S., Yoshimoto, T., Tsuru, D., 1980. Purification and comparison of two types of β-galactosidases from *Aspergillus oryzae*. Journal of Fermentation Technology 58, 115–122.

Ohtsuka, K., Tanoh, A., Ozawa, O., Kanematsu, T., Uchida, T., Shinke, R., 1990. Purification and properties of a β-galactosidase with high galactosyl transfer activity from *Cryptococcus laurentii* OKN-4. Journal of Fermentation and Bioengineering 70, 301–307.

Oliveira, C., Guimarães, P.M.R., Domingues, L., 2011. Recombinant microbial systems for improved β-galactosidase production and biotechnological applications. Biotechnology Advances 29, 600–609.

Onishi, N., Yamashiro, A., Yokozeki, K., 1995. Production of galacto-oligosaccharide from lactose by *Sterigmatomyces elviae* CBS8119. Environmental Microbiology 11, 4022–4025.

Onishi, N., Tanaka, T., 1996. Purification and properties of a galacto- and gluco-oligosaccharide-producing β-glycosidase from *Rhodotorula minuta* IFO879. Journal of Fermentation and Bioengineering 82, 439–443.

Onishi, N., Yokozeki, K., 1996. Gluco-oligosaccharide and galacto-oligosaccharide production by *Rhodotorula minuta* IFO879. Journal of Fermentation and Bioengineering 82, 124–127.

Onishi, N., Kira, I., Yokozeki, K., 1996. Galacto-oligosaccharide production from lactose by *Sirobasidium magnum* CBS6803. Letters in Applied Microbiology 23, 253–256.

Onishi, N., Tanaka, T., 1997. Purification and characterization of galacto-oligosaccharide-producing β-galactosidase from *Sirobasidium magnum*. Letters in Applied Microbiology 24, 82–86.

Onishi, N., Tanaka, T., 1998. Galacto-oligosaccharide production using a recycling cell culture of *Sterigmatomyces elviae* CBS8119. Letters in Applied Microbiology 26, 136–139.

Osman, A., Tzortzis, G., Rastall, R.A., Charalampopoulos, D., 2010. A comprehensive investigation of the synthesis of prebiotic galactooligosaccharides by whole cells of *Bifidobacterium bifidum* NCIMB 41171. Journal of Biotechnology 150, 140–148.

Osman, A., Tzortzis, G., Rastall, R.A., Charalampopoulos, D., 2012. BbgIV is an important *Bifidobacterium* β-galactosidase for the synthesis of prebiotic galactooligosaccharides at high temperatures. Journal of Agricultural and Food Chemistry 60, 740–748.

Osman, A., Symeou, S., Trisse, V., Watson, K.A., Tzortzis, G., Charalampopoulos, D., 2014. Synthesis of prebiotic galactooligosaccharides from lactose using bifidobacterial β-galactosidase (BbgIV) immobilised on DEAE-Cellulose, Q-Sepharose and amino-ethyl agarose. Biochemical Engineering Journal 82, 188–199.

Otieno, D.O., 2010. Synthesis of β-galactooligosaccharides from lactose using microbial β-galactosidases. Comprehensive Reviews in Food Science and Food Safety 9, 471–482.

Ouwehand, A.C., Tiihonen, K., Lahtinen, S., 2010. The potential of probiotics and prebiotics for skin health. In: Farage, M.A., Miller, K.W., Maibach, H.I. (Eds.), Textbook of Aging Skin. Springer, New York, pp. 799–809.

Ozmihci, S., Kargi, F., 2007. Effects of feed sugar concentration on continuous ethanol fermentation of cheese whey powder solution (CWP). Enzyme and Microbial Technology 41, 876–880.

Padilla, B., Ruiz-Matute, A.I., Belloch, C., Cardelle-Cobas, A., Corzo, N., Manzanares, P., 2012. Evaluation of oligosaccharide synthesis from lactose and lactulose using β-galactosidases from *Kluyveromyces* isolated from artisanal cheeses. Journal of Agricultural and Food Chemistry 60, 5134–5141.

Palai, T., Bhattacharya, P.K., 2013. Kinetics of lactose conversion to galacto-oligosaccharides by β-galactosidase immobilized on PVDF membrane. Journal of Bioscience and Bioengineering 115, 668–673.

Palai, T., Singh, A.K., Bhattacharya, P.K., 2014. Enzyme, β-galactosidase immobilized on membrane surface for galacto-oligosaccharides formation from lactose: kinetic study with feed flow under recirculation loop. Biochemical Engineering Journal 88, 68–76.

Palframan, R., Gibson, G.R., Rastall, R.A., 2003. Development of a quantitative tool for the comparison of the prebiotic effect of dietary oligosaccharides. Letters in Applied Microbiology 37, 281–284.

Pan, C., Hu, B., Li, W., Sun, Y.I., Ye, H., Zeng, X., 2009. Novel and efficient method for immobilization and stabilization of β-D-galactosidase by covalent attachment onto magnetic Fe_3O_4-chitosan nanoparticles. Journal of Molecular Catalysis B: Enzymatic 61, 208–215.

Panesar, P.S., Panesar, R., Singh, R.S., Kennedy, J.F., Kumar, H., 2006. Microbial production, immobilization and application of β-D-galactosidase. Journal of Chemical Technology and Biotechnology 81, 530–543.

Panesar, P.S., Kumari, S., Panesar, R., 2011. Prebiotics: current status and perspectives. International Journal of Food and Fermentation Technology 1, 149–159.

Panesar, P.S., Kumari, S., Panesar, R., 2013. Biotechnological approaches for the production of prebiotics and their potential applications. Critical Reviews in Biotechnology 33, 345–364.

Park, N.H., Yoshida, S., Takakashi, A., Kawabata, Y., Sun, H.J., Kusakabe, I., 2001. A new method for the preparation of crystalline L-arabinose from arabinoxylan by enzymatic hydrolysis and selective fermentation with yeast. Biotechnology Letters 23, 411–416.

Park, H.Y., Kim, H.J., Lee, J.K., Kim, D., Oh, D.K., 2008. Galactooligosaccharide production by a thermostable β-galactosidase from *Sulfolobus solfataricus*. World Journal of Microbiology and Biotechnology 24, 1553–1558.

Park, A.R., Oh, D.K., 2010. Galacto-oligosaccharide production using microbial β-galactosidase: current state and perspectives. Applied Microbiology and Biotechnology 85, 1279–1286.

Patel, S., Goyal, A., 2011. Functional oligosaccharides: production, properties and applications. World Journal of Microbiology and Biotechnology 27, 1119–1128.

Patil, N.V., Janssen, A.E.M., Boom, R.M., 2014. The potential impact of membrane cascading on downstream processing of oligosaccharides. Chemical Engineering Science 106, 86–98.

Patterson, J.A., Burkholder, K.M., 2003. Application of prebiotics and probiotics in poultry production. Poultry Science 82, 627–631.

Pawlak-Szukalska, A., Wanarska, M., Popinigis, A.T., Kur, J., 2014. A novel cold-active β-D-galactosidase with transglycosylation activity from the Antarctic *Arthrobacter* sp. 32cB – gene cloning, purification and characterization. Process Biochemistry 49, 2122–2133.

Pereira-Rodríguez, A., Fernández-Leiro, R., González-Siso, M.I., Cerdán, M.E., Becerra, M., Sanz-Aparicio, J., 2012. Structural basis of specificity in tetrameric *Kluyveromyces lactis* β-galactosidase. Journal of Structural Biology 177, 392–401.

Pérez-Conesa, D., López, G., Ros, G., 2007. Effects of probiotic, prebiotic and symbiotic follow-up infant formulas on large intestine morphology and bone mineralization in rats. Journal of the Science of Food and Agriculture 87, 1059–1068.

Perotti, M.C., Wolf, V., Vénica, I., Bergamini, V., 2012. Dairy products modified in their lactose content. Current Nutrition and Food Science 8, 8–18.

Pessela, B.C.C., Mateo, C., Fuentes, M., Vian, A., García, J.L., Carrascosa, A.V., Guisán, J.M., Fernández-Lafuente, R., 2003. The immobilization of a thermophilic β-galactosidase on Sepabeads supports decreases product inhibition. Complete hydrolysis of lactose in dairy products. Enzyme and Microbial Technology 33, 199–205.

Pessela, B.C., Mateo, C., Fuentes, M., Vian, A., Garcia, J.L., Carrascosa, A.V., Guisán, J.M., Fernández-Lafuente, R., 2004. Stabilization of a multimeric β-galactosidase from *Thermus* sp. strain T2 by immobilization on novel heterofunctional epoxy supports plus aldehyde-dextran cross-linking. Biotechnology Progress 20, 388–392.

Pessela, B.C.C., Betancor, L., Lopez-Gallego, F., Torres, R., Dellamora-Ortiz, G.M., Alonso-Morales, N., Fuentes, M., Fernández-Lafuente, R., Guisán, J.M., Mateo, C., 2005. Increasing the binding strength of proteins to PEI coated supports by immobilizing at high ionic strength. Enzyme and Microbial Technology 37, 295–299.

Petrova, S.D., Bakalova, N.G., Atev, A.P., Kolev, D.N., 2009. Production of galacto-oligosaccharides from lactose and whey permeate by β-glycosidase from *Sporotrichum thermophile* ATCC 346. Biotechnology and Biotechnological Equipment 23, 672–675.

Petrova, V.Y., Kujumdzieva, A.V., 2010. Thermotolerant yeast strains producers of galacto-oligosaccharides. Biotechnology and Biotechnological Equipment 24, 1612–1619.

Petzelbauer, I., Spletchna, B., Nidetzky, B., 2002. Development of ultra-high temperature process for the enzymatic hydrolysis of lactose. III. Utilization of two thermostable β-glycosidases in a continuous ultrafiltration membrane reactor and galacto-oligosaccharide formation under steady-state conditions. Biotechnology and Bioengineering 77, 394–404.

Piirainen, L., Kekkonen, R.A., Kajander, K., Ahlroos, T., Tynkkynen, S., Nevala, R., Korpela, R., 2008. In school-aged children a combination of galacto-oligosaccharides and *Lactobacillus* GG increases bifidobacteria more than *Lactobacillus* GG on its own. Annals of Nutrition and Metabolism 52, 204–208.

Pineiro, M., Asp, N.G., Reid, G., Macfarlane, S., Morelli, L., Brunser, O., Tuohy, K., 2008. FAO technical meeting on prebiotics. Journal of Clinical Gastroenterology 42, S156–S159.

Pinelo, M., Jonsson, G., Meyer, A., 2009. Membrane technology for purification of enzymatically produced oligosaccharides: molecular and operational features affecting performance. Separation and Purification Technology 70, 1–11.

Placier, G., Watzlawick, H., Rabiller, C., Mattes, R., 2009. Evolved β-galactosidases from *Geobacillus stearothermophilus* with improved transgalactosylation yield for galacto-oligosaccharide production. Applied and Environmental Microbiology 75, 6312–6321.

Planas, A., Faijes, M., 2002. Glycosidases and glycol synthases in enzymatic synthesis of oligosaccharides. An overview. Afinidad LIX 500, 295–313.

Playne, M.J., Crittenden, R.G., 2009. Galacto-oligosaccharides and other products derived from lactose. In: Fox, P.F., McSweeney, P.L.H. (Eds.), Advanced Dairy Chemistry. Lactose, Water, Salts and Minor Constituents, vol. 3. Springer, New York, pp. 121–202.

Pocedičová, K., Čurda, L., Mišun, D., Dryáková, A., Diblíková, L., 2010. Preparation of galacto-oligosaccharides using membrane reactor. Journal of Food Engineering 99, 479–484.

Pospiskova, K., Safarik, I., 2013. Low-cost, easy-to-prepare magnetic chitosan microparticles for enzymes immobilization. Carbohydrate Polymers 96, 545–548.

Prenosil, J.E., Stuker, E., Bourne, J.R., 1987. Formation of oligosaccharides during enzymatic lactose hydrolysis and their importance in a whey hydrolysis process: Part I: state of art. Biotechnology and Bioengineering 30, 1019–1025.

Propst, L., Flickinger, E.A., Bauer, L.L., Merchen, N.R., Fahey, G.C., 2003. A dose-response experiment evaluating the effects of oligofructose and inulin on nutrient digestibility, stool quality, and fecal protein catabolites in healthy adult dogs. Journal of Animal Science 81, 3057–3066.

Pruksasri, S., 2007. Production and Separation of Galacto-Oligosaccharides From Lactose by β-Galactosidase Immobilized on Nanofiltration Membranes (Ph.D. thesis). Ohio State University.

Pruksasri, S., Nguyen, T.H., Haltrich, D., Novalin, S., 2015. Fractionation of a galacto-oligosaccharides solution at low and high temperature using nanofiltration. Separation and Purification Technology 151, 124–130.

Qiang, X., YongLie, C., QianBing, W., 2009. Health benefit application of functional oligosaccharides. Carbohydrate Polymers 77, 435–441.

Rabiu, B.A., Jay, A.J., Gibson, G.R., Rastall, R.A., 2001. Synthesis and fermentation properties of novel galacto-oligosaccharides by β-galactosidases from *Bifidobacterium* species. Applied and Environmental Microbiology 67, 2526–2530.

Rajendran, A., Paredes, G., Mazzotti, M., 2009. Simulated moving bed chromatography for the separation of enantiomers. Journal of Chromatography A 1216, 709–738.

Rastall, R.A., 2006. Galacto-oligosaccharides as prebiotic food ingredients. In: Gibson, G.R., Rastall, R.A. (Eds.), Prebiotics: Development and Applications. John Wiley, New York, pp. 101–109.

Redmond, J.W., Packer, N.H., 1999. The use of solid-phase extraction with graphitised carbon for the fractionation and purification of sugars. Carbohydrate Research 319, 74–79.

Ren, H., Fei, J., Shi, X., Zhao, T., Cheng, H., Zhao, N., Chen, Y., Ying, H., 2015. Continuous ultrafiltration membrane reactor coupled with nanofiltration for the enzymatic synthesis and purification of galactosyl-oligosaccharides. Separation and Purification Technology 144, 70–79.

Reuter, S., Nygaard, A.R., Zimmermann, W., 1999. β-Galactooligosaccharide synthesis with β-galactosidases from *Sulfolobus solfataricus, Aspergillus oryzae*, and *Escherichia coli*. Enzyme and Microbial Technology 25, 509–516.

Riva, E., Verduci, E., Agostoni, C., Giovannini, M., 2005. Closer to the gold standard: an appraisal of formulae available in Italy for use in formula-fed infants. The Journal of International Medical Research 33, 595–611.

Rivero-Urgell, M., Santamaria-Orleans, A., 2001. Oligosaccharides: application in infant food. Early Human Development 65, S43–S52.

Roberts, H.R., McFarren, E.F., 1953. The chromatographic observation of oligosaccharides formed during the lactase hydrolysis of lactose. Journal of Dairy Science 36, 620–632.

Rodriguez-Colinas, B., de Abreu, M.A., Fernandez-Arrojo, L., de Beer, R., Poveda, A., Jimenez-Barbero, J., Haltrich, D., Ballesteros, A.O., Fernandez-Lobato, M., Plou, F.J., 2011. Production of galacto-oligosaccharides by the β-galactosidase from Kluyveromyces lactis: comparative analysis of permeabilized cells versus soluble enzyme. Journal of Agricultural and Food Chemistry 59, 10477–10484.

Rodriguez-Colinas, B., Fernandez-Arrojo, L., Ballesteros, A.O., Plou, F.J., 2014. Galactooligosaccharides formation during enzymatic hydrolysis of lactose: towards a prebiotic enriched milk. Food Chemistry 145, 388–394.

Rodriguez-Fernandez, M., Cardelle-Cobas, A., Villamiel, M., Banga, J.R., 2011. Detailed kinetic model describing new oligosaccharides synthesis using different β-galactosidases. Journal of Biotechnology 153, 116–124.

Rojas, A.L., Nagem, R.A.P., Neustroev, K.N., Arand, M., Adamska, M., Eneyskaya, E.V., Kulminskaya, A.A., Garratt, R.C., Golubev, A.M., Polikarpov, I., 2004. Crystal structures of β-galactosidase from *Penicillium* sp. and its complex with galactose. Journal of Molecular Biology 5, 1281–1292.

Rosatella, A.A., Branco, L.C., Afonso, C.A.M., 2009. Studies on dissolution of carbohydrates in ionic liquids and extraction from aqueous phase. Green Chemistry 11, 1406–1413.

Roy, G.M., 1995. Activated Carbon Applications in the Food and Pharmaceutical Industries. Technomic Publishing Company, Lancaster.

Roy, D., Daoudi, L., Azaola, A., 2002. Optimization of galacto-oligosaccharide production by *Bifidobacterium infantis* RW-8120 using response surface methodology. Journal of Industrial Microbiology and Biotechnology 29, 281–285.

Sairanen, U., Piirainen, L., Nevala, R., Korpela, R., 2007. Yoghurt containing galacto-oligosaccharides, prunes and linseed reduces the severity of mild constipation in elderly subjects. European Journal of Clinical Nutrition 61, 1423–1428.

Sakai, T., Tsuji, H., Shibata, S., Hayakawa, K., Matsumoto, K., 2008. Repeated-batch production of galacto-oligosaccharides from lactose at high concentration by using alginate-immobilized cells of *Sporobolomyces singularis* YIT 10047. Journal of General and Applied Microbiology 54, 285–293.

Sako, T., Matsumoto, K., Tanaka, R., 1999. Recent progress on research and applications of non-digestible galacto-oligosaccharides. International Dairy Journal 9, 69–80.

Sangeetha, K., Abraham, T.E., 2008. Preparation and characterization of cross-linked enzyme aggregates (CLEA) of subtilisin for controlled release applications. International Journal of Biological Macromolecules 43, 314–319.

Sangwan, V., Tomar, S.K., Singh, R.R.B., Singh, A.K., Ali, B., 2011. Galactooligosaccharides: novel components of designer foods. Journal of Food Science 76, R103–R111.

Sangwan, V., Tomar, S.K., Ali, B., Singh, R.R.B., Singh, A.K., Mandal, S., 2014. Galactooligosaccharides purification using microbial fermentation and assessment of its prebiotic potential by *in vitro* method. International Journal of Current Microbiology and Applied Sciences 3, 573–585.

Santoso, B., Mwenya, B., Sar, C., Gamo, Y., Kobayashi, T., Morikawa, R., Kimura, K., Mizukoshi, H., Takahashi, J., 2004. Effects of supplementing galacto-oligosaccharides, *Yucca schidigera* or nisin on rumen methanogenesis, nitrogen and energy metabolism in sheep. Livestock Production Science 91, 209–217.

Sanz-Valero, J.I., 2009. Production of Galacto-Oligosaccharides From Lactose by Immobilized β-Galactosidase and Posterior Chromatographic Separation (Ph.D. thesis). Ohio State University.

Saudagar, P.S., Singhal, R.S., 2004. Curdlan as a support matrix for immobilization of enzyme. Carbohydrate Polymers 56, 483–488.

Savaiano, D.A., Ritter, A.J., Klaenhammer, T.R., James, G.M., Longcore, A.T., Chandler, J.R., Walker, W.A., Foyt, H.L., 2013. Improving lactose digestion and symptoms of lactose intolerance with a novel galacto-oligosaccharide (RP-G28): a randomized, double-blind clinical trial. Nutrition Journal. 12, 160. http://www.nutritionj.com/content/12/1/160.

Scalabrin, D.M.F., Mitmesser, S.H., Welling, G.W., Harris, C.L., Marunycz, J.D., Carey Walker, D., Bos, N.A., Tölkko, S., Salminen, S., Vanderhoof, J.A., 2012. New prebiotic blend of polydextrose and galacto-oligosaccharides has a bifidogenic effect in young infants. Journal of Pediatric Gastroenterology and Nutrition 54, 343–352.

Schmidt, M., Stougaard, P., 2010. Identification, cloning and expression of a cold-active β-galactosidase from a novel Arctic bacterium, *Alkalilactibacillus ikkense*. Environmental Technology 31, 1107–1114.

Scholtens, P.A.M.J., Alles, M.S., Bindels, J.G., van der Linde, E.G.M., Tolboom, J.J.M., Knol, K., 2006. Bifidogenic effects of solid weaning foods with added prebiotic oligosaccharides: a randomised controlled clinical trial. Journal of Pediatric Gastroenterology and Nutrition 42, 553–559.

Schouten, B., Van Esch, B.C.A.M., Kormelink, T.G., Moro, G.E., Arslanoglu, S., Boehm, G., Knippels, L.M.J., Redegeld, F.A., Willemsen, L.E.M., Garssen, J., 2011. Non-digestible oligosaccharides reduce immunoglobulin free light-chain concentrations in infants at risk for allergy. Pediatric Allergy and Immunology 22, 537–542.

Searle, L.E.J., Best, A., Nunez, A., Salguero, F.J., Johnson, L., Weyer, U., Dugdale, A.H., Cooley, W.A., Carter, B., Jones, G., Tzortzis, G., Woodward, M.J., La Ragione, R.M., 2009. A mixture containing galactooligosaccharide, produced by the enzymic activity of *Bifidobacterium bifidum*, reduces *Salmonella enterica* serovar *Typhimurium* infection in mice. Journal of Medical Microbiology 58, 37–48.

Sears, P., Wong, C.H., 2001. Toward automated synthesis of oligosaccharides and glycoproteins. Science 291, 2344–2350.

Seeberger, P.H., Werz, D.B., 2005. Automated synthesis of oligosaccharides as a basis for drug discovery. Nature Reviews Drug Discovery 4, 751–763.

Sen, D., Sarkar, A., Gosling, A., Gras, S.L., Stevens, G.W., Kentish, S.E., Bhattacharya, P.K., Barber, A.R., Bhattacharjee, C., 2011a. Feasibility study of enzyme immobilization on polymeric membrane: a case study with enzymatically galacto-oligosaccharides production from lactose. Journal of Membrane Science 378, 471–478.

Sen, D., Gosling, A., Stevens, G.W., Bhattacharya, P.K., Barber, A.R., Kentish, S.E., Bhattacharjee, C., Gras, S.L., 2011b. Galactosyl oligosaccharide purification by ethanol precipitation. Food Chemistry 128, 773–777.

Sen, D., Sarkar, A., Das, S., Chowdhury, R., Bhattacharjee, C., 2012. Batch hydrolysis and rotating disk membrane bioreactor for the production of galacto-oligosaccharides: a comparative study. Industrial and Engineering Chemistry Research 51, 10671–10681.

Sen, P., Nath, A., Bhattacharjee, C., Chowdhury, R., Bhattacharya, P., 2014. Process engineering studies of free and micro-encapsulated β-galactosidase in batch and packed bed bioreactors for production of galactooligosaccharides. Biochemical Engineering Journal 90, 59–72.

Shadid, R., Haarman, M., Knol, J., Theis, W., Beermann, C., Rjosk-Dendorfer, D., Schendel, D.J., Koletzko, B.V., Krauss-Etschmann, S., 2007. Effects of galactooligosaccharide and long-chain fructooligosaccharide supplementation during pregnancy on maternal and neonatal microbiota and immunity-a randomized, double-blind, placebo-controlled study. The American Journal of Clinical Nutrition 86, 1426–1437.

Sheik Asraf, S., Gunasekaran, P., 2010. Current trends of β-galactosidase research and application. In: Mendez-Vilas, A. (Ed.), Current Research, Technology and Education Topics in Applied Microbiology and Microbial Biotechnology. vol. 2. Formatex Research Center, Badajoz, pp. 880–890.

Sheldon, R.A., 2007. Enzyme immobilization: the quest for optimum performance. Advanced Synthesis and Catalysis 349, 1289–1307.

Sheldon, R.A., 2011a. Characteristic features and biotechnological applications of cross-linked enzyme aggregates (CLEAs). Applied Microbiology and Biotechnology 92, 467–477.

Sheldon, R.A., 2011b. Cross-linked enzyme aggregates as industrial biocatalysts. Organic Process Research and Development 15, 213–223.

Sheldon, R.A., van Pelt, S., 2013. Enzyme immobilisation in biocatalysis: why, what and how. Chemical Society Reviews 42, 6223–6235.

Sheu, D.C., Li, S.Y., Duan, K.J., Chen, C.W., 1998. Production of galactooligosaccharides by β-galactosidase immobilized on glutaraldehyde-treated chitosan beads. Biotechnology Techniques 12, 273–276.

Shin, H.J., Park, J.M., Yang, J.W., 1998. Continuous production of galacto-oligosaccharides from lactose by *Bullera singularis* β-galactosidase immobilized in chitosan beads. Process Biochemistry 33, 787–792.

Shioiri, T., Yahagi, K., Nakayama, S., Asahara, T., Yuki, N., Kawakami, K., Yamaoka, Y., Sakai, Y., Nomoto, K., Totani, M., 2006. The effects of a synbiotic fermented milk beverage containing *Lactobacillus casei* strain Shirota and transgalactosylated oligosaccharides on defecation frequency, intestinal microflora, organic acid concentrations, and putrefactive metabolites of sub-optimal health state volunteers: a randomized placebo-controlled cross-over study. Bioscience and Microflora 25, 137–146.

Shoaf, K., Mulvey, G.L., Armstrong, G.D., Hutkins, R.W., 2006. Prebiotic galactooligosaccharides reduce adherence of enteropathogenic *Escherichia coli* to tissue culture cells. Infection and Immunity 74, 6920–6928.

Sinnott, M.L., 1990. Catalytic mechanisms of enzymic glycosyl transfer. Chemical Review 90, 1171–1202.

Song, J., Imanaka, H., Imamura, K., Minoda, M., Katase, T., Hoshi, Y., Yamaguchi, S., Nakanishi, K., 2011. Cloning and expression of a β-galactosidase gene of *Bacillus circulans*. Bioscience. Biotechnology and Biochemistry 75, 1194–1197.

Splechtna, B., Petzelbauer, I., Baminger, U., Haltrich, D., Kulbe, K.D., Nidetzky, B., 2001. Production of a lactose-free galacto-oligosaccharide mixture by using selective enzymatic oxidation of lactose into lactobionic acid. Enzyme and Microbial Technology 29, 434–440.

Splechtna, B., Nguyen, T.H., Steinbo, M., Klaus, C.K., Kulbe, D., Lorenz, W., Haltrich, D., 2006. Production of prebiotic galacto-oligosaccharides from lactose using β-galactosidases from *Lactobacillus reuteri*. Journal of Agricultural and Food Chemistry 54, 4999–5006.

Splechtna, B., Nguyen, T.H., Haltrich, D., 2007a. Comparison between discontinuous and continuous lactose conversion processes for the production of prebiotic galacto-oligosaccharides using β-galactosidase from *Lactobacillus reuteri*. Journal of Agricultural and Food Chemistry 55, 6772–6777.

Splechtna, B., Nguyen, T.H., Zehetner, R., Lettner, H.P., Lorenz, W., Haltrich, D., 2007b. Process development for the production of prebiotic galacto-oligosaccharides from lactose using β-galactosidase from *Lactobacillus* sp. Biotechnology Journal 2, 480–485.

Sriphannam, W., Lumyong, S., Niumsap, P., Ashida, H., Yamamoto, K., Khanongnuch, C., 2012. A selected probiotic strain of *Lactobacillus fermentum* CM33 isolated from breast-fed infants as a potential source of β-galactosidase for prebiotic oligosaccharide synthesis. The Journal of Microbiology 50, 119–126.

Srisimarat, W., Pongsawasdi, P., 2008. Enhancement of the oligosaccharide synthetic activity of β-galactosidase in organic solvents by cyclodextrin. Enzyme and Microbial Technology 43, 436–444.

Srivastava, A., Mishra, S., Chand, S., 2015. Transgalactosylation of lactose for synthesis of galacto-oligosaccharides using *Kluyveromyces marxianus* NCIM 3551. New Biotechnology 32, 412–418.

Srivastava, A., Mishra, S., Chand, S., 2016. Synthesis of galacto-oligosaccharides from lactose using immobilized cells of *Kluyveromyces marxianus* NCIM 3551. Journal of Molecular Catalysis B: Enzymatic 123, 147–153.

Sun, X., Duan, X., Wu, D., Chen, J., Wu, J., 2014. Characterization of *Sulfolobus solfataricus* β-galactosidase mutant F441Y expressed in *Pichia pastoris*. Journal of the Science of Food and Agriculture 94, 1359–1365.

Surakka, A., Kajander, K., Rajilic, M., Karjalainen, H., Hatakka, K., Vapaatalo, H., Zoetendal, E.G., de Vos, W.M., Korpela, R., Tynkkynen, S., 2009. Yoghurt containing galactooligosaccharides facilitates defecation among elderly subjects and selectively increases the number of bifidobacteria. International Journal of Probiotics and Prebiotics 4, 65–74.

Swennen, K., Courtin, C., van der Bruggen, B., Vandecasteele, C., Delcour, J., 2005. Ultrafiltration and ethanol precipitation for isolation of arabinoxylooligosaccharide with different structures. Carbohydrate Polymers 62, 283–292.

Takahashi, Y., Goto, S., 1994. Continuous separation of fructooligosaccharides using an annular chromatograph. Separation Science and Technology 10, 1311–1318.

Tanaka, Y., Kagamiishi, A., Kiuchi, A., Horiuchi, T., 1975. Purification and properties of β-galactosidase from *Aspergillus oryzae*. Journal of Biochemistry 77, 241–247.

Taniguchi, H., 2004. Carbohydrate research and industry in Japan and the Japanese society of applied glycoscience. Starch/Stärke 56, 1–5.

Teng, W., 2012. From chitin to chitosan. In: Yao, K., Li, J., Yao, F., Yin, Y. (Eds.), Chitosan-Based Hydrogels. Functions and Applications. CRC Press, Boca Raton, pp. 1–38.

Toba, T., Tomita, Y., Itoh, T., Adachi, S., 1981. β-Galactosidases of lactic acid bacteria: characterization by oligosaccharides formed during hydrolysis of lactose. Journal of Dairy Science 64, 185–192.

Toba, T., Yokota, A., Adachi, S., 1985. Oligosaccharide structures formed during the hydrolysis of lactose by *Aspergillus oryzae* β-galactosidase. Food Chemistry 16, 147–162.

Todorova-Balvay, D., Stoilova, I., Gargova, S., Vijayalakshmi, M.A., 2006. An efficient two step purification and molecular characterization of β-galactosidases from *Aspergillus oryzae*. Journal of Molecular Recognition 19, 299–304.

Torres, D.P.M., Gonçalves, M.P.F., Teixeira, J.A., Rodrigues, L.R., 2010. Galacto-oligosaccharides production, properties, applications, and significance as prebiotics. Comprehensive Reviews in Food Science and Food Safety 9, 438–454.

Torres, D.P.M., Bastos, M., Gonçalves, M.P.F., Teixeira, J.A., Rodrigues, L.R., 2011. Water sorption and plasticization of an amorphous galacto-oligosaccharide mixture. Carbohydrate Polymers 83, 831–835.

Tran, C.T.H., Nosworthy, N., Bilek, M.M.M., McKenzie, D.R., 2015. Covalent immobilization of enzymes and yeast: towards a continuous simultaneous saccharification and fermentation process for cellulosic ethanol. Biomass and Bioenergy 81, 234–241.

Tymczyszyn, E.E., Gerbino, E., Illanes, A., Gómez-Zavaglia, A., 2011. Galacto-oligosaccharides as protective molecules in the preservation of *Lactobacillus delbrueckii* subsp. *bulgaricus*. Cryobiology 62, 123–129.

Tymczyszyn, E.E., Santos, M.I., Costa, M.C., Illanes, A., Gómez-Zavaglia, A., 2014. History, synthesis, properties, applications and regulatory issues of probiotic oligosaccharides. In: Gil, M.H. (Ed.), Carbohydrates Applications in Medicine. Research Signpost, Kerala, pp. 127–154.

Tzortzis, G., Goulas, A.K., Gee, J.M., Gibson, G.R., 2005a. A novel galactooligosaccharide mixture increases the bifidobacterial population numbers in a continuous in vitro fermentation system and in the proximal colonic contents of pigs in vivo. Journal of Nutrition 135, 1726–1731.

Tzortzis, G., Goulas, A.K., Gibson, G.R., 2005b. Synthesis of prebiotic galactooligosaccharides using whole cells of a novel strain, *Bifidobacterium bifidum* NCIMB 41171. Applied Microbiology and Biotechnology 68, 412–416.

Tzortzis, G., Vulevic, J., 2009. Galacto-oligosaccharide prebiotics. In: Charalampopoulos, D., Rastall, R.A. (Eds.), Prebiotics and Probiotics Science and Technology. Springer, New York, pp. 207–244.

Urrutia, P., Mateo, C., Guisan, J.M., Wilson, L., Illanes, A., 2013a. Immobilization of *Bacillus circulans* β-galactosidase and its application in the synthesis of galacto-oligosaccharides under repeated-batch operation. Biochemical Engineering Journal 77, 41–48.

Urrutia, P., Rodriguez-Colinas, B., Fernandez-Arrojo, L., Ballesteros, A.O., Wilson, L., Illanes, A., Plou, F.J., 2013b. Detailed analysis of galactooligosaccharides synthesis with β-galactosidase from *Aspergillus oryzae*. Journal of Agricultural and Food Chemistry 61, 1081–1087.

Urrutia, P., Bernal, C., Wilson, L., Illanes, A., 2014. Improvement of chitosan derivatization for the immobilization of *Bacillus circulans* β-galactosidase and its further application in galacto-oligosaccharide synthesis. Journal of Agricultural and Food Chemistry 62, 10126–10135.

Vaghjiani, J.D., Lee, T.S., Lye, G.J., Turner, M.K., 2000. Production and characterisation of cross-linked enzyme crystals (CLECs) for application as process scale biocatalysts. Biocatalysis and Biotransformation 18, 151–175.

van den Heuvel, E.G., Schoterman, M.H., Muijs, T., 2000. Transgalacto-oligosaccharides stimulate calcium absorption in postmenopausal women. The Journal of Nutrition 130, 2938–2942.

Vandenplas, Y., Devreker, T., Salvatore, S., 2008. Prebiotics and infant nutritiom. In: Gibson, G.R., Roberfroid, M.B. (Eds.), Handbook of Prebiotics. Taylor & Francis, Boca Raton, pp. 393–403.

van Hoffen, E., Ruiter, B., Faber, J., M'Rabet, L., Knol, E.F., Stahl, B., Arslanoglu, S., Moro, G., Boehm, G., Garssen, J., 2009. A specific mixture of short-chain galacto-oligosaccharides and long-chain fructo-oligosaccharides induces a beneficial immunoglobulin profile in infants at high risk for allergy. Allergy 64, 484–487.

Vaňková, K., Onderková, Z., Antošová, M., Polakovič, M., 2008. Design and economics of industrial production of fructooligosaccharides. Chemical Papers 62, 375–381.

Van Loo, J., Vancraeynest, D., 2008. Prebiotics and animal nutrition. In: Gibson, G.R., Roberfroid, M. (Eds.), Handbook of Prebiotics. Taylor & Francis, Boca Raton, pp. 421–436.

Vanneste, J., De Ron, S., Vandecruys, S., Soare, S.A., Darvishmanesh, S., Van der Bruggen, B., 2011. Techno-economic evaluation of membrane cascades relative to simulated moving bed chromatography for the purification of mono- and oligosaccharides. Separation and Purification Technology 80, 600–609.

Varzakas, T.H., Arvanitoyannis, I.S., 2010. Bioactive dairy ingredients. In: Yildiz, F. (Ed.), Development and Manufacture of Yogurt and Other Functional Dairy Products. CRC Press, Boca Raton, pp. 197–227.

Veereman-Wauters, G., 2005. Application of prebiotics in infant foods. British Journal of Nutrition 93, S57–S60.

Vegas, R., Moure, A., Domínguez, H., Parajó, J.C., Alvarez, J.R., Luque, S., 2008. Evaluation of ultra- and nanofiltration for refining soluble products from rice husk xylan. Bioresource Technology 99, 5341–5351.

Velagapudi, V.R., Wittmann, C., Schneider, K., Heinzle, E., 2007. Metabolic flux screening of *Saccharomyces cerevisiae* single knockout strains on glucose and galactose supports elucidation of gene function. Journal of Biotechnology 132, 395–404.

Vendrig, J.C., Coffeng, L.E., Fink-Gremmels, J., 2014. Effects of orally administered galactooligosaccharides on immunological parameters in foals: a pilot study. BMC Veterinary Research. 10, 278. http://www.biomedcentral.com/1746-6148/10/278.

Venema, K., 2012. Intestinal fermentation of lactose and prebiotic lactose derivatives, including human milk oligosaccharides. International Dairy Journal 22, 123–140.

Vénica, C.I., Bergamini, C.V., Rebechi, S.R., Perotti, M.C., 2015. Galacto-oligosaccharides formation during manufacture of different varieties of yogurt. Stability through storage. LWT – Food Science and Technology 63, 198–205.

Vera, C., Guerrero, C., Illanes, A., 2011. Determination of the transgalactosylation activity of *Aspergillus oryzae* β-galactosidase: effect of pH, temperature and galactose and glucose concentrations. Carbohydrate Research 346, 745–752.

Vera, C., Guerrero, C., Conejeros, R., Illanes, A., 2012. Synthesis of galacto-oligosaccharides by -galactosidase from *Aspergillus oryzae* using partially dissolved and supersaturated solution of lactose. Enzyme and Microbial Technology 50, 188–194.

Vera, C., Guerrero, C., Illanes, A., Conejeros, R., 2014. Fed-batch synthesis of galacto-oligosaccharides with *Aspergillus oryzae* β-galactosidase using optimal control strategy. Biotechnology Progress 30, 59–67.

Vetere, A., Paoletti, S., 1998. Separation and characterization of three β-galactosidases from *Bacillus circulans*. Biochimica et Biophysica Acta 1380, 223–231.

Voragen, A.G.J., 1998. Technological aspects of functional food related carbohydrates. Trends in Food Science and Technology 9, 328–335.

Vulevic, J., Rastall, R.A., Gibson, G.R., 2004. Developing a quantitative approach for determining the in vitro prebiotic potential of dietary oligosaccharides. FEMS Microbiology Letters 236, 153–159.

Vulevic, J., Drakoularakou, A., Yaqoob, P., Tzortzis, G., Gibson, G.R., 2008. Modulation of the fecal microflora profile and immune function by a novel trans-galactooligosaccharide mixture (B-GOS) in healthy elderly volunteers. The American Journal of Clinical Nutrition 88, 1438–1446.

Walton, G.E., van den Heuvel, E.G.H.M., Kosters, M.H.W., Rastall, R.A., Tuohy, K.M., Gibson, G.R., 2012. A randomised crossover study investigating the effects of galacto-oligosaccharides on the faecal microbiota in men and women over 50 years of age. British Journal of Nutrition 107, 1466–1475.

Wanarska, M., Kur, J., Pladzyk, R., Turkiewicz, M., 2005. Thermostable *Pyrococcus woesei* β-D-galactosidase – high level expression, purification and biochemical properties. Acta Biochimica Polonica 52, 781–787.

Wang, K., 2012. Enzyme immobilization on chitosan-based supports. In: Yao, K., Li, J., Yao, F., Yin, Y. (Eds.), Chitosan-Based Hydrogels. Functions and Applications. CRC Press, Boca Raton, pp. 339–406.

Wang, K., Lu, Y., Liang, W.Q., Wang, S.D., Jiang, Y., Huang, R., Liu, Y.H., 2012. Enzymatic synthesis of galacto-oligosaccharides in an organic–aqueous biphasic system by a novel β-galactosidase from a metagenomic library. Journal of Agricultural and Food Chemistry 60, 3940–3946.

Wang, T.H., Lu, S., 2013. Production of xylooligosaccharide from wheat bran by microwave assisted enzymatic hydrolysis. Food Chemistry 138, 1531–1535.

Wang, K., Gao, Y., Wang, Z., Meng, G., 2014. Cross linked enzyme aggregates of β-galactosidase from different source by dialdehyde starch as cross-linker. Lecture Notes in Electrical Engineering 251, 1733–1739.

Warmerdam, A., Paudel, E., Jia, W., Boom, R.M., Janssen, A.E.M., 2013a. Characterization of β-galactosidase isoforms from *Bacillus circulans* and their contribution to GOS production. Applied Biochemistry and Biotechnology 170, 340–358.

Warmerdam, A., Boom, R.M., Janssen, A.E.M., 2013b. β-Galactosidase stability at high substrate concentrations. SpringerPlus. 2, 402. http://www.springerplus.com/content/2/1/402.

Warmerdam, A., Benjamins, E., De Leeuw, T.F., Broekhuis, T.A., Boom, R.M., Janssen, A.E.M., 2014. Galacto-oligosaccharide production with immobilized β-galactosidase in a packed-bed reactor vs. free β-galactosidase in a batch reactor. Food and Bioproducts Processing 92, 383–392.

Weaver, C.M., Martin, B.R., Nakatsu, C.H., Armstrong, A.P., Clavijo, A., McCabe, L.D., McCabe, G.P., Duignan, S., Schoterman, M.H.C., van den Heuvel, E.G.H.M., 2011. Galactooligosaccharides improve mineral absorption and bone properties in growing rats through gut fermentation. Journal of Agricultural and Food Chemistry 59, 6501–6510.

Whisner, C.M., Martin, B.R., Schoterman, M.H.C., Nakatsu, C.H., McCabe, L.D., McCabe, G.P., Wastney, M.E., van den Heuvel, E.G.H.M., Weaver, C.M., 2013. Galacto-oligosaccharides increase calcium absorption and gut bifidobacteria in young girls: a double-blind cross-over trial. British Journal of Nutrition 110, 1292–1303.

Widmer, F., Leuba, J.L., 1979. β-Galactosidase from *Aspergillus niger*. Separation and characterization of three multiple forms. European Journal of Biochemistry 100, 559–567.

Wilson, L., Illanes, A., Pessela, B.C.C., Abian, O., Fernández-Lafuente, R., Guisán, J.M., 2004a. Encapsulation of crosslinked penicillin G acylase aggregates in Lentikats: evaluation of a novel biocatalyst in organic media. Biotechnology and Bioengineering 86, 558–562.

Wilson, L., Betancor, L., Fernández-Lorente, G., Fuentes, M., Hidalgo, A., Guisán, J.M., Pessela, B.C.C., Fernández-Lafuente, R., 2004b. Cross-linked aggregates of multimeric enzymes: a simple and efficient methodology to stabilize their quaternary structure. Biomacromolecules 5, 814–817.

Wiśniewski, Ł., Antošová, M., Polakovič, M., 2013. Simulated moving bed chromatography separation of galacto-oligosaccharides. Acta Chimica Slovaca 6, 206–210.

Wiśniewski, Ł., Pereira, C.S.M., Polaković, M., Rodrigues, A.E., 2014. Chromatographic separation of prebiotic oligosaccharides. Case study: separation of galacto-oligosaccharides on a cation exchanger. Adsorption 20, 483–492.

Woodley, J.M., 2006. Choice of biocatalyst form for scalable processes. Biochemical Society Transactions 34, 301–303.

Woychik, J.H., Wondolowski, M.V., 1973. Lactose hydrolysis in milk and milk products by bound fungal beta-galactosidase. Journal of Milk and Food Technology 36, 31–33.

Wu, Y., Yuan, S., Chen, S., Wu, D., Chen, J., Wu, J., 2013. Enhancing the production of galacto-oligosaccharides by mutagenesis of *Sulfolobus solfataricus* β-galactosidase. Food Chemistry 138, 1588–1595.

Xia, Y., Zhao, J., Chen, H., Liu, X., Wang, Y., Tian, F., Zhang, H.P., Zhang, H., Chen, W., 2010. Extracellular secretion in *Bacillus subtilis* of a cytoplasmic thermostable β-galactosidase from *Geobacillus stearothermophilus*. Journal of Dairy Science 93, 2838–2845.

Xie, W., Wang, J., 2012. Immobilized lipase on magnetic chitosan microspheres for transesterification of soybean oil. Biomass and Bioenergy 36, 373–380.

Yanahira, S., Kobayashi, T., Suguri, T., Nakakoshi, M., Miura, S., Ishikawa, H., Nakajima, I., 1995. Formation of oligosaccharides from lactose by *Bacillus circulans* β-galactosidase. Bioscience Biotechnology and Biochemistry 59, 1021–1026.

Yang, Y.L., Wang, J.H., Teng, D., Zhang, F., 2008. Preparation of high-purity fructo-oligosaccharides by *Aspergillus japonicus* β-fructofuranosidase and successive cultivation with yeast. Journal of Agricultural and Food Chemistry 56, 2805–2809.

Yoon, S.H., Mukerjea, R., Robyt, J.F., 2003. Specificity of yeast (*Saccharomyces cerevisiae*) in removing carbohydrates by fermentation. Carbohydrate Research 338, 1127–1132.

Yoshida, E., Hidaka, M., Fushinobu, S., Koyanagi, T., Minami, H., Tamaki, H., Kitaoka, M., Katayama, T., Kumagai, H., 2010. Role of a Pa14 domain in determining substrate specificity of a glycoside hydrolase family 3 β-glucosidase from *Kluyveromyces marxianus*. Biochemical Journal 431, 39–49.

Zárate, S., López-Leiva, M.H., 1990. Oligosaccharide formation during enzymatic lactose hydrolysis: a literature review. Journal of Food Protection 3, 262–274.

Zhang, H., Li, W., Rui, X., Sun, X., Dong, M., 2013. *Lactobacillus plantarum* 70810 from Chinese paocai as a potential source of β-galactosidase for prebiotic galactooligosaccharides synthesis. European Food Research and Technology 236, 817–826.

Zheng, P., Yu, H., Sun, Z., Ni, Y., Zhang, W., Fan, Y., Xu, Y., 2006. Production of galacto-oligosaccharides by immobilized recombinant β-galactosidase from *Aspergillus candidus*. Biotechnology Journal 1, 1464–1470.

Zucca, P., Sanjust, E., 2014. Inorganic materials as supports for covalent enzyme immobilization: methods and mechanisms. Molecules 19, 14139–14194.

Chapter 5

Enzymatic Production of Lactulose

C. Guerrero and L. Wilson

5.1 LACTULOSE PRODUCTION: CHEMICAL OR ENZYMATIC

Lactulose (4-O-β-D-galactopyranosyl-D-fructose) is a nondigestible synthetic disaccharide in which both monomers are linked by a β-glycosidic bond (Schumann, 2002). The β 1–4 linkage between both sugars allows the existence of five anomeric forms of lactulose (Kim et al., 2006; Aider and de Halleux, 2007).

Lactulose is hygroscopic and solubility in water is 76.4% w/w at 30°C (Oosten, 1967). Solubility of lactulose in propanol, isopropanol, ethanol, and methanol is more than 15 times higher than lactose at 30°C, which allows it to be recovered from alcohol solutions (Montañés et al., 2007a). Lactulose is 50% sweeter than lactose and 60% as sweet as sucrose (Schumann, 2002; Panesar and Kumari, 2011).

Lactulose has received much attention in recent years because of its therapeutic and health-promoting properties, being considered as one of the most valuable compounds derived from lactose (Wang et al., 2013).

Initially, the International Dairy Federation and the European Union defined lactulose as a chemical indicator of milk quality by identifying the thermal treatment to which milk has been subjected. In fact, depending on the lactulose content in heat-treated milk, the thermal treatment used can be established, be it direct ultra-high temperature (UHT), indirect UHT, sterilization, or pasteurization (Adachi, 1958, 1965, Adhikari et al., 1991; Marconi et al., 2004). Before assessing its prebiotic properties (Mizota et al., 2002), lactulose applications referred to the pharmaceutical industry where it was produced mostly as a mild laxative against constipation (Wesselius-De Casparis et al., 1968; Tamura et al., 1993), and also as a drug for the treatment of hepatic encephalopathy (Elkington et al., 1969; Als-Nielsen et al., 2004), a condition that relates to liver failure by the accumulation of toxic substances (Riordan and Williams, 1997). Then, applications in the food sector appeared as a sugar substitute and, more interestingly, as a health promoter (Schuster-Wolff-Bühring et al., 2010; Panesar and Kumari, 2011; Song et al., 2013a; Wang et al., 2013; Nahla and Musa, 2015). Current

Lactose-Derived Prebiotics. http://dx.doi.org/10.1016/B978-0-12-802724-0.00005-6

interest lies mostly in its prebiotic condition, which is now well sustained with scientific evidence (De Souza Oliveira et al., 2011; Förster-Fromme et al., 2011).

Lactulose can be produced by chemical isomerization with alkaline catalysts or by transgalactosylation of fructose with lactose catalyzed by β-galactosidases (Hicks and Parrish, 1980; Schuster-Wolff-Bühring et al., 2010; Aït-Aissa and Aïder, 2014). However, at the industrial level it is exclusively produced by chemical synthesis (Aider and de Halleux, 2007; Panesar and Kumari, 2011; Wang et al., 2013).

Being a nonnatural sugar, early research on lactulose was focused mostly on its production during milk or whey processing (Montgomery and Hudson, 1930; Adachi, 1958; Andrews, 1986). Later on, reports referred to its purposely chemical synthesis and purification and, more recently, to its physiological effects on human health (Aider and de Halleux, 2007; Wang et al., 2013).

5.2 CHEMICAL SYNTHESIS OF LACTULOSE

Chemical synthesis of lactulose is now performed by two routes. The first one is the Lobry de Bruyn–van Ekenstein reaction, which is based on the formation of an enolic intermediate (lactose and epilactose) in an alkaline medium where the glucose residue of lactose is isomerized to fructose yielding lactulose (Hajek et al., 2013). The second route is based on the Amadori rearrangement (Wrodnigg and Eder, 2001); in this case, lactose reacts with ammonia and amines and a lactosyl–amine complex is formed that is then hydrolyzed to yield lactulose. In practical terms, the most used strategy for lactulose production is alkaline isomerization (Hicks and Parrish, 1980; Montilla et al., 2005; Aider and de Halleux, 2007; Wang et al., 2013). Electrophysical methods also have been proposed for lactulose synthesis from lactose in whey (Bologa et al., 2009).

Since the first article published by Montgomery and Hudson (1930), where the synthesis of lactulose from lactose was described using calcium hydroxide as catalyst, several authors have synthesized lactulose based on the same principle, but using different chemical catalysts, like sodium hydroxide, potassium hydroxide, alkaline organic compounds, and tertiary amines.

Zokaee et al. (2002) reported a yield of $0.3\,g_{lactulose}/g_{lactose}$ when using sodium hydroxide as catalyst, which is higher than the values reported by Nagasawa et al. (1974) and Hashemi and Ashtiani (2010) with the same catalyst. The former also used potassium hydroxide and sodium carbonate as catalysts obtaining a yield around $0.11\,g_{lactulose}/g_{lactose}$. Primary and secondary amines are not well suited as catalysts since side reactions occurs with the formation of glycosamines and Amadori compounds. However, these compounds are not formed when using a tertiary amine (triethylamine), and yields attained are similar to those obtained with alkaline catalysts (Hicks and Parrish, 1980; Aider and de Halleux, 2007), but a high catalyst concentration is required for obtaining high yields. In most cases, lactulose synthesis from lactose is accompanied by degradation reactions leading to the formation of several compounds that complicate product purification, which is a critical issue in its industrial production

(Hicks and Parrish, 1980; Dendene et al., 1994; Aider and de Halleux, 2007). With the purpose of increasing lactulose yields, complexing reagents, such as borates or aluminates, were added to the reaction mixture that allowed shifting the isomerization equilibrium toward the formation of lactulose, greatly reducing the secondary irreversible reactions that occur when using alkaline catalysts (Olano and Corzo, 2009). Yields higher than 0.7 $g_{lactulose}/g_{lactose}$ were obtained when using these reagents (Table 5.1). Using a mixture of sodium hydroxide, boric acid, and sodium aluminate, a yield of 0.7 $g_{lactulose}/g_{lactose}$ was obtained when using pure lactose as substrate (Zokaee et al., 2002), while a yield of 0.66 $g_{lactulose}/g_{lactose}$ was obtained when using whey as substrate (Nahla and Musa, 2015). However, aluminates and borates are hard to remove from the final product, hampering its use at the industrial level (Kozempel and Kurantz, 1994). Using an equimolar mixture of lactose-boric acid with tertiary amines in alkaline medium allowed an increase in lactulose yield and a reduction in boric acid, reducing the number of purification steps required (Hicks and Parrish, 1980) (Table 5.1). Other less-conventional catalysts have also been used for lactulose chemical synthesis, aiming to increase product yield and avoid secondary reactions. In this regard, sepiolite (a complex magnesium silicate) was used as a green catalyst (Troyano et al., 1996); yield was low but increased when using alkaline-substituted sepiolites (de la Fuente et al., 1999; Villamiel et al., 2002) (Table 5.1). Egg-shell powder was also used as catalyst, and even though lactulose yield was low, secondary products (epilactose, galactose, and organic acids) were at very low levels (Montilla et al., 2005); yields were higher when using calcium carbonate as catalyst (Paseephol et al., 2008) and lactose and whey as substrates (Seo et al., 2015). Lastly, yields of 0.25 and 0.12 $g_{lactulose}/g_{lactose}$ were obtained by electro-isomerization using lactose and whey as substrates, respectively (Aider and Gimenez-Vidal, 2012); this strategy is quite appealing since it operates at low temperature and no side reactions occur, so at the end of reaction only lactulose and residual lactose are present, which simplifies product purification.

Table 5.1 summarizes the reaction conditions and the lactulose yields obtained with different catalysts in the chemical isomerization of lactose into lactulose. The wide variety of reagents used indicates that an ideal chemical catalyst has not been developed yet (Montilla et al., 2005; Aider and de Halleux, 2007; Aider and Gimenez-Vidal, 2012). Such ideal chemical catalyst should produce lactulose at high yield with minimum product degradation, should be nontoxic and environmentally sound, easy to remove, cheap, and readily available with no restrictions of use (Aider and de Halleux, 2007; Panesar and Kumari, 2011). This is certainly challenging and has opened the option of using enzymes as catalysts that, in principle, comply with some of these requirements (see Section 5.3).

Chemical synthesis of lactulose is certainly complex, high concentrations of catalyst are required, and reaction is poorly specific generating unwanted side products like epilactose, galactose, tagatose, isosaccharic acids, and colored compounds so that several steps of downstream processing are required to attain

TABLE 5.1 Production of Lactulose Using Chemical Catalysts

Mechanism	Catalyst/Substrate	Reaction Time (h)	pH	Temperature (°C)	$Y_{lactulose}$ ($g_{lactulose}/g_{lactose}$)	References
Heat treatment	Buffer system/milk	8	–	120	–	Adachi (1958)
Alkaline agents	Calcium hydroxide/lactose	36	–	35	0.30	Montgomery and Hudson (1930)
	Potassium hydroxide/lactose	24	–	37	0.11	Nagasawa et al. (1974)
	Sodium carbonate/lactose	0.08	–	90	0.12	Nagasawa et al. (1974)
	Calcium hydroxide/lactose	0.5	–	85	0.213	Nagasawa et al. (1974)
	Sodium hydroxide/lactose	0.08	–	90	0.214	Nagasawa et al. (1974)
	Sodium hydroxide/lactose	0.25	11	70	0.27	Zokaee et al. (2002)
	Sepiolite with alkaline ions/milk permeate	2.5	–	90	0.20	Villamiel et al. (2002)
	Eggshell powder/milk permeate	1	–	98	0.0118	Montilla et al. (2005)
	Calcium carbonate/milk permeate	2	–	96	0.18–0.21	Paseephol et al. (2008)
	Sodium hydroxide/lactose	1	11	70	0.23	Hashemi and Ashtiani (2010)
	Calcium carbonate/whey	0.33	–	90	0.296	Seo et al. (2015)
	Eggshell powder/milk permeate	1	–	97	0.17	Nooshkam and Madadlou (2016)

Complexing agents	Boric acid with trimethylamine/lactose	4	11	70	0.87	Hicks and Parrish (1980)
	Magnesium oxide/lactose	–	–	100	0.35–0.40	Carobbi et al. (1985)
	Sodium hydroxide, sodium sulfite/lactose	–	–	75–80	0.3	Carobbi et al. (1985)
	Boric acid/lactose	2	11	70	0.75	Kozempel and Kurantz (1994)
	Sodium hydroxide and boric acid/lactose	0.25	11	70	0.77–0.80	Zokaee et al. (2002)
	Sodium aluminate/lactose	0.5	12	70	0.78	Zokaee et al. (2002)
	Sodium aluminate/whey	1	–	70	0.66	Nahla and Musa (2015)
Ion-mediated	Electro isomerization/lactose	1	–	23	0.25	Aider and Gimenez-Vidal (2012)
	Electro isomerization/whey	1	–	23	0.12	Aider and Gimenez-Vidal (2012)

the required purity (Zokaee et al., 2002; Aider and de Halleux, 2007; Hashemi and Ashtiani, 2010; Panesar and Kumari, 2011; Nooshkam and Madadlou, 2016). Pure lactose is required for the chemical synthesis of lactose because impurities in whey or whey permeate lead to unwanted side reactions reducing lactulose yield; this makes the process more expensive as the cost of raw material is a significant part of the processing cost (Panesar and Kumari, 2011); besides, the low specificity of the reaction makes downstream operations cumbersome and costly (Mayer et al., 2004; Mayer et al., 2010; Tang et al., 2011; Song et al., 2013a).

5.3 ENZYMATIC SYNTHESIS OF LACTULOSE

Based on the previous considerations, the synthesis of lactulose by enzyme catalysis appears as an interesting technological option to overcome the limitations inherent to chemical synthesis. Enzyme-catalyzed synthesis of lactulose has the advantage of not requiring pure substrates; whey or whey permeate, that are frequently underutilized side products from cheese manufacturing (see Section 1.1), can be used as lactose sources (Schuster-Wolff-Bühring et al., 2010). On the other hand, selectivity and mild reaction conditions make the process more compliant with green chemistry principles and put less burden on downstream operations, which are advantages that may make biocatalysis competitive with the ongoing chemical process (Illanes, 2011).

Formation of lactulose by enzyme catalysis can occur by a molecular rearrangement of lactose or by the formation of a β-glycosidic bond between galactose and fructose (Mayer et al., 2004; Panesar et al., 2013; Wang et al., 2013). The latter can be catalyzed by β-galactosidases (also by some β-glycosidases) or glycosyl transferases (Schuster-Wolff-Bühring et al., 2010; Panesar et al., 2013). Galactosidases can catalyze the synthesis of transgalactosylated oligosaccharide (TOS) by two different mechanisms: the first one is the thermodynamically controlled synthesis (TCS), which is essentially a reverse hydrolysis reaction where both sugars act as substrates; the second one is the kinetically controlled synthesis (KCS), where the enzyme transfers one monosaccharide unit from the donor saccharide to another saccharide, which is the acceptor of the transferred monosaccharide (Mayer et al., 2004; Plou et al., 2007; Wang et al., 2013; Sitanggang et al., 2014a). Higher yields are usually obtained in KCS than in TCS (Plou et al., 2007). From a technological point of view, glycosyl transferases have the drawbacks of requiring activated substrates and not being commercially available. They also require specific cofactors according to the transglycosylation reaction catalyzed (Schuster-Wolff-Bühring et al., 2010). Therefore, β-galactosidases (EC 3.2.1.23) are the catalysts to be chosen, being robust enzymes readily available and with a long record of use as catalyst in the food industry for the hydrolysis of lactose in the production of low-lactose milk and dairy products (because of this, the enzyme is sometimes referred to as lactase). However, under appropriate conditions β-galactosidases can catalyze transgalactosylation reactions (Sanz Valero, 2009), leading to the synthesis of galacto-oligosaccharide (GOS) (see Chapter 4: Enzymatic Production of Galacto-Oligosaccharides) and other lactose-derived compounds (see Chapter 6: Enzymatic Production of Other Lactose-Derived Prebiotic Candidates). In this way, in the presence of lactose and fructose, the enzyme will catalyze the synthesis of both lactulose and GOS, as well as the hydrolysis of lactose (Plou et al., 2007; Panesar and Kumari, 2011).

As mentioned in "Chapter 4, Enzymatic Production of Galacto-Oligosaccharides", β-galactosidase is a rather ubiquitous enzyme, whose activity has been detected in several organisms, but only a few can be considered as a source for technological applications. Most of the β-galactosidases currently used come from yeasts of the genus *Kluyveromyces* and filamentous fungi from the genus *Aspergillus*. These enzymes are readily available and have a generally recognized as safe (GRAS) (or equivalent) status allowing its unrestricted use in foods and pharmaceuticals (Nakayama and Amachi, 1999).

5.3.1 Synthesis With Free Enzyme

Vaheri and Kaupinnen (1978) were the first to report the lactulose synthesis by enzymatic transgalactosylation of fructose with lactose using β-galactosidase as catalyst (see Table 5.2). Twenty-five years elapsed before a comprehensive study was done to evaluate the potential of β-galactosidases from different origins for the synthesis of lactulose (Lee et al., 2004); enzyme preparations from *Escherichia coli*, *Aspergillus oryzae*, *Kluyveromyces lactis*, and *Kluyveromyces fragilis* were tested, all of them being able to use fructose as acceptor of the galactosyl–enzyme complex formed in the presence of lactose, so synthesizing lactulose; according to their results, the enzyme from *K. lactis* was the one producing the highest lactulose yield under the conditions tested (see Table 5.2). On the other hand, Mayer et al. (2004) working with *A. oryzae* β-galactosidase obtained a lactulose yield almost five times higher than with β-galactosidase-containing ethanol-permeabilized cells of *K. lactis* (Lee et al., 2004) and with a commercial soluble *K. lactis* β-galactosidase (Fattahi et al., 2010). Kim et al. (2006) working with a recombinant β-galactosidase from *Sulfolobus solfataricus* expressed in *E. coli* obtained a lactulose yield three times higher than reported by Lee et al. (2004) and Fattahi et al. (2010), although lower than reported by Mayer et al. (2004) and Guerrero et al. (2011) with *A. oryzae* β-galactosidase.

Lactulose synthesis was mostly performed using commercial β-galactosidase preparations, since most of them have a GRAS status or equivalent and are sold as commodities (Sanz Valero, 2009). Eleven commercial β-galactosidase preparations of different origin have been recently evaluated in the synthesis of lactulose (see Table 5.2). Best results in terms of lactulose yield were obtained with an *A. oryzae* β-galactosidase; second best was a preparation from *K. lactis* but it produced a complex mixture of TOS that is difficult to remove from the reaction medium (Guerrero et al., 2015a). With the purpose of increasing lactulose yield,

TABLE 5.2 Production of Lactulose Using β-Glycosidase and β-Galactosidase Catalysts

Mechanism	Biocatalyst	pH	Temperature (°C)	Lactose/Fructose Concentration (% w/w)	Free or Immobilized	$Y_{lactulose}$ ($g_{lactulose}$/$g_{lactose}$)	References
β-Glycosidases	*Pyrococcus furiosus*	5	75	2.62/20.7	Free/recombinant enzyme	0.44	Mayer et al. (2004)
	P. furiosus	5	75	2.62/20.7	Immobilized/ recombinant enzyme	0.43	Mayer et al. (2010)
β-Galactosidase	*Kluyveromyces fragilis*	7.2	37	12/20	Free	0.075	Vaheri and Kauppinen (1978)
	Aspergillus oryzae	5	37	2.62/20.7	Free	0.3	Mayer et al. (2004)
	Escherichia coli	7.3	37	15/5	Free	0.0113	Lee et al. (2004)
	A. oryzae	4.5	30	15/5	Free	0.017	Lee et al. (2004)
	K. fragilis	7.3	30	15/5	Free	0.035	Lee et al. (2004)
	Kluyveromyces lactis	6.5	37	15/5	Free	0.061	Lee et al. (2004)
	K. lactis	7	60	40/20	Permeabilized cells	0.05	Lee et al. (2004)
	Sulfolobus solfataricus	6	80	40/20	Free/recombinant enzyme	0.125	Kim et al. (2006)
	K. lactis	6.7	40	10/30	Free	0.122	Fattahi et al. (2010)
	Arthrobacter sp.	6	20	40/20	Free	–	Tang et al. (2011)

	A. oryzae	4.5	40	3.1/46.9	Free	0.282	Guerrero et al. (2011)
	K. lactis	7.5	47	20/20	Immobilized	0.006	Song et al. (2012)
	Lactobacillus acidophilus NRRL 4495	6.6	40	40/20	Free	0.063	Hashem et al. (2013)
	K. lactis	7.5	47	20/20	Immobilized	0.096	Song et al. (2013a)
	K. lactis	7.5	47	40/20	Immobilized	0.0395	Song et al. (2013b)
	K. lactis	7.5	40	40/20	Free	0.032	Hua et al. (2013)
	K. lactis	6.7	40	25–30/9–12	Free/ nonconventional medium	0.2	Khatami et al. (2014)
	K. lactis	6.8	40	16.7/33.3	Free	0.0685	Sitanggang et al. (2014b)
	Aspergillus aculeatus	3.5	50	1.68/38.2	Free	0.2	Guerrero et al. (2015a)
	Aspergillus niger	3.5	50	1.68/38.2	Free	0.23	Guerrero et al. (2015a)
	K. lactis	7	50	1.68/38.2	Free	0.24	Guerrero et al. (2015a)
	Bacillus circulans	6	50	1.68/38.2	Free	0.09	Guerrero et al. (2015a)
	A. oryzae	4.5	50	2.1/47.9	Free	0.31	Guerrero et al. (2015c)
	A. oryzae	4.5	50	2.1/47.9	Immobilized	0.254	Guerrero et al. (2015c)

the effect of several operational variables (temperature, pH, enzyme–substrate ratio, initial concentrations of sugars) have been studied; these variables only have effect on the reaction velocity (therefore on lactulose productivity), but not on lactulose yield (Lee et al., 2004; Kim et al., 2006; Guerrero et al., 2011, 2015b). However, fructose–lactose ratio has a strong influence on lactulose yield, with a threefold increase when increasing that ratio from 1 to 8 (Guerrero et al., 2011). Besides, by manipulating such ratio, synthesis can be driven to obtain a product with a determined lactulose–TOS ratio, which is an advantage over the chemical synthesis in the sense that a product with maximum prebiotic effect (optimum lactulose–TOS ratio) can be obtained without further fractionation to obtain the compounds separately (Guerrero et al., 2015b).

Another alternative in which much effort has been paid to increase lactulose yield with *K. lactis* β-galactosidase is the use of different reactor configurations and biocatalyst engineering (Song et al., 2012, 2013a,b; Hua et al., 2013; Sitanggang et al., 2014b, 2015; Khatami et al., 2014). The enzyme was immobilized in multiwalled carbon nanotubes and synthesis was conducted in a microreactor, but lactulose yield was low (Song et al., 2012). Later on the same authors conducted the synthesis of lactulose in batch and continuous mode of operation to determine if catalyst reuse could increase lactulose yield (Song et al., 2013a,b). Lactulose concentration increased 3.3 times when operating a packed-bed reactor in continuous mode with respect to repeated-batch operation, even though in the latter the catalyst could be reused 10 times (Song et al., 2013a). Working with membrane reactors, Sitanggang et al. (2014b) obtained lactulose yields similar to those obtained by Lee et al. (2004) and Fattahi et al. (2010) with the soluble enzyme. Recently, cross-linked aggregates (CLEAs) of *A. oryzae* β-galactosidase were used for lactulose synthesis, and despite that 17% reduction in yield was obtained with respect to the soluble enzyme, the catalyst could be reused 100 times so that the accumulated mass of product per unit mass of catalyst used increased 12 times (Guerrero et al., 2015b). Nonconventional media have also been used for lactulose synthesis; acetone had a negative effect on lactulose synthesis, but a yield 3.3 times higher than reported by Lee et al. (2004) and Fattahi et al. (2010) was obtained in triethyl phosphate medium (Khatami et al., 2014).

Table 5.2 summarizes the yields obtained in the synthesis of lactulose from lactose and fructose with the different β-galactosidases used. As can be appreciated, yields are always lower than obtained by chemical synthesis and vary considerably according to the enzyme catalyst used, which can be attributed to differences in structure and/or enzyme-substrate coupling mechanisms (Gosling et al., 2010). The ability of β-galactosidase to accept nucleophiles other than water in the active site affects its capacity of properly transferring the galactose molecule and producing lactulose or TOS. This capacity is determined by the tertiary structure of the enzyme and the amino acids forming the active site, by the concentration of lactose, which is the donor for the formation of the galactosyl–enzyme complex, and by the concentration of fructose, which is the galactosyl acceptor (Schuster-Wolff-Bühring et al., 2010). Large fructose excess is required to drive the reaction to lactulose synthesis rather than GOS synthesis and lactose hydrolysis, which generates a high proportion of unreacted fructose, being a major drawback of this route to lactulose synthesis (Guerrero et al., 2015c). Even though the formation of GOS during lactulose synthesis may be interesting in terms of prebiotic effect when used in functional foods, high residual concentrations of lactose and fructose may impair the sensory and nutritional quality of the product. An additional operation to separate and recycle both lactose and fructose may be a significant asset in economic terms and quality of the prebiotic mix (Schuster-Wolff-Bühring et al., 2010).

With the purpose of reducing fructose concentration, a process for lactulose production was proposed in which the synthesis of lactulose was coupled to the isomerization of glucose into fructose with glucose isomerase (Yang and Liu, 2008). This system was tested in conventional aqueous medium and in aqueous-organic two-phase system with cyclohexane, ethyl acetate, and n-butanol, obtaining a yield of $0.076\, g_{lactulose}/g_{lactose}$ in water and $0.19\, g_{lactulose}/g_{lactose}$ in water-cyclohexane (Hua et al., 2010), which is higher than the value reported by Khatami et al. (2014) with β-galactosidases in nonconventional medium (see Table 5.2). This dual enzymatic system was used for lactulose production directly from whey without external fructose addition but lactulose yield obtained was low (Song et al., 2013c). Similar results were reported by Lorenzen et al. (2013) using lactose and whey permeate as substrates, obtaining about 1% w/w of lactulose, but the primary product in this case was GOS. It is concluded that the dual enzymatic system cannot increase lactulose concentration with respect to the one-enzyme system, making the strategy economically unattractive (Schuster-Wolff-Bühring et al., 2010; Panesar and Kumari, 2011).

It is to be expected that novel more selective and specific β-galactosidases will be generated by protein engineering and metagenomic strategies that will allow increasing lactulose yield, which is the major constraint for making this technology competitive with ongoing chemical synthesis (Gosling et al., 2010; Schuster-Wolff-Bühring et al., 2010; Wang et al., 2013).

Another route proposed for the synthesis of lactulose is the two-step redox isomerization of lactose with pyranose oxidase and aldose reductase, which has been applied to the synthesis of other related disaccharides (Leitner et al., 2001; Peterbauer and Volc, 2010). However, there is no information available on the actual production of lactulose by this route.

A quite promising strategy for lactulose synthesis is the direct isomerization of lactose catalyzed by a specific isomerase. This may be the most obvious enzymatic route for lactulose production, but glucose (xylose) isomerase, which catalyzes the isomerization of glucose into fructose, is not active on lactose and searching such specific isomerase has been yet unsuccessful. However, the direct isomerization of lactose into lactulose was recently proven feasible with cellobiose 2-epimerases from thermophilic bacteria like *Dictyoglomus turgidum* (Kim et al., 2012) and *Caldicellulosiruptor saccharolyticus* (Kim and Oh, 2012) (see Table 5.3). The latter produced a lactulose-epilactose mixture with yields of

TABLE 5.3 Production of Lactulose Using Other Enzymatic Catalysts

Synthesis Route	Enzyme Source	pH	Temperature (°C)	Lactose/Fructose Concentration (% w/w)	Free or Immobilized Enzyme	$Y_{lactulose}$ ($g_{lactulose}/g_{lactose}$)	References
β-Galactosidase/glucose isomerase	*Kluyveromyces lactis/Streptomyces murinus*	8	30	80/10	Immobilized	0.19	Hua et al. (2010)
	K. lactis/S. murinus	8	30	80/10	Immobilized/non-conventional media	0.19	Hua et al. (2010)
	K. lactis/Streptomyces rubiginosus	7.5	53.5	20/–	Immobilized	0.0384	Song et al. (2013c)
	K. lactis/S. rubiginosus	7.8	45	40/–	Free/immobilized	0.011	Lorenzen et al. (2013)
Cellobiose 2-epimerase	*Caldicellulosiruptor rubiginosus*	7.5	80	70	Free	0.58	Kim and Oh (2012)
	C. rubiginosus with the addition of boric acid	7.5	80	70	Free	0.88	Kim et al. (2013)
	C. rubiginosus	7.5	87.5	70	Free/mutant	0.75	Shen et al. (2015)
	C. rubiginosus	7.5	80	60	Immobilized/recombinant enzyme	0.651	Wang et al. (2015)
	C. rubiginosus with milk	7.5	50	48.5	Free	0.577	Rentschler et al. (2015)
	C. rubiginosus with milk	7.5	8	48.5	Free	0.567	Rentschler et al. (2015)

0.58 $g_{lactulose}/g_{lactose}$ and 0.15 $g_{epilactose}/g_{lactose}$, respectively. Such lactulose yield is higher than obtained with β-glycosidases and β-galactosidases, though still lower than obtained by chemical synthesis. Trying to further increase lactulose yields, the direct isomerization of lactose into lactulose was carried out with the thermostable cellobiose 2-epimerase from *C. saccharolyticus* in the presence of borate (Kim et al., 2013); yield increased to 51%, which is close to the values obtained by chemical synthesis with complexing agents. However, borate removal from the reacted medium is complex and requires several purification steps, and also some enzyme inactivation occurs, increasing processing cost. In an effort to avoid the addition of complexing agents, a thermostable cellobiose 2-epimerase from *C. saccharolyticus* was obtained by site-directed mutagenesis, allowing an increase in operating temperature from 80 to 87.5°C, which resulted in a 29% increase in the concentration of lactulose (Shen et al., 2015). A higher lactulose yield (65%) was reported with ethanol-permeabilized cells of recombinant *E. coli* hosting the *C. saccharolyticus* cellobiose 2-epimerase gene when reaction was performed at high lactose concentration in phosphate buffer; at such conditions, lactose into epilactose conversion was less than 2%, so this strategy is quite promising for the enzymatic production of high purity lactulose (Wang et al., 2015). Recently the isomerization of lactose into lactulose in milk was studied at high (50°C) and low (8°C) temperatures with a recombinant *C. saccharolyticus* cellobiose 2-epimerase produced in *E. coli*; lactulose yield obtained at both temperatures was similar to that reported by Kim and Oh (2012), but productivity at 8°C was much lower since reaction time was three times the one required at 50°C (Rentschler et al., 2015).

Lactulose production by direct lactose isomerization with cellobiose 2-epimerase will probably be the way to go for the industrial production of lactulose in the near future. To the best of our knowledge, this enzyme is not available for use at industrial level and has no GRAS status (or equivalent) precluding for the moment its use in the food and pharmaceutical industries, which are aspects yet to be dealt with (Schuster-Wolff-Bühring et al., 2010; Wang et al., 2013).

5.3.2 Synthesis With Immobilized Enzyme

Few studies have reported the synthesis of lactulose with immobilized enzymes and most of them are recent (see Table 5.2). Table 5.4 summarizes the support materials and type of enzyme–support interaction used for the immobilization of β-galactosidases for lactulose synthesis.

A continuous enzymatic process for the production of lactulose by transgalactosylation of fructose with lactose was developed using free and immobilized β-glycosidase from *Pyrococcus furiosus* (Mayer et al., 2010). The hyperthermostable enzyme was immobilized onto an anion-exchange resin (Amberlite IRA-93) and onto Eupergit C with immobilization yields of 72% and 83% and specific activities of 55 and 90 IU/g dry support, respectively measured at 75°C with *p*-nitrophenyl-D-galactopyranoside as substrate. Yields of lactulose synthesis at 75°C were

TABLE 5.4 Carrier-Bound Immobilized β-Galactosidases Used for Lactulose Synthesis

Carrier	Enzyme-Support Interaction	Enzyme Source	References
Anion-exchange resin (Amberlite IRA-93)	Ionic	*Pyrococcus furiosus*	Mayer et al. (2010)
Eupergit	Covalent	*P. furiosus*	Mayer et al. (2010)
Silica	Covalent	*Kluyveromyces lactis*	Song et al. (2012)
Silica	Covalent	*K. lactis*	Song et al. (2013a)
Silica	Covalent	*K. lactis*	Song et al. (2013b)
CLEAs	Covalent	*Aspergillus oryzae*	Guerrero et al. (2015c)
Permeabilized cells	–	*K. lactis*	Lee et al. (2004)
Silica	Covalent	*K. lactis/ Streptomyces rubiginosus*	Song et al. (2013c)
Magnetic chitosan microspheres for β-galactosidase	Covalent	β-galactosidase from *K. lactis*; glucose isomerase from *Streptomyces murinus*	Hua et al. (2010)

CLEAs, cross-linked aggregates.

similar to the ones obtained with the free enzyme (43% and 41%, respectively), but productivity and stability was much higher with the immobilized catalysts. Productivities were 52, 15, and 12 $g_{lactulose}$/L/h for the Amberlite-immobilized enzyme, the Eupergit-immobilized enzyme, and the free enzyme, respectively. Both immobilized catalysts remained fully active after 14 days, while the half-life of the free enzyme in a membrane reactor was only 1.5 days.

Song et al. have published a series of works of lactulose synthesis with immobilized β-galactosidases. In the first one (Song et al., 2012), several surface functionalization techniques were used to immobilize β-galactosidase in a microreactor. β-Galactosidase was pretreated with lactose before immobilization, and functionalized multiwalled carbon nanotubes (MWNTs),

DNA-wrapped single-walled carbon nanotubes, and glutaraldehyde were used as linkers to immobilize the enzyme on a microchannel surface. In the case of MWNTs immobilization, N-(3-dimethylaminopropyl)-N′-ethylcarbodiimide hydrochloride (EDC·HCl) was used as the coupling agent, catalyzing the formation of amide bonds between the carboxyl groups of MWNTs and the amino groups of the enzyme. Immobilization yields were in all cases lower than 8.6%. Best microreactor performance in the synthesis of lactulose was obtained when functionalized MWNTs were used as linkers for the immobilization of pretreated β-galactosidase. In a microreactor continuous operation, synthesis of lactulose was maintained at a concentration of about 1.3 g/L for 48 h, indicating that enzyme immobilization can protect the active conformation of the enzyme under operating conditions. Later on, the same authors focused on the use of glutaraldehyde-activated silica as support for β-galactosidase immobilization and on enzyme pretreatment by incubation with lactose (Song et al., 2013a). Pretreatment allowed an increase in immobilization yield of expressed activity from 33% to 50%. The authors suggested that the effect of enzyme pretreatment prior to immobilization was due to a steric effect at the enzyme active site by the substrate so that interaction with the solid surface of the support occurred at enzyme regions far from the active site. Immobilized β-galactosidase was used in consecutive batches of lactulose synthesis to assess its reusability. Reactions were conducted at 47°C using 20% (w/v) lactose, 20% (w/v) fructose, and 12 IU/mL of immobilized β-galactosidase in 50 mM sodium phosphate buffer pH 7.5. Lactulose concentration obtained in the first batch was 10.8 g/L, and after 10 cycles of use lactulose concentration had dropped to 5.7 g/L, the immobilized enzyme still retaining 53% of its initial catalytic activity. In another study, Song et al. (2013b) used the same immobilization strategy to covalently immobilize β-galactosidase in activated silica gels, using this catalyst for the synthesis of lactulose at lactose concentrations higher than 40% (w/v). To assess its reusability, the catalyst was used in consecutive batches of lactulose synthesis; lactulose concentration in the first batch was 15.8 g/L and after 10 cycles of use lactulose concentration had dropped to 9.6 g/L, the immobilized enzyme still retaining 61% of its initial catalytic activity.

Carrier-free immobilization of β-galactosidase for the synthesis of lactulose has been recently reported by Guerrero et al. (2015c). Synthesis of lactulose under repeated-batch operation was done with cross-linked aggregates of β-galactosidase from *A. oryzae*, using lactose and fructose as substrates. In the preparation of the catalyst, the effect of the cross-linking agent to enzyme mass ratio and cross-linking time was studied, determining that the best conditions were 5.5 g glutaraldehyde/g enzyme and 5 h of cross-linking; at such conditions, 30% immobilization yield of expressed activity and a catalyst with a specific activity of 15,000 IU/g were obtained. The catalyst was much more stable than the free enzyme with a half-life of 123 h under nonreactive conditions at 50°C; when used in repeated-batch operation for the synthesis of lactulose, yield and productivity were 3.8 and 4.3 times higher than with the free enzyme, respectively.

Lactulose synthesis with permeabilized yeast cells was reported by Lee et al. (2004). Reaction conditions for lactulose production were optimized using *K. lactis* cells that had been permeabilized by treatment with 50% (v/v) ethanol, being cell concentration 10.4 g/L, 40% (w/v) lactose and 20% (w/v) fructose, 60°C, and pH 7.0. Under these conditions, the permeabilized cells produced approximately 20 g/L of lactulose in 3 h, corresponding to a productivity of 6.7 g/L/h. These results represent 1.3- and 2.1-fold increase in lactulose concentration and productivity, respectively, compared with untreated washed cells. Operational stability of the enzyme was not reported.

Synthesis of lactulose has also been conducted with a dual system with immobilized β-galactosidase and glucose isomerase. Hua et al. (2010) used a commercial immobilized glucose isomerase (Novozymes) and *K. lactis* β-galactosidase immobilized in magnetic chitosan microspheres cross-linked with glutaraldehyde, obtaining an immobilization yield of expressed activity of 26%. Optimum reaction temperature for lactulose synthesis was 5°C higher, and the optimum pH was 1.5 units higher than those of the system using free β-galactosidase (30°C and pH 6.5). Operational stability of the enzymes was assessed in repeated-batch operation. After 10 cycles of lactulose production, 65% of the initial β-galactosidase activity still remained; operational stability of glucose isomerase was not reported but probably remained mostly active during that period. Song et al. (2013c) conducted the synthesis of lactulose in a dual system using *K. lactis* β-galactosidase and *Streptomyces rubiginosus* glucose isomerase covalently immobilized to glutaraldehyde-activated silica, using the strategy of enzyme preincubation, in lactose in the case of β-galactosidase and in xylose in the case of glucose (xylose) isomerase. Protein immobilization yield and immobilization yield of expressed activity were 32.8% and 49.2%, respectively, and the specific activity of the immobilized β-galactosidase was 776.1 IU/$g_{support}$; the corresponding values for glucose isomerase were 52.3%, 47.1%, and 246.7 IU/$g_{support}$. Immobilized β-galactosidase and glucose isomerase were used in consecutive batches of lactulose synthesis at optimized conditions (53.5°C, 20% (w/v) substrates concentrations, pH 7, 12 IU/mL of immobilized β-galactosidase, and 60 IU/mL of immobilized glucose isomerase). The activity of the immobilized enzymes decreased with the number of reuses. After seven reuses, the synthesized lactulose concentration had dropped from 7.7 g/L (initial value) to 4.3 g/L and the catalytic activity of the immobilized enzymes was 57.1% of the initial.

5.4 MECHANISMS OF LACTULOSE SYNTHESIS AND OPTIMIZATION

Independent of the route used for lactulose synthesis, the kinetic mechanism is not completely elucidated, because of the several reactions involved. Some approximations described in the literature help in understanding the process of synthesis. The formulation of a kinetic mechanism allows the construction of mathematical models where the kinetic parameters of each of the reactions involved can be determined and in this way predict the behavior and optimize the process of lactulose synthesis. Mechanisms for lactulose synthesis by chemical and enzymatic routes now will be described.

5.4.1 Chemical Synthesis

Industrial production of lactulose is done exclusively by the chemical isomerization of lactose produced by the rearrangement of the glucose residues into fructose, so converting an aldose into a ketose (Fig. 5.1). The reaction mechanism has not been fully elucidated and the information about it is scarce, but it is assumed that transformation occurs by the Lobry de Bruyn–van Ekenstein reaction, which consists of the formation in alkaline medium of an enolic intermediate in the form of lactose and epilactose, where the glucose residue in the lactose moiety is converted into fructose, yielding lactulose (Fig. 5.1). This reaction is rapidly followed by the degradation of the lactulose formed into galactose and isosaccharinic acids, which needs to be arrested or significantly reduced (Verhaar et al., 1978; Dendene et al., 1994; Montilla et al., 2005).

Dendene et al. (1994) proposed a simplified mechanism for this reaction, where first-order kinetics was assumed for the isomerization reaction, while epilactose and tagatose were not considered in the kinetic mechanism and the model derived from it for considering them negligible (Fig. 5.2); it also considered that the concentration of hydroxyl ion was constant throughout the reaction, which is not so at all sugar concentrations. Even so, the experimental data

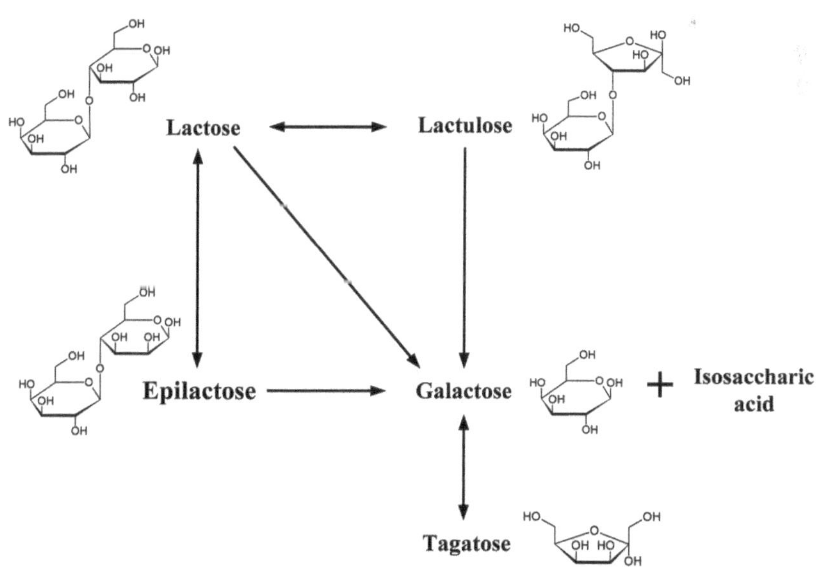

FIGURE 5.1 Proposed scheme for the alkaline isomerization of lactose into lactulose.

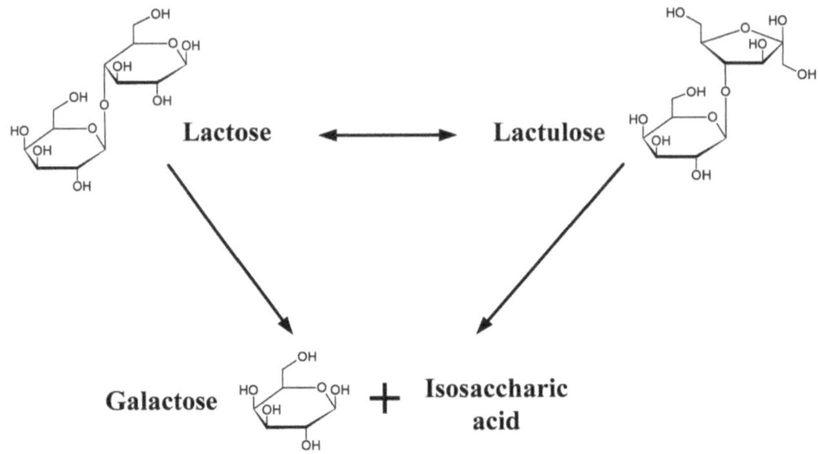

FIGURE 5.2 Simplified scheme for the mechanism of alkaline isomerization of lactose into lactulose, proposed by Dendene et al. (1994).

were in good agreement with the model, allowing the simulation of the isomerization reaction and optimization of the operation conditions in terms of product yield and degradation.

On the other hand, a kinetic model was proposed for the reactions occurring by heating disaccharide–casein mixtures, considering two degradation routes: the isomerization of lactose (aldose) into lactulose (ketose) and the subsequent degradation to galactose and formic acid, and the Maillard reactions. This model was developed in connection with the purpose of avoiding lactulose formation during heating of products containing lactose and protein, and it could predict and optimize the disaccharide content in foods (Brands and van Boekel, 2003).

5.4.2 Enzymatic Synthesis

As in the case of chemical synthesis, the mechanism of enzymatic synthesis of lactulose from lactose and fructose with β-galactosidase has not been completely elucidated. However, it is well known that β-galactosidases can catalyze both hydrolysis and transgalactosylation reactions synthesizing GOS from lactose (Prenosil et al., 1987; Gänzle, 2012; Lamsal, 2012; Wang et al., 2013). When the enzyme is in a lactose medium, hydrolysis and transgalactosylation are competing reactions, the latter prevailing at the beginning of the reaction when lactose concentration is high, being progressively displaced by hydrolysis as lactose conversion increases so that glucose and galactose are the final products. Hydrolysis of lactose follows Michaelis–Menten kinetics with galactose competitive inhibition (Prenosil et al., 1987; Vera et al., 2011a).

β-Galactosidases are rather nonspecific with respect to the galactose acceptor molecule; therefore other sugars present in the reaction medium, ie, fructose or sucrose, can act as acceptors (Kim et al., 2006; Li et al., 2009; Gänzle, 2012).

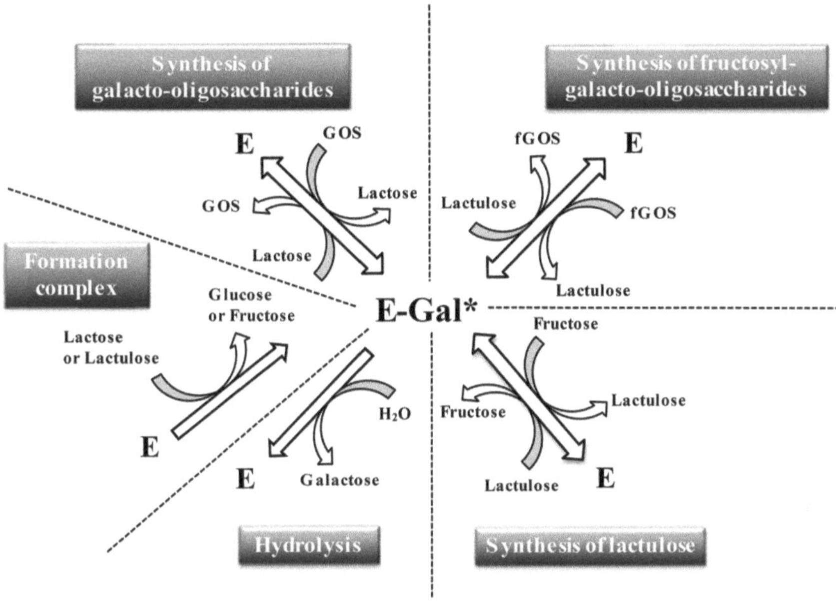

FIGURE 5.3 Simplified scheme for the mechanism for the enzymatic synthesis of lactulose and galacto-oligosaccharides (GOS) or fructosyl-galacto-oligosaccharides (fGOS) from lactose and fructose with β-galactosidase (E).

Not only sugars can act as acceptors but also other hydroxyl-containing compounds, like alcohols (Stevenson et al., 1993; Klewicki, 2000; Irazoqui et al., 2009). Therefore, the synthesis of other transgalactosylated compounds is possible (Wang et al., 2013; Guerrero et al., 2015c).

The medium for lactulose synthesis contains both lactose and fructose so that both sugars can act as galactose acceptors leading to GOS and lactulose synthesis, respectively. The mechanism for the synthesis of lactulose with β-galactosidase is similar to the one described for GOS (see Section 4.3) and a scheme is shown in Fig. 5.3 (Vera et al., 2011b, 2013; Guerrero et al., 2011).

Lactulose synthesis from lactose and fructose is a kinetically controlled reaction where the enzyme first acts on lactose forming the galactosyl–enzyme complex with liberation of one molecule of glucose per molecule of lactose reacted. The galactosyl–enzyme complex is then reacted with a molecule of fructose to yield lactulose, or with lactose molecules to consecutively form GOS of different chain lengths (Vera et al., 2011b, 2013), both acceptors competing for the galactose from the galactosyl–enzyme complex. Alternatively, the complex can react with water to yield galactose. As long as lactose and fructose are in high concentration, transgalactosylation reactions will prevail over hydrolysis; however, as they decrease as a consequence of reaction, hydrolysis will take over, water being then the main acceptor, and therefore galactose will be released. Hydrolysis of the formed products (lactulose and GOS) will also occur and some authors have reported that lactulose can also act as acceptor producing oligosaccharides that have a terminal fructose instead of glucose, designated as fructosyl-galacto-oligosaccharides (fGOS) (Martínez-Villaluenga et al., 2008; Olano and Corzo, 2009; Rodriguez-Fernandez et al., 2011). Maximum lactulose concentration is obtained at a time after which it is reduced due to the formation of fGOS (di-, tri-, and tetrasaccharides) (Lee et al., 2004; Kim et al., 2006; Guerrero et al., 2011).

The unavoidable presence of TOS in the synthesis of lactulose complicates its purification since separation is difficult and costly, but TOSs are prebiotic compounds so that lactulose-TOS mixtures are valuable by themselves (Palframan et al., 2002; Ghoddusi et al., 2007). However, if pure lactulose is the objective product, TOS synthesis can be minimized displacing the reaction toward lactulose synthesis by using kinetically controlled strategies where the use of high fructose–lactose ratios has been quite effective, as described in Section 5.3 (Mayer et al., 2004; Guerrero et al., 2011, 2015b). The mechanism of lactulose synthesis is quite complex so that no models have been developed allowing to predict it. There are, however, several models describing the synthesis of GOS from lactose (Boon et al., 1999; Kim et al., 2004; Vera et al., 2011b) and one for the synthesis of fGOS from lactulose (Rodriguez-Fernandez et al., 2011). Most of them consider the hydrolysis of the galactosyl–enzyme complex and the competitive inhibition by galactose, but a model considering the reactions of hydrolysis and transgalactosylation of lactulose, GOS, and fGOS altogether remains to be developed. This is why optimization of lactulose synthesis has been based solely on surface-response experimental designs. Khatami et al. (2014) used a two-factor three-level factorial design to determine the effect of lactose and fructose concentrations on lactulose yield, defining the range of substrate concentrations that maximized it both in aqueous and nonconventional (aqueous-organic) media. Likewise, Guerrero et al. (2015b) used a Box–Behnken surface response experimental design for optimizing lactulose synthesis in terms of yield and selectivity considering sugar concentrations, fructose–lactose molar ratio, and temperature as operational variables.

As in the case of lactulose synthesis from lactose and fructose with β-galactosidase, the mechanism of lactulose synthesis from lactose with cellobiose 2-epimerase is not yet elucidated. This enzyme was initially reported to be able to catalyze the epimerization of lactose into epilactose (see Section 6.5); however, Kim and Oh (2012) reported that the enzyme from *Caldicellulosiruptor rubiginosus* could also catalyze the isomerization of lactose into lactulose producing a mixture of lactulose and epilactose. Rentschler et al. (2015) on examining that epilactose was produced in the early stage of reaction before lactulose formation, proposed that it is possible that some epilactose was isomerized into lactulose during the course of reaction so that no more lactulose could be produced in addition to that produced by lactose isomerization once the concentration of epilactose has dropped to certain level, so precluding obtaining 100% lactulose yield by this route; at the end of reaction there will always be a lactulose–epilactose mixture.

A simplified scheme for the concomitant synthesis of lactulose and epilactose from lactose catalyzed by cellobiose-2 epimerase (Wang et al., 2013; Rentschler et al., 2015) is represented in Fig. 5.4.

The kinetic mechanism of this synthetic route is mostly unknown so that no mathematical models have been developed to optimize lactulose synthesis by lactose epimerization with cellobiose 2-epimerase. However, this route has the obvious advantage over transgalactosylation with β-galactosidase of using a single substrate. Increasing the affinity for lactulose is a goal that should be met by screening of new cellobiose 2-epimerase and protein engineering of existing ones.

5.5 DOWNSTREAM PROCESSING FOR LACTULOSE PURIFICATION

All the processes for lactulose synthesis described in the previous sections produce significant amounts of undesirable side products and unreacted substrates, which are hard to remove from the reaction medium (Hernández et al., 2009; Panesar and Kumari, 2011). For the industrial production of lactulose, these contaminants should be removed or kept at a minimum in the final product, since the presence of monosaccharides, lactose, and colored compounds is not tolerable, particularly in the pharmaceutical applications of lactulose. This may be not so for food applications where residual amounts of other sugar components can be considered acceptable (Montilla et al., 2005; Hernández et al., 2009).

In the chemical synthesis of lactulose, the reactor output contains mainly lactulose, lactose, and epilactose besides the chemical catalyst, so that several steps of

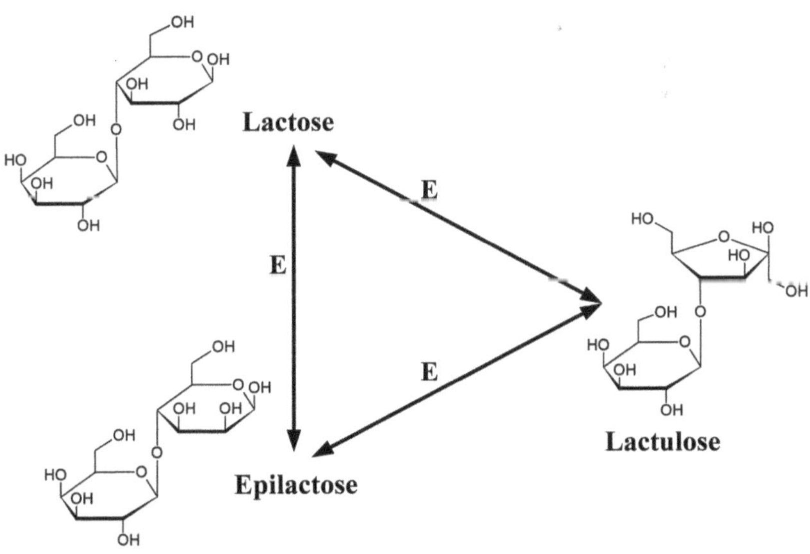

FIGURE 5.4 Simplified scheme for the mechanism of the concomitant lactulose synthesis by lactose isomerization and epilactose synthesis by lactose epimerization with cellobiose 2-epimerase (E).

purification are required (Panesar and Kumari, 2011). In the enzymatic synthesis by lactose transgalactosylation, contaminants to be removed are glucose, galactose, and residual fructose and lactose. Removal of unreacted fructose is mostly important since fructose alters the sweetness and caloric value of the product and induces Maillard reactions (Panesar and Kumari, 2011; Schuster-Wolff-Bühring et al., 2010). The enzyme catalyst can be easily removed if immobilized; if soluble enzyme is used, removal from the product stream may be required for certain applications, but if properly inactivated, its presence will be acceptable at least in food applications; this represents a clear advantage of enzymatic over chemical synthesis.

Downstream operations may well represent up to 90% of the operating costs, and the increasing pressure for developing environmentally sustainable processes and more stringent regulations on waste management have been powerful driving forces for developing several strategies for the effective purification of prebiotic carbohydrates (Feng et al., 2009; Hernández et al., 2009; Pinelo et al., 2009; Montañés et al., 2012; Nath et al., 2013; Guerrero et al., 2014). This certainly applies to lactulose production, which is still conducted by chemical synthesis.

Purification of lactulose at production level is done by ion-exchange or adsorption liquid chromatography (Carobbi et al., 1990; Feng et al., 2009). Ion-exchange resins and activated carbon are the most used chromatographic matrices, having a higher affinity for di- and oligosaccharides than for monosaccharides so that the latter elute first from the chromatographic column; this operation also allows the removal of the chemical catalyst (Carobbi et al., 1990).

Hicks et al. (1984) evaluated five purification strategies for lactulose produced by chemical synthesis with borate as catalyst. The selected purification scheme considered the treatment with strong acids followed by adsorption, and finally ion-exchange chromatography in which borate was separated from the synthesized product. Dendene et al. (1995) evaluated different cations (K^+, Na^+ and Ca^{++}) in ion-exchange resins, with the best separation of lactulose from lactose and galactose being obtained with calcium resins that performed nicely, even though high purification yields were not attained due to the high sugar concentration and the presence of traces of the chemical catalyst. The operation is time-consuming and significant amounts of eluent (usually water) and energy are required (Feng et al., 2009). In this way, the operation is complex, difficult to scale up, costly, and not efficient enough for the complete removal of lactose and other oligosaccharides from the reacted medium (Li et al., 2008; Hernández et al., 2009; Duarte et al., 2010; Torres et al., 2010).

Another strategy for lactulose purification is membrane fractionation by ultrafiltration and nanofiltration. Both inorganic (ceramic) and organic (cellulose, polysulfone, polyamide) membranes are available, the latter being the most used at industrial scale (Vanneste et al., 2011; Zhang et al., 2011). Potential for purification is high and significant advances have been experienced in recent years (Li et al., 2008; Pinelo et al., 2009; Feng et al., 2009); however, membrane fractionation of carbohydrate mixtures is not yet a mature technology and even though catalyst can be effectively removed, lactose cannot be separated from lactulose by this strategy (Pinelo et al., 2009; Feng et al., 2009; Nath et al., 2013). Nanofiltration was used at pilot scale level for the removal of the chemical catalyst (NaCl and H_3BO_3) from a reacted mixture containing mostly lactulose and lactose. Operating at 25 bar, more than 95% of the catalyst was removed with 90% recovery of the disaccharides (Zhang et al., 2011). Even though lactose could not be removed, this strategy can be used to separate monosaccharides from oligosaccharides, as reported for the case of GOS, where a purification yield of 70% was obtained delivering a product with 55% GOS (Feng et al., 2009). Using this strategy, monosaccharides can be efficiently removed from lactulose–GOS mixtures. Time, eluent, and energy savings are assets of membrane technology when compared to chromatography (Goulas et al., 2002; Feng et al., 2009).

The use of borate catalysts for lactose isomerization, which are stable at alkaline conditions, allows the selective separation of lactulose from the reaction medium with yields of recovery higher than 75% (Kozempel et al., 1995; Montilla et al., 2005; Aider and de Halleux, 2007).

Other strategies for lactulose purification are based on the different solubility of lactulose and lactose in organic solvents. Solubility of both disaccharides in alcoholic solutions increased with the water content and decreased with the molecular mass of the alcohol. However, at all conditions evaluated, solubility of lactulose was more than 10 times higher than lactose, so that starting from equimolar mixtures of them lactulose could be obtained with purity higher than 90% (Montañés et al., 2007a; Ruiz-Matute et al., 2007; Panesar and Kumari, 2011).

Different strategies for lactulose recovery have been developed, like extraction with supercritical fluids, solid-phase extraction, and pressurized liquid extraction. Ruiz-Matute et al. (2007) evaluated the recovery of lactulose from a mixture containing 70% w/w of lactose and 30% w/w of lactulose, using pressurized liquid extraction with a 70/30 ethanol–water mixture obtaining a product with 94% lactulose. This strategy has the advantages of lower extraction time and solvent consumption than with conventional extraction. Using supercritical carbon dioxide and ethanol as cosolvent at 100 bar and 100°C, a product with 95% lactulose was obtained from a mixture containing 70% w/w of lactose and 30% w/w of lactulose, although lactulose recovery was only 45% (Montañés et al., 2007b). Supercritical carbon dioxide extraction is an appealing technology, being efficient, productive, flexible, gentle enough for biologically active products, and compliant with green chemistry principles. One disadvantage is the low solubility of polar compounds in carbon dioxide so that cosolvents must be used, as stated earlier. The same strategy was used for the purification of lactulose from a carbohydrate mixture with 74% w/w lactulose, 12.2% w/w galactose, 6.6% w/w lactose, 5.2% w/w epilactose, and 2% w/w tagatose used as a simulated product from the chemical synthesis of lactulose; a product with 84% lactulose was obtained with a yield of recovery of 67% (Montañés et al., 2008). In order to evaluate the technological meaning of this operation, the extraction of lactulose from the reacted mixture produced by alkaline isomerization, containing lactulose, tagatose, and fGOS, was scaled up. Three extraction steps with supercritical carbon dioxide were required to fractionate the mixture, different operation conditions being required at each step; the process was considered viable from an economic assessment with a payout time of 10 years (Montañés et al., 2012).

Other purification strategies that have been applied for GOS purification (see Section 4.4) are in principle applicable to the purification of enzymatically synthesized lactulose. Among them, enzymatic oxidation has been used for the removal of glucose and lactose by conversion into gluconic acid and lactobionic acid with glucose oxidase and cellobiose dehydrogenase, respectively. Even though removal of the acids by ion-exchange chromatography is feasible, the cost of the oxidases and the complexity of the system make the process unrealistic at production level (Splechtna et al., 2001). Selective fermentation of monosaccharides with *Saccharomyces cerevisiae* and mono and di-saccharides with *K. lactis* is also a feasible strategy for lactulose purification that has been used for the purification of oligosaccharides; the efficient removal of contaminant sugars and low cost are their main advantages (Rabiu et al., 2001; Yoon et al., 2003; Cheng et al., 2006; Li et al., 2008; Hernández et al., 2009; Duarte et al., 2010). Selective fermentation was successfully used in the purification of GOS with *S. cerevisiae* and *K. lactis* obtaining a product with very high purity (Guerrero et al., 2014). A variation of this strategy was implemented by Rada et al. (2008) utilizing *Lactobacillus* strains for the removal of nonprebiotic sugars. Selective fermentation is attractive for lactulose purification since it allows the complete removal of monosaccharides without consumption of the products of synthesis (lactulose and GOS). In the case of lactulose purification, only *S. cerevisiae* can be used since *K. lactis* consumes lactose and lactulose as well.

It can be concluded that most of the techniques described herein are not efficient enough for lactulose purification, since separation from lactose is a major problem because of the similar structure and physicochemical properties of both. Even so, separation of monosaccharides is important for making low-calorie and poorly sweet products amenable for consumption by diabetic persons and weight watchers (Li et al., 2008; Hernández et al., 2009).

Purification of lactulose at industrial scale is then a complex task that may involve several steps. A process has been described considering a first step of spent chemical catalyst (aluminum hydroxide) removal by centrifugation, which is then mixed with sodium hydroxide and calcined at 750°C to recover the catalyst in its active form (sodium aluminate). The supernatant containing lactulose and unreacted lactose besides some other minor components is subjected to several steps of ultrafiltration and nanofiltration, removal of ions with ion-exchange resins, and then concentration; unreacted lactose is then removed by crystallization obtaining a highly purified lactulose with small amounts of contaminants (residual lactose, epilactose, and galactose) (Carobbi et al., 1990; Panesar and Kumari, 2011).

5.6 APPLICATIONS IN FOODS AND PHARMACEUTICALS

Lactulose importance has increased considerably in recent years because of its multiple applications in the food and pharma sectors (Schumann, 2002; Panesar and Kumari, 2011).

Lactulose has been used since 1950 as a drug for the treatment of specific medical conditions. In 1957 lactulose was considered as having bifidus factor, its use as prebiotic being much more recent; this evolution explains why lactulose was initially classified as a medicinal product rather than as a food additive (Schumann, 2002; Olano and Corzo, 2009). Potential applications of lactulose are presented in Fig. 5.5.

5.6.1 Lactulose in Food Applications

Use of NDOs as health-promoting ingredients has increased considerably in recent years paralleling the development of functional foods. Lactulose has been scientifically proven as a prebiotic, conferring also excellent functional

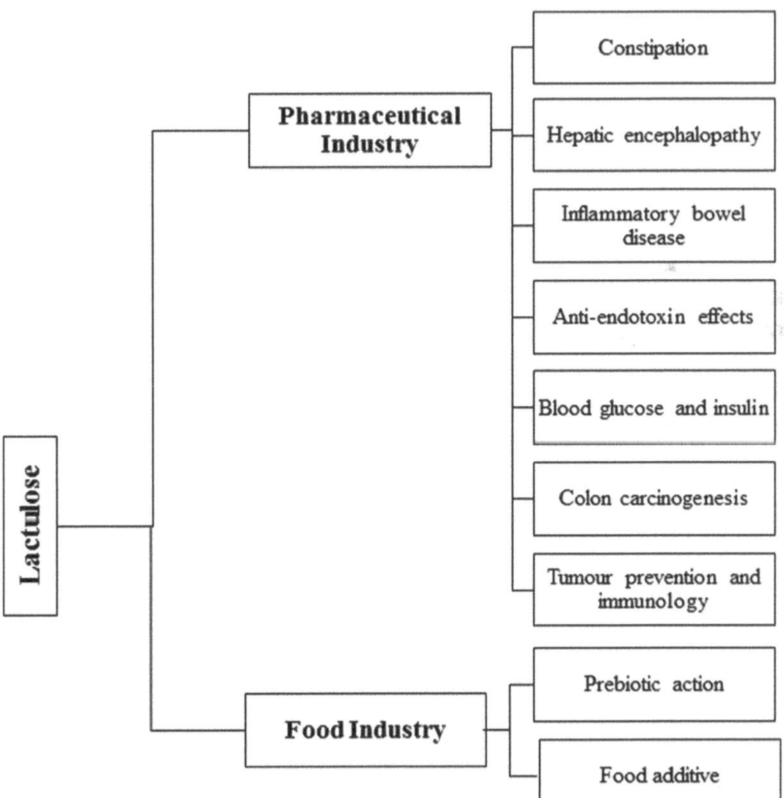

FIGURE 5.5 Applications of lactulose in the food and pharmaceutical sectors.

properties to the food matrices that contain it (Schumann, 2002; Olano and Corzo, 2009; Seki and Saito, 2012), so that the food industry uses lactulose both as a prebiotic and as functional additive. Main uses for lactulose in the food industry are presented in the following sections.

5.6.1.1 Lactulose as Prebiotic

Many food applications of lactulose are based on its prebiotic condition. At low doses, from 0.04 to 0.2 g/kg of body weight, lactulose contributes to increase mineral absorption; at between 0.07 and 0.2 g/kg of body weight it promotes the growth of probiotic bacteria, becoming an important health promoter component in functional foods, food supplements, and nutraceuticals. However, the use of lactulose as prebiotic is limited since at daily doses higher than 0.25 g/kg of body weight it promotes bowel movement, acting as a laxative (Schuster-Wolff-Bühring et al., 2010; Panesar and Kumari, 2011).

Beneficial effect of lactulose ingestion is associated with the relief of health conditions related to an altered intestinal microbiota. Preclinical studies have demonstrated the effect of lactulose in the treatment of intestinal infections and the prevention of gallbladder stones, but controlled clinical studies with patients are still lacking. There is also evidence of its effect against the formation of colonic carcinomas (Moore and Moore, 1995; Macfarlane et al., 2008).

Lactulose is not digested in the upper intestinal tract so that it produces an osmotic effect promoting the displacement of water in the intestinal lumen and intestinal movement, accelerating the intestinal passage and increasing fecal volume. Therefore, the main use of lactulose actually refers to its pharmaceutical use in the treatment of chronic constipation. A negative effect of lactulose ingestion is the increase in intestinal gas evolution and the advent of diarrhea when ingested in high doses. Patients with diarrheic syndrome do not tolerate prebiotics well, but administered in adequate doses no adverse effects are produced and, in fact, lactulose at low doses is prescribed for infants and pregnant women (Macfarlane et al., 2008; Schuster-Wolff-Bühring et al., 2010).

Lactulose degradation to short-chain fatty acids (SCFAs) produced in the colon by *Bifidobacterium* and *Lactobacillus* lowers the intestinal pH, reducing ammonia evolution by displacing the equilibrium to the formation of ammonium ion, while the proliferation of the probiotic strains inhibits the growth of ammonia-producing bacteria. Since ammonia is toxic at brain level, lactulose is effective in the treatment of hepatic encephalopathy (Bruzzese et al., 2006; Macfarlane et al., 2008) (see Section 5.1). Lowering pH reduces the survival of *Salmonella* (Panesar and Kumari, 2011) and the prevalence of urinary and respiratory tract infections (Liao et al., 1994). Lactulose also has antidiabetic effect by reducing blood glucose levels and pancreatic insulin production (Panesar and Kumari, 2011).

Lactulose has been used as prebiotic also in animal feed, reducing the amount of antibiotic supplementation, improving intestinal motility and proliferation of

Bifidobacteria and *Lactobacilli* while reducing intestinal pathogens; the effect of lactulose as animal growth promoter has also been reported (Schumann, 2002; Aït-Aissa and Aïder, 2014).

Fig. 5.6 summarizes the mechanisms of action of lactulose and their main consequences in bacterial metabolism.

5.6.1.2 Lactulose as a Food Additive

Lactulose is used as a functional ingredient in a wide variety of food products. In 1957 its stimulating effect on intestinal *Bifidobacteria* was reported for the first time so that it is also considered as a bifidus factor for the regulation of intestinal function, and more recently, as a prebiotic (Olano and Corzo, 2009; Seki and Saito, 2012). Since then, significant information has accumulated proving that lactulose ingestion provides multiple health benefits to the consumer (Olano and Corzo, 2009), suggesting that the incorporation of lactulose-containing products to the regular diet is a step forward to healthy and equilibrated nutrition (Schuster-Wolff-Bühring et al., 2010). Its use as

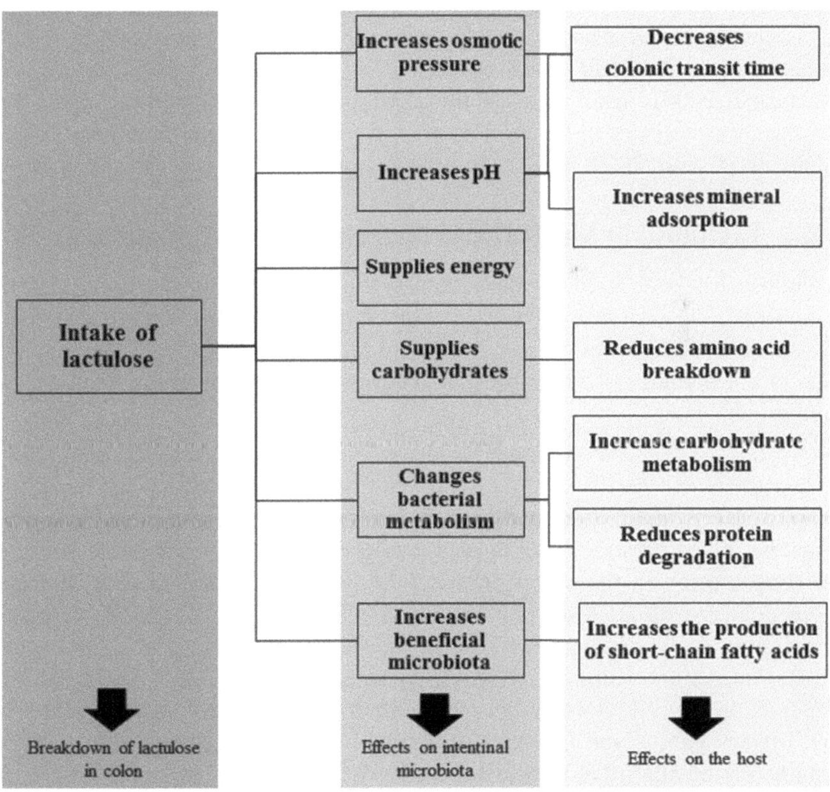

FIGURE 5.6 Main effects of lactulose consumption on gut microbiota and on the host.

prebiotic in milk formulas and different dairy products is preponderant (Lee et al., 2004; Kim et al., 2006; Seki and Saito, 2012). It was early reported that the incorporation of 0.5% w/w lactulose in powder stimulates the healthy intestinal microbiota, while at a level of 1% it is mildly laxative (Nagendra et al., 1995). Lactulose has been also reported as a stimulant of calcium and magnesium absorption (Olano and Corzo, 2009).

As a functional food ingredient, lactulose has been reported as flavor enhancer improving the organoleptic quality of the food matrix to which it is added (Schuster-Wolff-Bühring et al., 2010; Panesar and Kumari, 2011; Seki and Saito, 2012). Its addition to yogurt was as effective as inulin and soy fiber in the treatment of infant constipation (Schumann, 2002; Olano and Corzo, 2009; Seki and Saito, 2012; Aït-Aissa and Aïder, 2014). Because of its high thermal stability at low pH, lactulose has been used to fortify fruit juices and other acid foods (Seki and Saito, 2012). It is increasingly being used as a sweetener for diabetic persons and as a sugar substitute in confectionery, soft drinks, baby formulas, bakery products, yogurt and other dairy products, and as a fortifier in formulated foods for the elderly (Schumann, 2002; Seki and Saito, 2012).

Lactulose has been used as cryoprotectant of probiotic strains used in yogurt, increasing the shelf life of the product at 4°C in five weeks (Tabatabaie and Mortazavi, 2008). Addition of lactulose to infant formulas allowed maintaining the characteristics of the product after prolonged storage (Schumann, 2002; Olano and Corzo, 2009; Aït-Aissa and Aïder, 2014).

5.6.2 Lactulose in Medical and Pharmaceutical Applications

Lactulose has been used for decades now in the treatment of chronic constipation and hepatic encephalopathy; other minor applications refer to hepatic disorders, tumor prevention, immunostimulation, and anti-endotoxin effect, and also to the maintenance of blood sugar levels (Schumann, 2002; Panesar and Kumari, 2011).

5.6.2.1 Constipation

As said before, lactulose is mostly used as a laxative in the treatment of chronic constipation of persons of all ages. Lactulose is an osmotic laxative because its consumption in the colon produces SCFA with the consequent increase in osmolarity; liquid level is maintained, pH drops, and feces are softened and can be easily evacuated, reducing the intestinal transit time (Schumann, 2002; Panesar and Kumari, 2011; Aït-Aissa and Aïder, 2014). Laxative action depends on many factors including health condition, age, body weight, gender, dietary habits, and ingested dose (Schuster-Wolff-Bühring et al., 2010). An interesting quality of lactulose as laxative is that it can be used in long-term treatments not producing dependence and being notably effective (Schumann, 2002; Aït-Aissa and Aïder, 2014).

5.6.2.2 Hepatic Encephalopathy

Second to constipation, the use of lactulose in the prevention and treatment of hepatic encephalopathy (HE) is the most significant pharmaceutical application. HE is a neuropsychiatric syndrome that can progress form mental disorders to coma (Schumann, 2002; Bruzzese et al., 2006; Macfarlane et al., 2008). HE is induced by high concentrations of blood ammonia, which is the consequence of hepatic malfunctioning not allowing its proper removal. Ammonia is produced by protein degradation caused by intestinal bacteria and, when insufficiently removed at hepatic level, acts as a neurotoxic compound in the brain (Bruzzese et al., 2006; Macfarlane et al., 2008; Aït-Aissa and Aïder, 2014). Lactulose reduces the number of putrefactive ammonia-producing bacteria in the colon while pH reduction inhibits ammonia adsorption so that its blood level is reduced (Aït-Aissa and Aïder, 2014).

5.6.2.3 Inflammatory Bowel Disease and Anti-endotoxin Effects

Another major use of lactulose is in the treatment of inflammatory bowel disease, which consists of the inflammation of the colon and upper intestine; ulcerative colitis and Crohn's disease are the more severe manifestations (Liao et al., 1994; Talley et al., 2011). As mentioned before, lactulose, as a prebiotic, increases SCFA production in the colon with a significant decrease in fecal pH, so creating favorable conditions for the proliferation of *Lactobacilli* (mostly *Lactobacillus acidophilus*), which inhibits the growth of coliforms, *Bacteroides*, *Salmonella*, and *Shigella* (Schumann, 2002; Aït-Aissa and Aïder, 2014). This change in intestinal microbiota is associated with intestinal health (Paul et al., 2007), reduction of urinary and respiratory infections (Liao et al., 1994), and reduction in the production and absorption of endotoxins in the intestine, which is a key aspect in the intestinal inflammatory response (Panesar and Kumari, 2011). Lactulose fermentation by colonic bacteria produces considerable amounts of endogenous hydrogen, which aids in the prevention of colitis and reduces the symptoms of intestinal inflammation (Chen et al., 2011). Anti-endotoxin effect of lactulose is also applicable in the treatment of metabolic disorders, like hepatorenal syndrome (Schuster-Wolff-Bühring et al., 2010), exocrine pancreatic dysfunction (Mack et al., 1992), diabetes mellitus (Tabatabaie et al., 1997), and hypercholesterolemia (Liao and Florin, 1995).

5.6.2.4 Blood Glucose and Insulin

Lactulose is effective in reducing blood sugar and pancreatic insulin, so exerting an antidiabetic effect (Schumann, 2002; Aït-Aissa and Aïder, 2014).

5.6.2.5 Colon Carcinogenesis, Tumor Prevention, and Immunology

Colon cancer generally develops as a result of biochemical changes in the lumen, mucosa, and adjacent tissues of the large intestine. Colon microbiota and the metabolic products generated may influence colon cancer development. It has been claimed in several reports that the intake of live probiotic bacteria and prebiotics, such as lactulose, reduces colon cancer risk, but further evidence is required to confirm this statement (Moore and Moore, 1995; Macfarlane et al., 2008; Panesar and Kumari, 2011; Aït-Aissa and Aïder, 2014). *Bifidobacteria* play an important role in tumor prevention, and lactulose ingestion has been reported to improve the antitumoral and immunogenic effects of such bacteria. Efficacy of *Bifidobacteria* in the prevention of breast and liver cancer has been proved consistently, and since most *Bifidobacteria* readily metabolize lactulose, immunogenic and antitumoral activity can be considered an indirect effect of lactulose ingestion (Aït-Aissa and Aïder, 2014). Use of lactulose in preoperatory treatment allowed preventing surgical complications arising from obstructive jaundice (Greve et al., 1990; Panesar and Kumari, 2011; Aït-Aissa and Aïder, 2014).

As extensively described in this chapter, lactulose is, together with inulin, FOS, and GOS, one of the most important and widely used prebiotics (Wang, 2009). Lactulose and GOS are lactose-derived prebiotics, both representing an outstanding opportunity for lactose upgrading (Sako et al., 1999; Gänzle et al., 2008). GOS–lactulose mixtures can be produced by enzyme biocatalysis so its performance as a synbiotic NDO mixture is worthy of further research (Guerrero et al., 2015b). Lactulose is a quite versatile compound that has significant applications both in the food and pharmaceutical fields, medical applications being salient among lactose-derived compounds, which forecasts a promising future in terms of market growth, consumer acceptance, and social impact.

REFERENCES

Adachi, S., 1958. Formation of lactulose and tagatose from lactose in strongly heated milk. Nature 181, 840–841.

Adachi, S., 1965. A rapid method for the assay of lactulose. Analytical Biochemistry 12, 137–142.

Adhikari, A.K., Sahai, D., Mathur, O.N., 1991. A rapid spectrophotometric method for quantitative determination of lactulose in heated milk and milk products. Lait 71, 555–564.

Aider, M., de Halleux, D., 2007. Isomerization of lactose and lactulose production: review. Trends in Food Science and Technology 187, 356–364.

Aider, M., Gimenez-Vidal, M., 2012. Lactulose synthesis by electro-isomerization of lactose: effect of lactose concentration and electric current density. Innovative Food Science and Emerging Technologies 16, 163–170.

Aït-Aissa, A., Aïder, M., 2014. Lactulose: production and use in functional food, medical and pharmaceutical applications. Practical and critical review. Food Science and Technology 49, 1245–1253.

Als-Nielsen, B., Gluud, L.L., Gluud, C., 2004. Non-absorbable disaccharides for hepatic encephalopathy: systemic review of randomized trials. British Medical Journal 328, 1046–1050.

Andrews, G.R., 1986. Formation and occurrence of lactulose in heated milk. Journal of Dairy Research 53, 665–680.

Bologa, M.K., Stepurina, T.G., Bologa, A.M., Polikarpov, A.A., Sprinchan, E.G., 2009. Optimization of lactose isomerization into lactulose by the electrophysical method. Surface Engineering and Applied Electrochemistry 45, 415–419.

Boon, M.A., Janssen, A.E.M., van der Padt, A., 1999. Modelling and parameter estimation of the enzymatic synthesis of oligosaccharides by β-galactosidase from *Bacillus circulans*. Biotechnology and Bioengineering 64, 558–567.

Brands, C.M.J., van Boekel, M.A.J.S., 2003. Kinetic modeling of reaction in heated disaccharide-casein systems. Food Chemistry 83, 13–26.

Bruzzese, E., Volpicelli, M., Squaglia, M., Tartaglione, A., Guarino, A., 2006. Impact of prebiotics on human health. Digestive and Liver Disease 38, 283–287.

Carobbi, R., Miletti, S., Franci, V., 1985. Process for Preparing Lactulose from Lactose, in the Form of a Syrup or a Crystalline Product. United States Patent 4536221.

Carobbi, R., Innocenti, F., Ripoli, B.A., 1990. Process for Preparing Lactulose from Lactose by Epimerization with Sodium Aluminate. United States Patent 4957564.

Chen, X., Zuo, Q., Hai, Y., Sun, X.J., 2011. Lactulose: an indirect antioxidant ameliorating inflammatory bowel disease by increasing hydrogen production. Medical Hypotheses 76, 325–327.

Cheng, C.C., Yu, M.C., Cheng, T.C., Sheu, D.C., Duan, K.J., Tai, W.L., 2006. Production of high-content galacto-oligosaccharide by enzyme catalysis and fermentation with *Kluyveromyces marxianus*. Biotechnology Letters 28, 793–797.

de la Fuente, M.A., Juárez, M., de Rafael, D., Villamiel, M., Olano, A., 1999. Isomerization of lactose catalyzed by alkaline-substituted sepiolites. Food Chemistry 66, 301–306.

De Souza Oliveira, R.P., Rodrigues Florence, A.C., Perego, P., De Oliveira, M.N., Converti, A., 2011. Use of lactulose as prebiotic and its influence on the growth, acidification profile and viable counts of different probiotics in fermented skim milk. International Journal of Food Microbiology 145, 22–27.

Dendene, K., Guihard, L., Nicolas, S., Bariou, B., 1994. Kinetics of lactose isomerization to lactulose in an alkaline medium. Journal of Chemical Technology and Biotechnology 61, 37–42.

Dendene, K., Guihard, L., Balannec, B., Barjou, B., 1995. Study of the separation of lactose, lactulose and galactose by liquid chromatography using cationic ion-exchange resin columns. Chromatographia 41, 561–567.

Duarte, P.M., Goncalves, M.P., Teixeira, J.A., Rodrigues, L.R., 2010. Galacto-oligosaccharides: production, properties, applications, and significance as prebiotics. Comprehensive Reviews in Food Science and Food Safety 9, 438–454.

Elkington, S.G., Floch, M.H., Conn, H.O., 1969. Lactulose in the treatment of chronic portal-systemic encephalopathy – a double-blind clinical trial. New England Journal of Medicine 281, 408–412.

Fattahi, H., Zokaee, F., Bonakdarpour, B., Hashemi, S.A., Khatami, S.H., 2010. Enzymatic synthesis of lactulose by commercial β-galactosidase from *Kluyveromyces lactis*. Afinidad 546, 149–153.

Feng, Y.M., Chang, X.L., Wang, W.H., Ma, R.Y., 2009. Separation of galacto-oligosaccharides mixture by nanofiltration. Journal of the Taiwan Institute of Chemical Engineers 40, 326–332.

Förster-Fromme, K., Schuster-Wolff-Bühring, R., Hartwig, A., Holder, H., Schwiertz, A., Bischoff, S.C., Hinrichs, J., 2011. A new enzymatically produced 1-lactulose: a pilot study to test the bifidogenic effects. International Dairy Journal 21, 940–948.

Gänzle, M., 2012. Enzymatic synthesis of galacto-oligosaccharides and other lactose derivatives (hetero-oligosaccharides) from lactose. International Dairy Journal 22, 116–122.

Gänzle, M.G., Haase, G., Jelen, P., 2008. Lactose: crystallization, hydrolysis and value-added derivatives. International Dairy Journal 18, 685–694.

Ghoddusi, H.B., Grandison, M.A., Grandison, A.S., Tuohy, K.M., 2007. *In vitro* study on gas generation and prebiotic effects of some carbohydrates and their mixtures. Anaerobe 13, 193–199.

Gosling, A., Stevens, G.W., Barber, A.R., Kentish, S.E., Gras, S.L., 2010. Recent advances refining galactooligosaccharide production from lactose. Food Chemistry 121, 307–318.

Goulas, A.K., Kapasakalidis, P.G., Sinclair, H.R., Rastall, R.A., Grandison, A.S., 2002. Purification of oligosaccharides by nanofiltration. Journal of Membrane Science 209, 321–335.

Greve, J.W., Gouma, D.J., Leeuwen, P.A.M., Buurman, W.A., 1990. Lactulose inhibits endotoxin induced tumour necrosis factor production by monocytes. An in vitro study. Gut 31, 198–203.

Guerrero, C., Vera, C., Plou, F., Illanes, A., 2011. Influence of reaction conditions on the selectivity of the synthesis of lactulose with microbial β-galactosidases. Journal of Molecular Catalysis B: Enzymatic 72, 206–212.

Guerrero, C., Vera, C., Novoa, C., Dumont, J., Acevedo, F., Illanes, A., 2014. Purification of highly concentrated galacto-oligosaccharide preparations by selective fermentation with yeast. International Dairy Journal 39, 78–88.

Guerrero, C., Vera, C., Conejeros, R., Illanes, A., 2015a. Transgalactosylation and hydrolytic activities of commercial preparations of β-galactosidase for the synthesis of prebiotic carbohydrates. Enzyme and Microbial Technology 70, 9–17.

Guerrero, C., Vera, C., Conejeros, R., Illanes, A., 2015b. Simultaneous synthesis of mixtures of lactulose and galacto-oligosaccharides and their selective fermentation. Journal of Biotechnology 209, 31–40.

Guerrero, C., Vera, C., Conejeros, R., Illanes, A., 2015c. Repeated-batch operation for the synthesis of lactulose with β-galactosidase immobilized by aggregation and crosslinking. Bioresource Technology 190, 122–131.

Hajek, J., Murzin, D., Salmi, T., Mikkola, J.P., 2013. Interconversion of lactose to lactulose in alkaline environment: comparison of different catalysis concepts. Topics in Catalysis 56, 839–845.

Hashem, A.M., Ismail, S.A.E., Helmy, W.A., El-Mohamady, Y., Abou-Romia, R., 2013. Factors affecting the production of lactulose by *Lactobacillus acidophilus* NRRL 4495 β-galactosidase and its biological activity. Malaysian Journal of Microbiology 9, 1–6.

Hashemi, A.A., Ashtiani, F.Z., 2010. The isomerization kinetics of lactose to lactulose in the presence of sodium hydroxide at constant and variable pH. Food and Bioproducts Processing 88, 181–187.

Hernández, O., Ruiz-Matute, A.I., Olano, A., Moreno, F.J., Sanz, M.L., 2009. Comparison of fractionation techniques to obtain prebiotic galactooligosaccharides. International Dairy Journal 19, 531–536.

Hicks, K.B., Parrish, F.W., 1980. A new method for the preparation of lactulose from lactose. Carbohydrate Research 82, 393–397.

Hicks, K.B., Raupp, D.L., Smith, P.W., 1984. Preparation and purification of lactulose from sweet cheese whey ultrafiltrate. Journal of Agricultural and Food Chemistry 32, 288–292.

Hua, X., Yang, R., Zhang, W., Fei, Y., Jum, Z., Jiang, B., 2010. Dual-enzymatic synthesis of lactulose in organic-aqueous two-phase media. Food Research International 43, 716–722.

Hua, X., Yang, R., Shen, Q., Ye, F., Zhang, W., Zhao, W., 2013. Production of 1-lactulose and lactulose using commercial β-galactosidase from *Kluyveromyces lactis* in the presence of fructose. Food Chemistry 137, 1–7.

Illanes, A., 2011. Whey upgrading by enzyme biocatalysis. Electronic Journal of Biotechnology 14 (6). http://dx.doi.org/10.2225/vol14-issue6-fulltext-11.

Irazoqui, G., Giacomini, C., Batista-Viera, F., Brena, B.M., Cardelle-Cobas, A., Corzo, N., Jimeno, M.L., 2009. Characterization of galactosyl derivatives obtained by transgalactosylation of lactose and different polyols using immobilized β-galactosidase from *Aspergillus oryzae*. Journal of Agriculture and Food Chemistry 57, 11302–11307.

Khatami, S., Ashtiani, F.Z., Bonakdarpour, B., 2014. The enzymatic production of lactulose via transglycosylation in conventional and non-conventional media. International Dairy Journal 34, 74–79.

Kim, C.H., Ji, E., Oh, D., 2004. A new kinetic model of recombinant β-galactosidase from *Kluyveromyces lactis* for both hydrolysis and transgalactosylation reaction. Biochemical and Biophysical Research Communications 316, 738–743.

Kim, Y.S., Park, C.S., Oh, D.K., 2006. Lactulose production from lactose and fructose by a thermostable β-galactosidase from *Sulfolobus solfataricus*. Enzyme and Microbial Technology 39, 903–908.

Kim, J.E., Kim, Y.S., Kang, L.W., Oh, D.K., 2012. Characterization of a recombinant cellobiose 2-epimerase from *Dictyoglomus turgidum* that epimerizes and isomerizes β-1,4- and α-1,4-gluco-oligosaccharides. Biotechnology Letters 34, 2061–2068.

Kim, Y.S., Oh, D.K., 2012. Lactulose production from lactose as a single substrate by a thermostable cellobiose 2-epimerase from *Caldicellulosiruptor saccharolyticus*. Bioresource Technology 104, 668–672.

Kim, Y.S., Kim, J.E., Oh, D.K., 2013. Borate enhances the production of lactulose from lactose by cellobiose 2-epimerase from *Caldicellulosiruptor saccharolyticus*. Bioresource Technology 128, 809–812.

Klewicki, R., 2000. Transglycosylation of a β-galactosyl radical, in the course of enzymic hydrolysis of lactose, in the presence of selected polyhydroxyalcohols. Biotechnology Letters 22, 1063–1066.

Kozempel, M., Kurantz, M., 1994. The isomerization kinetics of lactose to lactulose in the presence of borate. Journal of Chemical Technology and Biotechnology 59, 25–29.

Kozempel, M.F., Kurantz, M.J., Craig, J.C., Hicks, K.B., 1995. Development of a continuous lactulose process: separation and purification. Biotechnology Progress 11, 592–595.

Lamsal, B.P., 2012. Production, health aspects and potential food uses of dairy prebiotic galactooligosaccharides. Journal of the Science of Food and Agriculture 92, 2020–2028.

Lee, Y.J., Kim, C.S., Oh, D.K., 2004. Lactulose production by β-galactosidase in permeabilized cells of *Kluyveromyces lactis*. Biotechnology Products and Process Engineering 64, 787–793.

Leitner, C., Mayr, P., Riva, S., Volc, J., Kulbe, K.D., Nidetzky, B., Haltrich, D., 2001. Enzymatic redox isomerization of 1,6-disaccharides by pyranose oxidase and NADH-dependent aldose reductase. Journal of Molecular Catalysis B: Enzymatic 11, 407–414.

Li, Z., Xiao, M., Lu, L., Li, Y., 2008. Production of non-monosaccharide and high-purity galactooligosaccharides by immobilized enzyme catalysis and fermentation with immobilized yeast cells. Process Biochemistry 43, 896–899.

Li, W., Xiang, X., Tang, S., Hu, B., Tian, L., Sun, Y., Ye, H., Zeng, Y., 2009. Effective enzymatic synthesis of lactosucrose and its analogues by β-galactosidase from *Bacillus circulans*. Journal of Agricultural and Food Chemistry 57, 3927–3933.

Liao, W., Cui, X.S., Jin, X.Y., Florén, C.H., 1994. Lactulose, a potential drug for the treatment of inflammatory bowel disease. Medical Hypotheses 43, 234–238.

Liao, W., Florin, C.H., 1995. Endotoxin, cytokines and hyperlipidemia. Scandinavian Journal of Gastroenterology 28, 97–103.

Lorenzen, P., Breiter, J., Clawin-Rädecker, I., Dau, A., 2013. A novel bi-enzymatic system for lactose conversion. Food Science and Technology 48, 1396–1403.

Mack, D.R., Flick, J.A., Durie, P.R., Rosentein, B.J., Ellis, L.E., Perman, J.A., 1992. Correlation of intestinal lactulose permeability with exocrine pancreatic dysfunction. Journal of Pediatrics 120, 696–701.

Macfarlane, G.T., Steed, H., Macfarlane, S., 2008. Bacterial metabolism and health-related effects of galacto-oligosaccharides and other prebiotics. Journal of Applied Microbiology 104, 305–344.

Marconi, E., Messia, M.C., Amine, A., Moscone, D., Vernazza, F., Stocchi, F., Palleschi, G., 2004. Heat-treated milk differentiation by a sensitive lactulose assay. Food Chemistry 84, 447–450.

Martínez-Villaluenga, C., Cardelle-Cobas, A., Corzo, N., Olano, A., Villamiel, M., Jimeno, M.L., 2008. Enzymatic synthesis and identification of two trisaccharides produced from lactulose by transgalactosylation. Journal of Agricultural and Food Chemistry 56, 557–563.

Mayer, J., Conrad, J., Klaiber, I., Lutz-Wahl, S., Beifuss, U., Fischer, L., 2004. Enzymatic production and complete nuclear magnetic resonance assignment of the sugar lactulose. Journal of Agricultural and Food Chemistry 52, 6983–6990.

Mayer, J., Kranz, B., Fischer, L., 2010. Continuous production of lactulose by immobilized thermostable β-glycosidase from *Pyrococcus furiosus*. Journal of Biotechnology 145, 387–393.

Mizota, T., Mori, T., Yaeshima, T., Yanagida, T., Iwatsuki, K., Ishibashi, N., Tamura, Y., Fukuwatari, Y., 2002. Effects of low dosages of lactulose on the intestinal function of healthy adults. Milchwissenschaft 57, 312–315.

Montañés, F., Olano, A., Ibáñez, E., 2007a. Modeling solubilities of sugars in alcohols based on original experimental data. Thermodynamics 53, 2411–2418.

Montañés, F., Fornari, T., Martín-Álvarez, P.J., Montilla, A., Corzo, N., Olano, A., Ibáñez, E., 2007b. Selective fractionation of disaccharide mixture by supercritical CO_2 with ethanol as co-solvent. The Journal of Supercritical Fluids 41, 61–67.

Montañés, F., Corzo, N., Olano, A., Reglero, G., Ibáñez, E., Fornari, T., 2008. Selective fractionation of carbohydrate complex mixture by supercritical extraction with CO_2 and different co-solvents. The Journal of Supercritical Fluids 45, 189–194.

Montañés, F., Fornari, T., Olano, A., Ibánez, E., 2012. Isolation of prebiotic carbohydrates by supercritical fluid extraction. Scaling-up and economical feasibility. Journal of Chromatography A 1250, 92–98.

Montilla, A., del Castillo, M.D., Sanz, M.L., Olano, A., 2005. Egg shell as catalyst of lactose isomerisation to lactulose. Food Chemistry 90, 883–890.

Montgomery, E.M., Hudson, C.S., 1930. Relations between rotatory power and structure in the sugar group XXVII. Synthesis of a new disaccharide ketose (lactulose) from lactose. Journal of the American Chemical Society 52, 2101–2106.

Moore, W.E., Moore, L.H., 1995. Intestinal floras of populations that have a high risk of colon cancer. Applied and Environmental Microbiology 61, 3202–3207.

Nagasawa, T., Tomita, M., Tamura, Y., Obayashi, T., Mizota, T., 1974. Process for Preparing a Lactulose Syrup. United States Patent 3814174.

Nagendra, R., Viswanatha, S., Kumar, S.A., Murthy, B.K., Rao, S.V., 1995. Effect of feeding milk formula containing lactulose to infants on fecal bifidobacterial flora. Nutrition Research 15, 15–24.

Nahla, T.K., Musa, T.N., 2015. Chemical isomerization of whey lactose to lactulose by using batch reaction. Pakistan Journal of Nutrition 14, 255–258.

Nakayama, T., Amachi, T., 1999. Bioprocess Technology Fermentation, Biocatalysis and Bioseparation. John Wiley & Sons, Inc., New York.

Nath, A., Bhattacharjee, C., Chowdhury, R., 2013. Synthesis and separation of galacto-oligosaccharides using membrane bioreactor. Desalination 316, 31–41.

Nooshkam, M., Madadlou, A., 2016. Maillard conjugation of lactulose with potentially bioactive peptides. Food Chemistry 192, 831–836.

Olano, A., Corzo, N., 2009. Lactulose as a food ingredient. Journal of the Science of Food and Agriculture 89, 1987–1990.

Oosten, B.J., 1967. Solubility diagram of lactose and lactulose in water. Recueil 86, 675–676.

Palframan, R.J., Gibson, G.R., Rastall, R.A., 2002. Effect of pH and dose on the growth of gut bacteria on prebiotic carbohydrates *in vitro*. Anaerobe 8, 287–292.

Panesar, P.S., Kumari, S., 2011. Lactulose: production, purification and potential applications. Biotechnology Advances 29, 940–948.

Panesar, P.S., Kumari, S., Panesar, R., 2013. Biotechnological approaches for the production of prebiotics and their potential applications. Critical Reviews in Biotechnology 33, 345–364.

Paseephol, T., Small, D.M., Sherkat, F., 2008. Lactulose production from milk concentration permeate using calcium carbonate-based catalysts. Food Chemistry 111, 283–290.

Paul, J., Kumar Verma, A., Verma, R., 2007. Role of gut flora in inflammatory bowel disease – a state of art. Communicating Current Research and Educational Topics and Trends in Applied Microbiology 705–718.

Peterbauer, C.K., Volc, J., 2010. Pyranose dehydrogenases: biochemical features and perspectives of technological applications. Applied Microbiology and Biotechnology 85, 837–848.

Pinelo, M., Jonsson, G., Meyer, A.S., 2009. Membrane technology for purification of enzymatically produced oligosaccharides: molecular and operational features affecting performance. Separation and Purification Technology 70, 1–11.

Plou, F.J., Gomez de Segura, A., Ballesteros, A., 2007. Application of glycosidases and transglycosidases in the synthesis of oligosaccharides. In: Polaina, J., MacCabe, A.P. (Eds.), Industrial Enzymes: Structure, Function and Applications. Springer, New York, pp. 141–157.

Prenosil, J.E., Stuker, E., Bourne, J.R., 1987. Formation of oligosaccharides during enzymatic lactose: part I: state of art. Biotechnology and Bioengineering 30, 1019–1025.

Rabiu, B.A., Jay, A.J., Gibson, G.R., Rastall, R.A., 2001. Synthesis and fermentation properties of novel galacto-oligosaccharides by β-galactosidases from *Bifidobacterium* species. Applied and Environmental Microbiology 67, 2526–2530.

Rada, V., Nevoral, J., Trojanová, I., Tománková, E., Smehilová, M., Killer, J., 2008. Growth of infant faecal bifidobacteria and clostridia on prebiotic oligosaccharides in *in vitro* conditions. Anaerobe 14, 205–208.

Rentschler, E., Schuh, K., Krewinkel, M., Baur, C., Claaben, W., Meyer, S., Kuschel, B., Stressler, T., Fischer, L., 2015. Enzymatic production of lactulose and epilactose in milk. Journal of Dairy Science 98, 1–9.

Riordan, S.M., Williams, R., 1997. Treatment of hepatic encephalopathy. New England Journal of Medicine 337, 473–479.

Rodriguez-Fernandez, M., Cardelle-Cobas, A., Villamiel, M., Banga, J.R., 2011. Detailed kinetic model describing new oligosaccharides synthesis using different β-galactosidases. Journal of Biotechnology 153, 116–124.

Ruiz-Matutte, A.I., Sanz, M.L., Corzo, N., Martín-Álvarez, P.J., Ibáñez, E., Martínez-Castro, I., Olano, A., 2007. Purification of lactulose from mixtures with lactose using pressurized liquid extraction with ethanol-water at different temperatures. Journal of Agricultural and Food Chemistry 55, 3346–3350.

Sako, T., Matsumoto, K., Tanaka, R., 1999. Recent progress on research and applications of non-digestible galacto-oligosaccharides. International Dairy Journal 9, 69–80.

Sanz Valero, J.I., 2009. Production of Galacto-oligosaccharides from Lactose by Immobilized β-galactosidase and Posterior Chromatographic Separation (Ph.D. thesis). Ohio State University, Ohio, USA.

Schumann, C., 2002. Medical, nutritional and technological properties of lactulose. An update. European Journal of Nutrition 41, 17–25.

Schuster-Wolff-Bühring, R., Fischer, L., Hinrichs, J., 2010. Production and physiological action of the disaccharide lactulose. International Dairy Journal 20, 731–741.

Seki, N., Saito, H., 2012. Lactose as a source for lactulose and other functional lactose derivatives. International Dairy Journal 22, 110–115.

Seo, Y.H., Park, G.W., Han, J.I., 2015. Efficient lactulose production from cheese whey using sodium carbonate. Food Chemistry 173, 1167–1171.

Shen, Q., Zhang, Y., Hua, X., Zhang, W., Zhao, W., 2015. Thermostability enhancement of cellobiose 2-epimerase from *Caldicellulosiruptor saccharolyticus* by site-directed mutagenesis. Journal of Molecular Catalysis B: Enzymatic 120, 158–164.

Sitanggang, A.B., Drews, A., Kraume, M., 2014a. Continuous synthesis of lactulose in an enzymatic membrane reactor reduces lactulose secondary hydrolysis. Bioresource Technology 167, 108–115.

Sitanggang, A.B., Drews, A., Kraume, M., 2014b. Rapid transgalactosylation towards lactulose synthesis in a small-scale enzymatic membrane reactor (EMR). Chemical Engineering Transactions 38, 19–24.

Sitanggang, A.B., Drews, A., Kraume, M., 2015. Influence of operating conditions on continuous lactulose synthesis in an enzymatic membrane reactor system: a basis prior to long-term operation. Journal of Biotechnology 203, 89–96.

Song, Y.S., Shin, H.Y., Lee, J.Y., Park, C., Kim, S.W., 2012. β-galactosidase immobilized microreactor fabricated using a novel technique for enzyme immobilization and its application for continuous synthesis of lactulose. Food Chemistry 133, 611–617.

Song, Y.S., Lee, H.U., Park, C., Kim, S.W., 2013a. Batch and continuous synthesis of lactulose from whey lactose by immobilized β-galactosidase. Food Chemistry 136, 689–694.

Song, Y.S., Suh, Y.J., Park, C., Kim, S.W., 2013b. Improvement of lactulose synthesis through optimization of reaction conditions with immobilized β-galactosidase. Korean Journal of Chemical Engineering 30, 160–165.

Song, Y.S., Lee, H.U., Park, C., Kim, S.W., 2013c. Optimization of lactulose synthesis from whey lactose by immobilized β-galactosidase and glucose isomerase. Carbohydrate Research 369, 1–5.

Splechtna, B., Petzelbauer, I., Baminger, U., Haltrich, D., Kulbe, K.D., Nidetzky, B., 2001. Production of a lactose-free galacto-oligosaccharide mixture by using selective enzymatic oxidation of lactose into lactobionic acid. Enzyme and Microbial Technology 29, 434–440.

Stevenson, D.E., Stanley, R.A., Furneaux, R.H., 1993. Optimization of alkyl-D galactopyranoside synthesis from lactose using commercially available β-galactosidases. Biotechnology and Bioengineering 42, 657–666.

Tabatabaie, T., Kotake, Y., Wallis, G., Jacob, J.M., Floyd, R.A., 1997. Spin trapping agent phenyl-N-tertbutylnitrone protects against the onset of drug-induced insulin-dependent diabetes mellitus. FEBS Letters 407, 148–152.

Tabatabaie, F., Mortazavi, A., 2008. Influence of lactulose on the survival of probiotic strains in yoghurt. World Applied Sciences Journal 3, 88–90.

Talley, N.J., Abreu, M.T., Achkar, J.P., Bernstein, C.N., Dubinsky, M.C., Hanauer, S.B., Kane, S.V., Sandborn, W.J., Ullman, T.A., Moayyedi, P., 2011. An evidence-based systematic review on medical therapies for inflammatory bowel disease. The American Journal of Gastroenterology 106, 2–25.

Tamura, Y., Mizota, T., Shimamura, S., Tomita, M., 1993. Lactulose and its application to the food and pharmaceutical industries. International Dairy Federation Bulletin 289, 43–53.

Tang, L., Li, Z., Dong, X., Yang, R., Zhang, J., Mao, Z., 2011. Lactulose biosynthesis by β-galactosidase from a newly isolated *Arthrobacter* sp. Journal of Industrial Microbiology and Biotechnology 38, 471–476.

Torres, D.P.M., Gonçalves, M.P., Teixeira, J.A., Rodrigues, L.R., 2010. Galacto-oligosaccharides: production, properties, applications, and significance as prebiotics. Comprehensive Reviews in Food Science and Food Safety 9, 438–454.

Troyano, E., de Rafael, D., Martínez-Castro, I., Olano, A., 1996. Isomerization of lactose over natural sepiolite. Journal of Chemical Technology and Biotechnology 65, 111–114.

Vaheri, M., Kaupinnen, V., 1978. The formation of lactulose (4-*O*-β-galactopyranosyl fructose) by β-galactosidase. Acta Pharmaceutica Fennica 87, 75–83.

Vanneste, J., De Ron, S., Vandecrys, S., Soare, S.A., Darvishmanesh, S., Van der Bruggen, B., 2011. Techno-economic evaluation of membrane cascades relative to simulated moving bed chromatography for the purification of mono- and oligosaccharides. Separation and Purification Technology 80, 600–609.

Vera, C., Guerrero, C., Illanes, A., 2011a. Determination of the transgalactosylation activity of *Aspergillus oryzae* β-galactosidase: effect of pH, temperature, and galactose and glucose concentrations. Carbohydrate Research 346, 745–752.

Vera, C., Guerrero, C., Illanes, A., Conejeros, R., 2011b. A pseudo steady-state model for galacto-oligosaccharides synthesis with β-galactosidase from *Aspergillus oryzae*. Biotechnology and Bioengineering 108, 2270–2279.

Vera, C., Guerrero, C., Illanes, A., Conejeros, R., 2013. Fed-batch synthesis of galacto-oligosaccharides with *Aspergillus oryzae* β-galactosidase using optimal control strategy. Biotechnology Progress 30, 59–67.

Verhaar, L.A.Th., van der Aalst, M.J.M., Beenackers, J.A.W.M., Kuster, B.F.M., 1978. Ion-exchange chromatography of lactose-lactulose isomerization mixtures using a boric acid-borate eluent. Journal of Chromatography 170, 363–370.

Villamiel, M., Corzo, N., Foda, M.I., Montes, F., Olano, A., 2002. Lactulose formation catalysed by alkaline-substituted sepiolites in milk permeate. Food Chemistry 76, 7–11.

Wang, H., Yang, R., Hua, X., Zhao, W., Zhang, W., 2013. Enzymatic production of lactulose and 1-lactulose: current state and perspectives. Applied Microbiology and Biotechnology 97, 6167–6180.

Wang, M., Yang, R., Hua, X., Shen, Q., Zhang, W., Zhao, W., 2015. Lactulose production from lactose by recombinant cellobiose 2-epimerase in permeabilised *Escherichia coli* cells. Food Science and Technology 50, 1625–1631.

Wang, Y., 2009. Prebiotics: present and future in food science and technology. Food Research International 42, 8–12.

Wesselius-De Casparis, A., Braadbaart, B., Bergh-Bohlken, G.E., Mimica, M., 1968. Treatment of chronic constipation with lactulose syrup: results of a double-blind study. Gut 9, 84–86.

Wrodnigg, T.M., Eder, B., 2001. The Amadori and Heyns rearrangements: landmark in the history of carbohydrate chemistry or unrecognized synthetic opportunities? Topics in Current Chemistry 215, 115–152.

Yang, R., Liu, F., 2008. Method for Preparing Lactulose by Using Co-immobilized Lactase and Glucose Isomerase. Chinese Patent CN 101182563 B.

Yoon, S.H., Mukerjea, R., Robyt, J.F., 2003. Specificity of yeast (*Saccharomyces cerevisiae*) in removing carbohydrates by fermentation. Carbohydrate Research 338, 1127–1132.

Zhang, Z., Yang, R., Zhang, S., Zhao, H., Hua, X., 2011. Purification of lactulose syrup by using nanofiltration in a diafiltration mode. Journal of Food Engineering 105, 112–118.

Zokaee, F., Kaghazchi, T., Zare, A., Soleimani, M., 2002. Isomerization of lactose to lactulose. Study and comparison of three catalytic systems. Process Biochemistry 37, 629–635.

Chapter 6

Enzymatic Production of Other Lactose-Derived Prebiotic Candidates

C. Guerrero and A. Illanes

Several nondigestible compounds, other than galacto-oligosaccharides (Chapter 4: Enzymatic Production of Galacto-Oligosaccharides) and lactulose (Chapter 5: Enzymatic Production of Lactulose), are derived from lactose, namely lactosucrose, lactitol, lactobionic acid, tagatose, epilactose, and fructosyl-galacto-oligosaccharide, as schematically presented in Fig. 6.1. Most of these compounds, though not being definitely established as prebiotics, display several health-promoting effects and also other functional properties of interest to the food and pharmaceutical industries.

6.1 LACTOSUCROSE

Lactosucrose (O-β-D-galactopyranosyl-(1-4)-O-α-D-glucopyranosyl-(1-2)-β-D-fructo-furanoside) is a mildly sweet (30% of sweetening power of sucrose) nondigestible oligosaccharide (NDO) that despite stimulating bifidobacterial growth is not properly considered a prebiotic (Fujita et al., 2009), although it has the status of a food for specified health use (FOSHU) ingredient in Japan where, in practical terms, it is used as a prebiotic (Seki and Saito, 2012). It has also been claimed that long-term lactosucrose ingestion enhances calcium absorption (Teramoto et al., 2006), prevents obesity (Kimura et al., 2002), and has immunoregulatory effects (Taniguchi et al., 2007). Lactosucrose is also used as a food ingredient by its functional properties, being mildly sweet, bacteriostatic, and noncariogenic (Lifran et al., 2009).

This rare trisaccharide is hardly produced by chemical synthesis and scarcely exists in nature so that enzymatic synthesis is the technologically favored option (Nath et al., 2015).

Lactosucrose can be produced by transfructosylation of lactose with sucrose in an enzyme-catalyzed reaction of synthesis with either levansucrase (sucrose 6-fructotransferase, E.C. 2.4.1.10) (Choi et al., 2004; Li et al., 2015) or β-fructofuranosidase (EC 3.2.1.26) (Pilgrim et al., 2001, 2006; Arakawa et al., 2002). Synthesis is conducted under kinetic control with the concomitant

Lactose-Derived Prebiotics. http://dx.doi.org/10.1016/B978-0-12-802724-0.00006-8

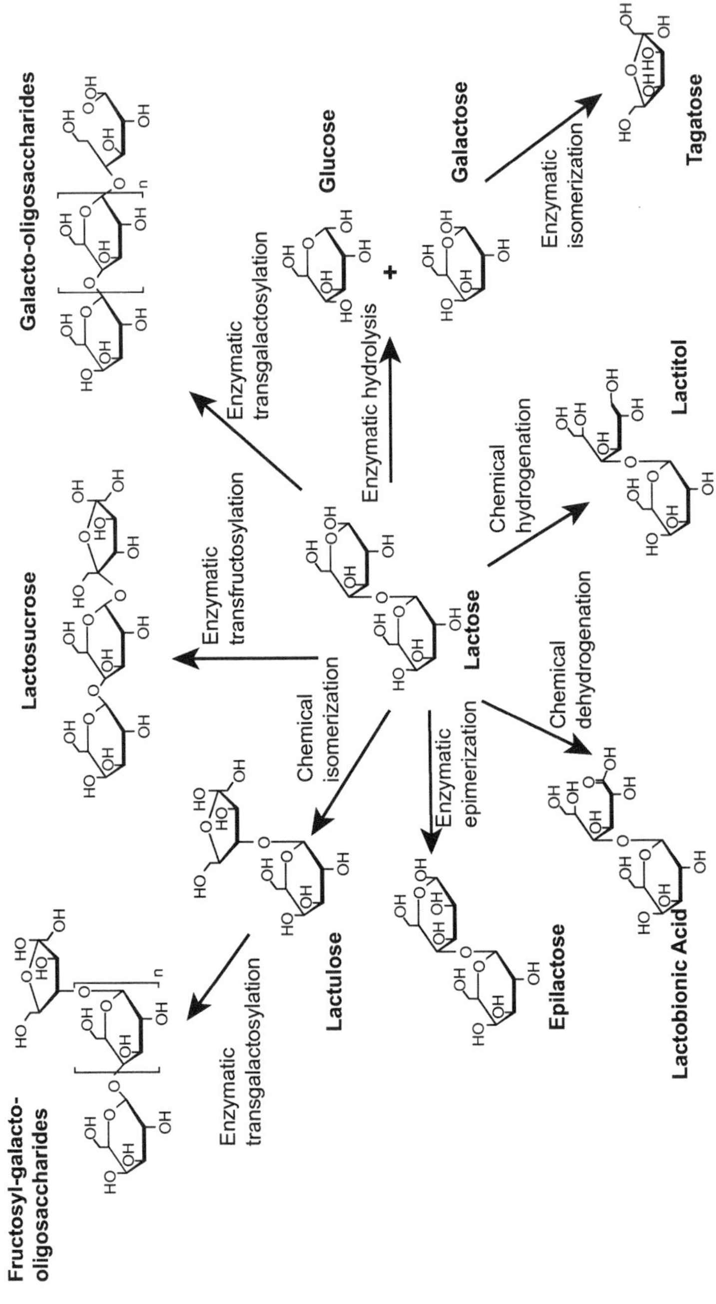

FIGURE 6.1 Lactose-derived nondigestible compounds.

reactions of hydrolysis of the substrates and the product lactosucrose so that yields are below 50% (Pilgrim et al., 2006).

Lactosucrose has also been produced with whole cells of *Bacillus subtilis*, *Geobacillus stearothermophilus*, *Paenibacillus polymyxa*, and *Sterigmatomyces elviae* with levansucrase activity. Best results were obtained with *B. subtilis*, 183 g/L of lactosucrose being produced from a mixture of 225 g/L each of lactose and sucrose (Park et al., 2005), which is a higher concentration than previously reported with β-fructofuranosidase from *Arthrobacter* sp (Pilgrim et al., 2001) and whole cells of *P. polymyxa* harboring levansucrase activity (Choi et al., 2004). A similar concentration of lactosucrose was obtained with a lysate from a recombinant *Escherichia coli* harboring a levansucrase from *Zymomonas mobilis*; in this case, a lactosucrose yield of 29% was obtained, which is lower than the 41% yield reported by Park et al. (2005); however, when using a mixed enzyme system in which glucose oxidase was added to eliminate the glucose formed, lactosucrose yield of 43% was obtained (Han et al., 2009). Lactosucrose was produced in continuous mode of operation with mutant cells of *Sterigmatomyces elviae* immobilized in alginate producing more than 180 g/L in a steady operation for 48 days, which is quite promising from an industrial perspective (Lee et al., 2007).

Synthesis of lactosucrose has also been performed by transgalactosylation of sucrose with lactose using a β-galactosidase from *Bacillus circulans* in a reaction similar to the synthesis of lactulose by transgalactosylation of fructose with lactose (Li et al., 2009); in this case, galacto-oligosaccharide (GOS) formation is inevitable since lactose acts both as galactosyl donor and acceptor so that a mixture of lactosucrose and GOS ensues (Díez-Municio et al., 2014).

Lactosucrose is produced at industrial level in Japan by an enzymatic process using β-fructofuranosidase from *Arthrobacter* sp (Fujita et al., 2009). The Japanese companies Ensuiko Sugar Refining Company and Hayashibara Shoji Inc. are reported producers (Playne and Crittenden, 2009) with many suppliers mostly in Asian countries. The estimated market in Japan was close to 1000 tons per year in 2002 (Nakakuki, 2002), and by 2009 more than 30 FOSHU products containing lactosucrose were in the Japanese market (Mu et al., 2013a). Lactosucrose is also marketed in the United States and to a lesser extent in the European Union where it has been considered as an emerging prebiotic candidate (Rastall, 2010). However, evidence for prebiotic status of lactosucrose is still insufficient, so it remains for the moment as a prebiotic candidate (Roberfroid, 2008). The current market of lactosucrose was estimated around 5000 tons per year (Lifran et al., 2009). Main challenges facing the production of lactosucrose are the increase in yield over 50% and a well-sustained prebiotic condition based on clinical studies in humans.

Updated reviews on lactosucrose production and applications have been published recently (Mu et al., 2013a; Silvério et al., 2015).

6.2 LACTITOL

Lactitol (4-O-β-D-galactopyranosyl-sorbitol) is a nonnatural sugar alcohol obtained from lactose that shares some of the properties of most NDOs since it is not metabolized nor absorbed in the small intestine and stimulates colonic *Bifidobacteria* and *Lactobacilli* (Ballongue et al., 1997; Finney et al., 2007); its effect on colonic bacteria has been considered similar to lactulose (Szilagyi, 2010), but prebiotic effects of lactitol have not been clearly disclosed (Majumder et al., 2011). However, lactitol is mostly used in the food and pharmaceutical industries because of its functional properties. Lactitol is a low-calorie sweetener with some laxative effect, being considered as dietary fiber (Kummel and Brokx, 2001). It has also been suggested as an appetite suppressor (Gee and Johnson, 2005). Traditional food applications refer mostly to its use as sucrose replacer in jams and marmalades, chocolates and candies, and in fine bakery crisp products because of its low hygroscopicity (van Velthuijsen, 1979).

Lactitol ingestion does not increase glucose level, so it is well suited as a food ingredient for diabetics (Drakoularakou et al., 2007). In the pharmaceutical field it has been used as a mild laxative and also in the treatment of hepatic encephalopathy as an alternative to lactulose (Cammà et al., 1993). It has also been used to prevent dental caries, although xylitol is used mostly to that purpose (Szilagyi, 2010). Many nonfood applications for lactitol have been proposed as humectant, plasticizer, lacquer, and hot melt additive (Audic et al., 2003).

Lactitol is produced chemically by hydrogenation of lactose with metal catalysts at high pressures and temperatures that cause the reduction of the glucose moiety in lactose to sorbitol (van Velthuijsen, 1979). Yields are typically higher than 90% and the process has been optimized obtaining yields close to 100% at high productivities when using ruthenium/carbon and sponge nickel catalysts at temperatures from 110°C to 120°C and hydrogen pressures over 50 bar (Kuusisto et al., 2007). The product is recovered by concentration and crystallization, very much as in the process for lactose production (Yang and Silva, 1995). Lactitol is marketed as a concentrated syrup and as a crystalline powder in the form of lactitol monohydrate (Seki and Saito, 2012). It is produced exclusively by chemical synthesis despite some attempts to produce it by biocatalysis (Klewicki, 2007).

Main producers of lactitol are Purac Biochem (The Netherlands), Danisco (Denmark), and Mitsubishi Shoji Foodtech Co (Japan), with a market size of 11,000 tons per year by 2009 (Playne and Crittenden, 2009) and an estimated present market size of 17,000 tons (Lifran et al., 2009). Its low price when compared with other prebiotics or prebiotic candidates makes it an appealing health-promoting lactose-derived product; however, extensive purification will be required by its most sophisticated uses, which may increase its price considerably. Even though it is considered safe and complies with some prebiotic requirements, very few studies have been done on in vitro and in vivo trials (Szilagyi, 2010), so that in the short term its uses will keep on being those described above.

Lactitol derivatives, like lactitol oligosaccharides (Yanahira et al., 1992) and lactitol fatty acid esters (van Velthuijsen, 1979; Drummond and Wells, 1998), have been proposed as valuable products for food and detergent applications, respectively, but its technological impact is doubtful.

6.3 LACTOBIONIC ACID

Lactobionic acid (4-O-β-galactopyranosil-D-gluconate) is a versatile polyhydroxy acid with several applications of industrial relevance (Alonso et al., 2013), which are a consequence of its structure and physicochemical properties. Many of such applications are unrelated to the field of functional foods. One of the most important uses of lactobionic acid refers to the preservation of organs for transplant where it acts as a chelating agent and impermeant reducing oxidative damage and cell swelling (Karam et al., 2005; Lindell et al., 2005); it is thought that lactobionic acid chelates iron and in such a way reduces the oxidative injury to organs during storage (Gänzle et al., 2008). Lactobionic acid is increasingly being used in the cosmetic industry as an antiaging agent in skin-care products (Green et al., 2009) and in the treatment of dermatitis and rosacea (Hatano et al., 2009). Lactobionic acid, being highly biocompatible and biodegradable, has excellent opportunities in the design of targetable drug delivery systems by using it as a ligand for cell surface receptors (Peng et al., 2007; Lin et al., 2009), and in tissue engineering where it favors the entrapment and attachment of cells and the structuration of biocompatible scaffolds (Alonso et al., 2013). Its use in the construction of biocompatible scaffolds for the attachment of hepatocytes to chitosan matrices has been highlighted; in this case, lactobionic acid is covalently coupled to hyaluronic acid to produce a glycosylated derivative that conforms a porous matrix with chitosan, improving the wettability and flexibility of the scaffold for the formation of hepatocyte aggregates with enhanced liver metabolic activities at high cell densities (Fan et al., 2010).

In the pharmaceutical area, lactobionic acid, in the form of calcium lactobionate, is used in calcium supplements (Rehman et al., 2011) and is being used in Japan as a component of new biopharmaceuticals, like equol, which is used in the treatment of osteoporosis and prevention of breast cancer and vascular diseases (Kenichi and Kimura, 2014).

In the chemical industry, it is used as a surfactant or co-builder in biodegradable detergents because of its chelating and emulsifying properties (Yoshimura et al., 2005). Lactobionic acid-based biodegradable surfactants have been developed with improved performance and reduced environmental threat (Bize et al., 2010). It is also being used as a functionalizing agent for novel biomaterials like carbon nanotubes (Feng et al., 2011).

The use of lactobionic acid as a food ingredient refers mostly to its antioxidant, gelling, and acidulant properties (Ruiz Matute et al., 2013). Lactobionic acid is mildly sweet, and being poorly metabolized in the gastrointestinal tract and fermentable by colonic bacteria, it is a potential prebiotic candidate (Saarela et al., 2003); however, its prebiotic condition is still unproven and safety issues have prevented its use as health-promoting additive in functional foods (Gänzle et al., 2008; Schaafsma, 2008). However, its presence has been detected in yogurts containing acetic acid bacteria (Kiryu et al., 2009).

Lactobionic acid is currently produced by the chemical oxidation of the aldehyde group in the glucose residue of lactose in an energy-intensive process requiring expensive catalysts and extensive downstreaming to remove unwanted side products (Kuusisto et al., 2007). Three technological alternatives have been proposed for lactobionic synthesis, namely heterogeneous chemical oxidation, electrochemical oxidation, and biocatalytic oxidation (Gutiérrez et al., 2012a). Dehydrogenation of lactose at high pH catalyzed by noble metals is used commercially (Chia et al., 2008).

Production of lactobionic acid by heterogeneous chemical oxidation of lactose with air or oxygen has been frequently reported using palladium and gold catalysts (Mirescu and Prüße, 2007; Tokarev et al., 2007). The former is prone to poisoning and deactivation by overoxidation, so that oxygen tension in the medium has to be carefully controlled, which is not the case with gold where high stability against overoxidation has been demonstrated. Supported gold catalysts were reported as successful for producing lactobionic acid from lactose with very high selectivity under mild conditions (Murzina et al., 2008), and stoichiometric conversion with high selectivity was reported for the synthesis of lactobionic acid using gold catalysts supported on mesoporous silica particles (Gutiérrez et al., 2012b). Production of lactobionic acid by electrocatalytic oxidation of lactose has also been proposed using noble metals as electrodes, but updated information is scarce and its industrial implementation doubtful (Kokoh and Alonso-Vante, 2006).

Lactobionic acid has also been produced by enzymatic or whole-cell microbial oxidation (Murakami et al., 2004, 2006; Alonso et al., 2011) and claimed to have reached industrial level (Climent et al., 2011). Several *Pseudomonas* strains have been used for lactobionic acid production since they can synthesize it via the lactose oxidation pathway, where a membrane-bound dehydrogenase catalyzes the oxidation of lactose into lactobiono-δ-lactone, which is converted into lactobionic acid by a lactonase (Miyamoto et al., 2000). Fermentation of whey lactose by *Pseudomonas taetrolens* has been suggested as a low-cost and eco-friendly alternative for producing lactobionic acid, but complex and costly downstream operations may render this option unattractive at least in the short term (Alonso et al., 2012). Several other bacterial strains have been proposed as potential lactobionic acid producers. Despite its present complexities, production of lactobionic acid by biocatalysis is considered to be a promising technological approach to meet its increasing demand, and production from whey-derived lactose has been considered attractive because of environmental considerations and cost-effectiveness (Alonso et al., 2013) within a technological platform of whey upgrading (see Chapter 1: Lactose: Production and Upgrading).

Enzymatic catalysis produces higher lactobionic acid titers, yields, and productivities than microbial fermentation, but enzymes are rather unstable under process conditions so the efficiency of catalyst use is low (Nordkvist et al., 2007). Lactose-oxidizing enzymes, like glucose-fructose dehydrogenase, cellobiose dehydrogenase (Van Hecke et al., 2009), and carbohydrate oxidase (Nordkvist et al., 2007), have been used for lactobionic acid production, but technological challenges are many since coenzyme regeneration is required and enzyme inactivation, mostly by the formation of H_2O_2, has to be reduced, which requires the addition of catalase (Hua et al., 2007); in addition, redox mediators required may be incompatible with the intended use of the product (Ludwig et al., 2004).

Cellobiose dehydrogenase has been the most used enzyme for lactobionic acid synthesis; it is an extracellular enzyme produced by several fungi, like *Sclerotium rolfsii*, *Phanerochaete chrysosporium*, and *Trametes versicolor* (Zamocky et al., 2006). The enzyme requires a redox mediator like ABTS (2,2'-azino-bis(3-ethylbenzothiazoline-6-sulphonic acid)) and DCIP (2,6-dichlorophenol indophenol) that can be regenerated by an auxiliary enzyme, like laccase; even though the system produces stoichiometric conversion of lactose into lactobionic acid at a high-volumetric productivity, the system is complex involving two enzymes and a redox mediator (Baminger et al., 2001; Van Hecke et al., 2009). Alternatively, the redox mediator (ABTS) can be regenerated electrochemically, which in principle is simpler and avoids the use of an auxiliary enzyme (Dhariwal et al., 2006). However, the enzymatic process is still complex and costly (Ludwig et al., 2004; Van Hecke et al., 2011), although this situation may change since the development of enzymatic cascade reactions is a technological goal that is intensively being pursued (Grondal et al., 2010; Ricca et al., 2011).

Purification of lactobionic acid has been attempted considering different unit operations, namely, extraction, solvent precipitation, crystallization, chromatography, and electrodialysis; some of them are considered amenable for large-scale operation (Nath et al., 2015).

The most important producers of lactobionic acid are Solvay (Germany) and Friesland Campina Domo (The Netherlands) (Playne and Crittenden, 2009). The market size for lactobionic acid was estimated in 15,000 tons at the end of 2010, with a projected annual increase of 5% (Lifran et al., 2009). The present market has been forecast close to 20,000 tons per year (Gutiérrez et al., 2012a).

6.4 TAGATOSE

D-Tagatose ((3S,4S,5R)-1,3,4,5,6-pentahydroxy-hexan-2-one) is a D-galactose isomer and a D-fructose epimer (Yoshida et al., 2007). It is a rare natural hexose that is found only in trace amounts in some fruits and exudates from plants like *Sterculia setigera*. It is found in small amounts in several processed dairy foods, like powdered milk (800 mg/kg), infant formulas (4–20 mg/kg), hot cocoa (140–1000 mg/kg), cheese (2–20 mg/kg), and yogurt (30 m/kg) (Mendoza et al., 2005; Skytte, 2006). Its striking similarity to D-fructose was on the basis of its development as a low-calorie sweetener, and its marketing as such has expanded considerably since receiving GRAS status in 2000. Tagatose has a sweetness profile similar to sucrose (92% of sweetening power in a 10% w/v solution), but a faster onset, like fructose. It has several beneficial properties over traditional sweeteners: low caloric value (estimated at 1.5 kcal/g), oral health compatibility, minimal glycemic response, and some health-promoting effects (Livesey and Brown, 1996; Levin, 2002; Lu et al., 2008; Vastenavond et al., 2012). GRAS status has allowed its incorporation in a variety of food matrices, like baked goods, ice cream, milk chocolate, yogurt, hard and chewy candies, instant cereals, top sweeteners, cereals, health bars, fruit juices, and soft drinks (Skytte, 2006), where it is used as sweetener, flavor enhancer, texturizer, and body builder. It is also used as flavor enhancer in several nondairy foods, like fruit juices and soft drinks, mint tablets and candies, and as healthy bulk sweetener in health bars and cereals. Allowed levels of use are in most cases below 100 mg/g (Bertelsen et al., 1999; Taylor, 2006; Taylor et al., 2008). Tagatose is particularly attractive as an ingredient in special foods for diabetics since its consumption does not increase glucose blood level (Patra et al., 2009; Roberts and Wright, 2012), and also for toothpaste, mouthwash, and flavored lipsticks for being compatible with oral hygiene (Lu and Levin, 2002).

Regarding nonfood uses, tagatose has been evaluated as cryoprotectant in organ preservation (Sui et al., 2007), in the treatment of obesity and type 2 diabetes (Moore, 2006); it is now in phase II clinical trials to assess its use as a medication for diabetic patients (Beerens et al., 2012) and as antibiofilm agent (Levin and Lu, 2007).

Health-promoting effects of tagatose have attracted much attention in recent years. Being poorly absorbable in the upper digestive tract (Lærke and Jensen, 1999; Buemann et al., 2000) and readily fermentable by colonic *Lactobacilli* producing short-chain fatty acids (SCFAs), it is a candidate prebiotic although not fully complying with all requirements (Patra et al., 2009; Fujimaru et al., 2012); for instance, stimulation of *Bifidobacteria* has not been conclusively proven (Bertelsen et al., 2001), and its use as prebiotic requires an improvement in its storage stability (Dobbs and Bell, 2010).

In this scenario, it is clear that market opportunities for tagatose are tightly related to their functional properties as a highly tolerable noncaloric sweetener; however, its health-promoting effects are indeed a bonus that will contribute to its consolidation within the functional food market (Skytte, 2006; Armstrong et al., 2009).

Being a rare natural hexose, tagatose cannot be produced by extraction from natural products, so it has to be synthesized from chemical precursors. The first economically viable process for tagatose production was patented by Biospherics Incorporated in 1992 based on the isomerization of galactose by chemical catalysts; the process considered the hydrolysis of whey-derived lactose, either by acid or β-galactosidase to yield a glucose–galactose mixture, glucose separation, and isomerization of galactose with alkaline calcium catalyst

(Beadle et al., 1991, 1992); alternatively, aluminates were used as catalysts (Ekeberg et al., 2002). However, the chemical isomerization of galactose is nonspecific leading to the formation of undesirable side products imparting color, off-flavor, and reduction in sweetness; this implies complex downstream processing and waste management (Oh, 2007). Therefore, production of tagatose by biotransformation appears as a sound technology to tackle the problems associated with the chemically catalyzed process.

A first step in this direction was taken by proposing a chemoenzymatic process for tagatose synthesis in which galactose was enzymatically oxidized to D-galactosone with immobilized pyranose 2-oxidase from *Peniophora gigantea* and then chemically reduced to tagatose by hydrogenation with a palladium catalyst; enzymatic oxidation produced hydrogen peroxide as a side product that had to be eliminated by catalase to avoid enzyme inactivation (Freimund et al., 1996). The system was rather complex and costly and yields were not comparable with the chemical synthesis, but the advantage of using enzymes in terms of specificity has been raised. Tagatose in the L configuration was synthesized from sorbose using the highly promiscuous D-tagatose 3-epimerase from *Pseudomonas cichorii*, but yield was only 20% (Itoh and Izumori, 1996; Ito, 2009). An engineered D-tagatose epimerase from *P. cichorii* obtained from a site-saturation mutagenesis library allowed obtaining a variant with significantly higher activity for the epimerization of sorbose into tagatose, but stability under reaction condition was impaired to some extent (Bosshart et al., 2015). Tagatose has been also synthesized by galactitol oxidation with galactose dehydrogenase from *Rhodobacter sphaeroides*; yield of 98% was obtained, but a high enzyme–substrate ratio was required and also a cofactor (NAD^+) regeneration system that used the reduction of pyruvate to lactate catalyzed by lactate dehydrogenase (Huwig et al., 1997). Different microorganisms were able to produce tagatose either from galactitol (Rollini and Manzoni, 2005), D-sorbose (Yoshida et al., 2007), D-psicose (Rao et al., 2008), or L-talitol (Poonperm et al., 2008). However, these are expensive substrates so their use for tagatose production is precluded, even though psicose production from fructose has become industrially feasible; therefore, production of tagatose from fructose via psicose is an interesting technological option (Kim et al., 2006a).

A major breakthrough in tagatose production came from the development of an enzymatic process based on the isomerization of galactose with L-arabinose isomerase (EC 5.3.1.4), which is more specific and environmentally sound than the chemical route (Kim et al., 2001). The enzyme actually catalyzes the isomerization of L-arabinose into L-ribulose, but due to structural similarity it can also act on D-galactose (Cheetham and Wootton, 1993). Several microorganisms have been used as L-arabinose isomerase source, as shown in Table 6.1.

Most of the arabinose isomerases from thermophilic organisms are produced as recombinant protein in *E. coli* hosts and require Co^{+2} as cofactor, which makes them unsuitable for food use, so that expression in GRAS hosts is needed (Kim, 2004). This has inspired the screening and selection of wild microbial strains

TABLE 6.1 L-Arabinose Isomerase–Producing Microorganisms

Microorganism	References
Acidothermus cellulolytics	Cheng et al. (2010a)
Alicyclobacillus acidocaldarius	Lee et al. (2005)
Anoxybacillus flavithermus	Li et al. (2011)
Bacillus halodurans	Lee et al. (2005)
Bacillus licheniformis	Prabhu et al. (2008)
Bacillus stearothermophilus	Cheng et al. (2010b)
Bacillus subtilis	Kim et al. (2010b)
Corynebacterium sp.	Kim et al. (2014a)
Enterococcus faecium	Torres et al. (2014)
Escherichia coli	Roh et al. (2000) Manjasetty and Chance (2006)
Geobacillus stearothermophilus	Ryu et al. (2003) Lee et al. (2005)
Geobacillus thermodenitrificans	Kim and Oh (2005)
Lactobacillus bifermentans	Givry and Duchiron (2008)
Lactobacillus diolivorans	Nath et al. (2015)
Lactobacillus fermentum	Xu et al. (2011)
Lactobacillus gayonii	Yamanaka (1975)
Lactobacillus plantarum	Zhang et al. (2007)
Lactobacillus reuteri	Staudigl et al. (2014)
Lactobacillus sakei	Rhimi et al. (2010)
Mycobacterium smegmatis	Takata et al. (2007)
Shewanella sp.	Rhimi et al. (2011)
Thermotoga neapolitana	Kim et al. (2002) Hong et al. (2007)
Thermotoga maritima	Lee et al. (2004)
Thermoanaerobacter mathranii	Jørgensen et al. (2004)
Thermus sp.	Kim et al. (2003b)

as arabinose isomerase sources; *Enterococcus faecium* appears as a promising strain for arabinose isomerase production since it belongs to the lactic acid bacteria, which are reputedly GRAS organisms (Manzo et al., 2013). *B. subtilis* has been used as a GRAS organism for hosting the arabinose isomerase gene from *G. stearothermophilus* (Cheon et al., 2009). Overexpression of *Bifidobacterium longum* L-arabinose isomerase in *Lactococcus lactis*, being both probiotic strains, is a promising alternative for producing a GRAS arabinose isomerase (Salonen et al., 2012). Tagatose at a yield of 40% was obtained from galactose with arabinose isomerase from the prebiotic strain *Lactobacillus plantarum*; the enzyme has a significantly lower affinity for galactose than for arabinose and the reaction was carried out at moderately low galactose concentration (0.56 M), but the use of a GRAS organism is a significant asset (Zhang et al., 2007).

A patent was issued to the Danish company Arla Foods Amba for the production of tagatose from lactose that considers the enzymatic hydrolysis of lactose into glucose and galactose, the recovery of galactose by chromatographic separation from glucose and unreacted lactose, the conversion of galactose into tagatose with a thermophilic L-arabinose isomerase, and the chromatographic purification of the product (Bertelsen et al., 2006). The patent covered the production using a thermophilic β-galactosidase and L-arabinose isomerases from several thermophilic organisms from the genera *Bacillus*, *Sulfolobus*, *Thermoanaerobacter*, *Thermotoga*, *Thermus*, and *Pyrococcus*.

A novel strategy for production of tagatose from lactose was developed in which recombinant *Pichia pastoris* β-galactosidase from *Arthrobacter chlorophenolicus* was secreted, hydrolyzing lactose with the simultaneous isomerization of galactose by a cold-adapted L-arabinose isomerase; while lactose conversion was 90%, 30% conversion of galactose into tagatose was obtained (Wanarska and Kur, 2012).

Technological challenges for arabinose isomerase use in tagatose industrial production are referred to increased yield (reaction is reversible), substrate specificity (the enzyme is properly an arabinose isomerase, with a higher affinity for arabinose than galactose) (Li et al., 2013), increased thermostability, lower pH optimum, and metal cofactors independency. Equilibrium of reaction has been modified in favor of arabinose by adding boric acid to the reaction medium taking advantage of the higher affinity of boric acid for D tagatose than for D-galactose (Kim et al., 2013a; Salonen et al., 2013). Using this strategy, unprecedented high tagatose conversion (74%) and concentration (370 g/L) were obtained (Lim et al., 2007). Directed evolution has been evaluated as a tool for obtaining improved enzymes for galactose isomerization into tagatose (Kim et al., 2010a).

Directed evolution of a *G. stearothermophilus* L-arabinose isomerase was conducted with the purpose of altering substrate specificity and enhancing activity toward galactose; a 50-fold increase in activity was obtained after two rounds of evolution. Site-directed mutagenesis of *G. stearothermophilus* L-arabinose isomerase cloned in *E. coli* allowed obtaining a threefold increase in catalytic efficiency and a tagatose yield 25% higher than the wild-type enzyme (Oh et al., 2006). A twofold increase in catalytic efficiency and 60% increase in yield with respect to the wild type was reported for a *G. stearothermophilus* mutant L-arabinose cloned in *E. coli* involving changes in three amino acids ($Me^{322}\rightarrow$Val; $Ser^{393}\rightarrow$Thr; $Val^{408}\rightarrow$Ala) (Kim et al., 2006b).

Site-directed mutagenesis of the arabinose isomerase gene from *Lactobacillus fermentum* was targeted to three aspartic acid residues and proved their key role in establishing the pH profile of the enzyme, which is relevant for tagatose production (Xu et al., 2014a). Significant improvement in arabinose isomerase stability at acidic pH was obtained with a recombinant *E. coli* expressing the genes araA from *L. plantarum* (Chouayekh et al., 2007) and *Lactobacillus sakei* (Rhimi et al., 2010) encoding L-arabinose isomerase. Improved activity for galactose isomerization and reduced pH optimum was obtained by site-directed mutagenesis of arabinose isomerase from *Alicyclobacillus hesperidum* (Fan et al., 2015). Independence on cation cofactors in arabinose isomerase thermophilic-producing organisms has been pursued by recombination of the N- and C-terminal regions of *Thermotoga maritima* and *G. stearothermophilus* with the homologous regions of the mesophilic noncation requiring arabinose isomerase from *Bacillus halodurans* (Hong et al., 2011). Arabinose isomerases have higher affinity for arabinose than galactose: highest catalytic efficiency for galactose is one order of magnitude lower than for arabinose (Xu et al., 2014b). However, recent studies based on a substrate-docking homology model allowed obtaining several variants by site-directed mutagenesis of the arabinose isomerase gene of *Geobacillus thermodenitrificans*; a triple-site variant enzyme was obtained having a catalytic efficiency for D-galactose two times higher than for L-arabinose, so this mutant enzyme can be properly designated as galactose epimerase (Kim et al., 2014b). This is undoubtedly a step forward for the enzymatic industrial production of tagatose.

Immobilized L-arabinose isomerases have been used in the synthesis of D-tagatose to increase operational stability and allow biocatalyst recovery. Moderately high yields and high productivities have been reported with a recombinant thermostable enzyme in continuous recycling operation in a packed-bed reactor with an alginate-immobilized thermostable L-arabinose isomerase (Kim et al., 2003a). A significant improvement in thermal stability (stabilization factor higher than 100) and catalytic efficiency was obtained by immobilizing *Bacillus licheniformis* L-arabinose isomerase in alginate (Zhang et al., 2009) and on aminopropyl glass (Zhang et al., 2011), with broader pH and temperature profiles, making them attractive catalysts for the large-scale production of tagatose. Immobilization on Chitopearl of a thermostable L-arabinose isomerase from *Thermotoga neapolitana* cloned in *E. coli* produced a 10-fold increase in enzyme half-life. A substantial improvement in tagatose production was obtained at 70°C when the pH was controlled at 7.5, obtaining tagatose yields of 52%, 47%, and 38% at 100, 300, and 500 g/L of galactose respectively (Lim et al., 2008).

An interesting scheme for tagatose production was proposed in which lactose was converted into tagatose in a one-pot reaction with a thermophilic β-glycosidase from *Sulfolobus solfataricus* and a recombinant thermophilic arabinose isomerase from *Thermoanaerobacter mathranii* expressed in *E. coli*; a further conversion of glucose into fructose was obtained when adding an immobilized commercial preparation of glucose isomerase from *Streptomyces murinus* (Jørgensen et al., 2004). After 24 h of reaction at 65°C at an initial lactose concentration of 0.875 M, a highly sweet product was obtained containing 0.8 M tagatose, 0.15 M galactose, 0.2 M fructose, and 0.2 M glucose, the rest being residual lactose and galacto-oligosaccharides; long-term stability of the enzymes at such high temperature and the use of GRAS microorganisms as hosts of arabinose isomerase were issues still to be addressed.

A process for producing tagatose from soy oligosaccharides was recently patented in which soy oligosaccharides are first hydrolyzed by a α-galactosidase from *Mortierella vinacea*, the hydrolysate is then isomerized with arabinose isomerase from *T. neapolitana* and the resulting sugar mixture is separated by chromatography to obtain purified tagatose; the patent also covered the one-pot conversion of soy oligosaccharides using both enzymes (Kim et al., 2013b).

Alternatively to enzyme biocatalysis, whole-cell biocatalysts containing arabinose isomerase activity have been used for tagatose production (Jung et al., 2005); however, *E.coli*, a frequently used host of arabinose isomerase, cannot be used for tagatose production as food additive where food-grade organisms are required, so that other microbial hosts have been used. The use of natural inhabitants of the colon, like *L. fermentum* (Xu et al., 2014a) and *Pediococcus pentosaceus* strains, are particularly interesting in this respect (Men et al., 2014).

The use of immobilized cells can increase operational stability considerably as demonstrated for a recombinant arabinose isomerase where a stabilization factor of 50 was obtained with respect to free cells in suspension (Zhang et al., 2010). A very promising strategy for a one-step production of tagatose from lactose recently has been reported in which a recombinant *E. coli* was constructed harboring the two vital genes lacZ and araA′ from two different *E. coli* strains (Zhan et al., 2014). At optimal conditions, the recombinant strain hydrolyzed almost all lactose and converted 43% of galactose into tagatose in borate buffer. To do so, optimal pH of arabinose isomerase (8.0) was altered to a value close to the optimal pH of β-galactosidase (6.5) by directed evolution of the araA′ gene. Conversion rate from galactose to tagatose needs to be further improved so that other original arabinose isomerases are being explored, and also enzyme engineering strategies are being conducted to increase the specific activity of the *E. coli* arabinose isomerase. Tagatose also has been produced with immobilized cells of the prebiotic bacterium *L. fermentum* in alginate beads (Xu et al., 2012).

Purification of tagatose is a critical step since conversion is modest because of the reversibility of the reaction so that a significant amount of residual galactose remains at the end of reaction that has to be removed. Purification from the sugar mixture was attained by selective precipitation with calcium hydroxide to form calcium tagatate, which was followed by neutralization and crystallization (Beadle et al., 1991, 1992). Galactose removal can also be achieved by chromatographic separation (Kim et al., 2002; Levin, 2002; Saari et al., 2010); a sustainable process for separation of galactose from tagatose has been devised using Amberlite ion-exchange columns, but even though almost pure tagatose was obtained, recovery was low (Hong et al., 2007). Simulated moving bed chromatography has also been proposed for tagatose purification (Hu et al., 2011; Wagner, 2014). Other strategies for tagatose purification have been proposed: the selective degradation of galactose by fermentation with *Saccharomyces cerevisiae* allowed obtaining a tagatose purity of 95% without any tagatose degradation (Liang et al., 2012); selective recovery of tagatose from a mixture with galactose was also achieved by supercritical carbon dioxide extraction, obtaining under optimized conditions a tagatose purity of 90% with 75% recovery (Montañés et al., 2007), this strategy being extended to the fractionation of complex carbohydrate mixtures (Montañés et al., 2008).

Industrial production of tagatose by galactose isomerization with L-arabinose isomerase is benefiting from the early developments in the mature technology of fructose (high-fructose syrup) production by glucose isomerization with glucose (xylose) isomerase, with which the analogy is clear (Kim, 2004). A review of recent patents for tagatose production was reported by Boudebbouze et al. (2011).

After GRAS status was given to tagatose by the US Food and Drug Administration, approval for its use as food ingredient in the European Union was granted in 2005 by the Food Standards Agency (http://acnfp.food.gov.uk/committee/acnfp/assess/fullapplics/tagatose). Other countries, like Australia, New Zealand, Mexico, Japan, Korea, South Africa, and Canada have also approved their use through their respective regulatory agencies (Dobbs and Bell, 2010). This indeed explains the blooming technological developments for tagatose production. The market is rapidly expanding; global production was estimated at 500 tons in 2009, and a conservative estimate of 3000 tons was forecast for 2015. Once the patent on tagatose owned by Spherix expired in 2007 and production by Arla Foods-Nordzucker joint venture stopped because of insufficient market volume, the Belgian company Nutrilab took the leadership and started producing it by enzymatic technology using spent galactose as raw material. Nutrilab is actively developing a market for tagatose and setting up a plant with a capacity of 10,000 tons per year. Significant cost reduction and compliance with green chemistry principles seem to underlie the commercial success of tagatose production by biocatalysis. More than 50 food products containing tagatose are now in the market (http://www.foodnavigator.com/Market-Trends/Nutrilab-works-towards-a-bright-future-for-tagatose).

6.5 EPILACTOSE

Epilactose (4-O-β-galactopyranosil-D-mannose) is a biologically active lactose epimer. This compound is chemically derived from lactose when milk or lactose are heat- or alkaline-treated (Martínez-Castro and Olano, 1980; de la Fuente et al., 1999). Epilactose formation is markedly increased by microwave heating of milk (Villamiel et al., 1996).

Like other NDOs, epilactose exhibits health-promoting effects including the selective stimulation of beneficial intestinal microbiota and improvement in mineral absorption (Nishimukai et al., 2008; Suzuki et al., 2010; Patel and Goyal, 2012), but most physiological tests have been done with laboratory animals, so it is premature to consider it as a prebiotic. However, in Japan it is considered a promising prebiotic (Watanabe et al., 2008).

Although epilactose can be chemically synthesized by epimerization of lactose, its polyhydroxyl nature makes it extremely difficult, inefficient, and costly, as in other lactose-derived NDOs. Enzymatic synthesis of epilactose from lactose has been conducted with cellobiose 2-epimerase from *Ruminococcus albus* (Ito et al., 2008). Similar activity has been found in other microorganisms, which are listed in Table 6.2, all of them being capable of catalyzing lactose epimerization. The enzyme is not yet available commercially and probably needs further improvement before being of technological significance (Sato et al., 2012), while epilactose needs to be validated regarding its health-promoting effects on humans (Seki and Saito, 2012).

Conversion of lactose into epilactose is low, usually below 30% (Sato et al., 2012; Mu et al., 2013b), so that lactose removal is a major issue in its production. A process was developed considering lactose fractional crystallization, hydrolysis of lactose with β-galactosidase, consumption of monosaccharides

TABLE 6.2 Cellobiose 2-Epimerase–Producing Microorganisms for Epilactose Synthesis

Microorganism	References
Bacteroides fragilis	Senoura et al. (2009)
Caldicellulosiruptor saccharolyticus	Park et al. (2011) Rentschler et al. (2015)
Cellulosilyticum lentocellum	Krewinkel et al. (2015)
Dictyoglomus turgidum	Kim et al. (2012)
Dysgonomonas gadei	Krewinkel et al. (2015)
Eubacterium cellulosolvens	Taguchi et al. (2008)
Flavobacterium johnsoniae	Krewinkel et al. (2014)
Rhodothermus marinus	Ojima et al. (2011)
Ruminococcus albus	Ito et al. (2008) Saburi et al. (2010)
Spirochaeta thermophila	Park et al. (2013)
Teredinibacter turnerae	Ojima et al. (2013)
Thermoanaerobacterium saccharolyticum	Chen et al. (2015)

by yeast fermentation, and column chromatography with cation-exchange resin (Saburi et al., 2010). Epilactose with purity higher than 90% was obtained with 43% yield of epilactose recovery; practical significance of this purification scheme remains to be assessed.

Improvement in cellobiose 2-epimarase is still required for epilactose industrial production, with increased thermal and acid pH stability and higher affinity for lactose traits to be targeted. Protein engineering strategies will be valuable tools once the three-dimensional structure of the enzyme is elucidated (Ito et al., 2009; Mu et al., 2013b).

6.6 FRUCTOSYL-GALACTO-OLIGOSACCHARIDES

Fructosyl-galacto-oligosaccharides (fGOSs) are NDOs produced by the β-galactosidase-catalyzed transgalactosylation of lactulose and are known also as lactulose-derived oligosaccharides (Cardelle-Cobas et al., 2008a; Guerrero et al., 2013). They were reported for the first time by Martínez-Villaluenga et al. (2008), who identified the structure of two trisaccharides derived from lactulose that were synthesized by *Kluyveromyces lactis* β-galactosidase. The structure of fGOS is similar to GOS, but having a fructose instead of glucose as sugar moiety at the reducing end. Due to this striking similarity, some authors have suggested to extend the concept of GOS to include fGOS (Cardelle-Cobas et al., 2008a; Martínez-Villaluenga et al., 2008; Olano and Corzo, 2009; Cardelle-Cobas et al., 2011a). As seen in Table 6.3, synthesis of fGOS has recently

TABLE 6.3 β-Galactosidase-Producing Microorganisms for Fructosyl-Galacto-Oligosaccharides Synthesis

Origin of β-Galactosidases	References
Kluyveromyces lactis	Martínez-Villaluenga et al. (2008) Hernández-Hernández et al. (2011) Cardelle-Cobas et al. (2011a) Padilla et al. (2012) Guerrero et al. (2013) Guerrero et al. (2015)
Aspergillus aculeatus	Cardelle-Cobas et al. (2008a) Hernández-Hernández et al. (2011) Guerrero et al. (2015)
Aspergillus oryzae	Hernández-Hernández et al. (2011) Guerrero et al. (2013) Guerrero et al. (2015)
Bacillus circulans	Guerrero et al. (2013) Guerrero et al. (2015)
Aspergillus niger	Guerrero et al. (2015)

received much attention, and different enzymes and reaction conditions have been tested to maximize fGOS yield.

During the synthesis of fGOS from lactulose, trisaccharides are mostly produced when *K. lactis* β-galactosidase is used (Martínez-Villaluenga et al., 2008), while di-, tri-, and higher oligosaccharides are produced with *Aspergillus aculeatus* and *Aspergillus oryzae* β-galactosidase (Cardelle-Cobas et al., 2008a; Guerrero et al., 2013). As previously mentioned, different types of linkages among the monosaccharide units can be formed according to the enzyme used; therefore, different synthetic products can be generated in the synthesis of lactulose-derived oligosaccharides (Martínez-Villaluenga et al., 2008). Synthesis of fGOS with *A. aculeatus* β-galactosidase produced mostly β 1–6 linkages, while with *K. lactis* β-galactosidase, β 1–6 and β 1–1 linkages were formed; fGOS yields were 0.21 and 0.28 g of fGOS/g of lactulose, respectively (Hernández-Hernández et al., 2011; Cardelle-Cobas et al., 2012).

Padilla et al. (2012) evaluated the synthesis of fGOS with *K. lactis* and *Kluyveromyces marxianus* β-galactosidases isolated from artisanal cheese, reporting yields of 0.35 and 0.42 g of fGOS/g of lactulose, respectively, being di-, tri-, and tetrasaccharides with β 1–6 and β 1–1 bonds the main products formed in both cases. Cardelle-Cobas et al. (2008a) studied the effect of operational parameters (temperature, pH, and lactulose initial concentration and enzyme concentration) on fGOS synthesis with *K. lactis* β-galactosidase; except for pH, all variables studied had a significant effect on fGOS yield. Guerrero et al. (2013) evaluated the synthesis of fGOS with three commercial preparations from *A. oryzae*, *B. circulans*, and *K. lactis* at different temperatures and lactulose initial concentrations; fGOS yield was not affected by temperature in the range studied, while yield slowly decreased at lactulose concentrations higher than 50% w/w; however, both variables had a significant effect on the specific productivity of fGOS synthesis. The origin of the enzyme had a strong effect on both yield and specific productivity; yields of 0.36, 0.14, and 0.192 g of fGOS/g of lactulose and specific productivities of 0.57, 0.055, and 0.011 g of fGOS/h/mg of protein, were obtained with *A. oryzae*, *B. circulans*, and *K. lactis* β-galactosidases, respectively. As seen, specific productivity obtained with the *A. oryzae* enzyme was 28 and 11 times higher than with *B. circulans* and *K. lactis* enzymes, respectively (Guerrero et al., 2013). The same authors compared the yields obtained with 11 commercial β-galactosidase preparations of different origin in the synthesis of lactulose, fGOS, and GOS, showing the strong effect of the enzyme origin on product yield. Based on their results, the β-galactosidase from *A. oryzae* was selected for fGOS synthesis; the yield obtained was 30% higher than in GOS synthesis at the same operational conditions. On the contrary, with the *B. circulans* enzyme fGOS yield was 61% lower than obtained in GOS synthesis (Guerrero et al., 2015). This means that the β-galactosidase from *A. oryzae* has a clear advantage over the others for fGOS synthesis, which is not the case for GOS synthesis where the β-galactosidases from *B. circulans* and *A. oryzae* are competitive with each other (see Chapter 4: Enzymatic Production of Galacto-Oligosaccharides).

With the purpose of gaining insight in the reaction of fGOS synthesis, Rodriguez-Fernandez et al. (2011) proposed a model describing the mechanism of synthesis of lactulose-derived oligosaccharides, which is quite similar to mechanisms previously reported for GOS synthesis from lactose.

It is known that the chemical structure of NDOs (number and type of hexoses, position and type of bonds between them, and polymerization degree) are important determinants with respect to their health-promoting effects and their fermentability by colonic bacteria (Martínez-Villaluenga et al., 2008; Olano and Corzo, 2009; Hernández-Hernández et al., 2011). Therefore, there is an intense search for new NDOs as potential prebiotics (Cardelle-Cobas et al., 2011a). As lactulose is a prebiotic NDO and a therapeutic drug (see Section 5.6 in Chapter 5, Enzymatic Production of Lactulose), molecules derived from it open up the possibility for obtaining products with improved properties or targeted functionality, useful both in the food and pharmaceutical sectors (Martínez-Villaluenga et al., 2008; Cardelle-Cobas et al., 2008a; Olano and Corzo, 2009).

López-Sanz et al. (2015) proved that fGOSs are stable during storage and processing conditions and can be incorporated in food matrices, like milk and fruit juices, in the pH range from 3.4 to 6.8. They can be incorporated in foods for diabetics and lactose-intolerant people, and their health-promoting effects make them interesting prebiotic candidates.

Prebiotic properties of fGOS have been assessed in anaerobic fermentations with probiotic strains in pure cultures and mixed fecal populations. Cardelle-Cobas et al. (2011b) compared the prebiotic effect of fGOS and GOS with similar patterns of β-glycosidic bonds, obtaining slightly higher cell growth of *Lactobacillus*, *Streptococcus*, and *Bifidobacterium* with fGOS. Likewise, the stimulatory effect of fGOS on probiotic *Bifidobacterium animalis* in cecum and colon has been reported, but no significant differences were observed with respect to GOS (Marín-Manzano et al., 2013). Cardelle-Cobas et al. (2012) evaluated the prebiotic effect of GOS, fGOS, lactulose, and Vivinal-GOS (commercial product with about 50% GOS) in pH-controlled anaerobic batch culture, determining the variations in the intestinal microbiota and in the production of SCFA. It was determined that fGOS were selectively consumed by *Bifidobacterium* and lactic acid bacterial populations, producing high amounts of SCFA, higher than with GOS, and presented a higher prebiotic score than GOS at 24 h of culture.

Increase in yields of fGOS synthesis has been pursued using a two-step chemoenzymatic process, in which GOS is enzymatically produced in the first step and then the reacted mixture is chemically isomerized using sodium aluminate as catalyst. In this way, GOS, lactose, glucose, and galactose yielded fGOS, lactulose, fructose, and tagatose with isomerization yield of 6′-galactosyl-lactose into 6′-galactosyl-lactulose higher than 60% (Cardelle-Cobas et al., 2008b). This product is certainly interesting as a potential health promoter and functional sweetener.

Another strategy proposed for the synthesis of fGOS considers a first step of alkaline isomerization of lactose in whey permeate, in which a mixture of lactulose and residual lactose is obtained with a yield of 0.161 g lactulose/g lactose, and a second step catalyzed by *B. circulans* β-galactosidase, in which both lactose and lactulose act as acceptors of transgalactosylated galactose yielding a mixture of GOS and fGOS with a 50% yield of transgalactosylated oligosaccharides (Corzo-Martínez et al., 2012).

Undoubtedly, fGOS is an interesting new family of candidate prebiotic NDOs whose application within the health-food sector is yet to be developed.

REFERENCES

Alonso, S., Rendueles, M., Díaz, M., 2011. Efficient lactobionic acid production from whey by *Pseudomonas taetrolens* under pH-shift conditions. Bioresource Technology 102, 9730–9736.

Alonso, S., Rendueles, M., Díaz, M., 2012. Role of dissolved oxygen availability on lactobionic acid production from whey by *Pseudomonas taetrolens*. Bioresource Technology 109, 140–147.

Alonso, S., Rendueles, M., Díaz, M., 2013. Bio-production of lactobionic acid: current status, applications and future prospects. Biotechnology Advances 31, 1275–1291.

Arakawa, K., Aoyama, Y., Ikeda, H., Mikuni, K., Fujita, K., Hara, K., 2002. The development of lactosucrose production and its applications in foods for specified health use. Journal of Applied Glycoscience 49, 63–72.

Armstrong, L.M., Luecke, K.J., Bell, L.N., 2009. Consumer evaluation of bakery product flavour as affected by incorporating the prebiotic tagatose. International Journal of Food Science and Technology 44, 815–819.

Audic, J.L., Chaufer, B., Dauphin, G., 2003. Non-food applications of milk components and dairy co-products: a review. Lait 83, 417–438.

Ballongue, J., Schumann, C., Quignon, P., 1997. Effects of lactulose and lactitol on colonic microflora and enzymatic activity. Scandinavian Journal of Gastroenterology 222, 41–44.

Baminger, U., Ludwig, R., Galhaup, C., Leitner, C., Kulbe, K.D., Haltrich, D., 2001. Continuous enzymatic regeneration of redox mediators used in biotransformation reactions employing flavoproteins. Journal of Molecular Catalysis B: Enzymatic 11, 541–550.

Beadle, J.R., Saunders, J.P., Wajda, T.J., 1991. Process for Manufacturing Tagatose. United States Patent 5002612.

Beadle, J.R., Saunders, J.P., Wajda, T.J., 1992. Process for Manufacturing Tagatose. United States Patent 5078796.

Beerens, K., Desmet, T., Soetaert, W., 2012. Enzymes for the biocatalytic production of rare sugars. Journal of Industrial Microbiology and Biotechnology 39, 823–834.

Bertelsen, H., Jensen, B.B., Buemann, B., 1999. D-Tagatose-a novel low-calorie bulk sweetener with prebiotic properties. World Review on Nutrition and Dietetics 85, 98–109.

Bertelsen, H., Andersen, H., Tvede, M., 2001. Fermentation of D-tagatose by human intestinal bacteria and dairy lactic acid bacteria. Microbial Ecology in Health and Disease 13, 87–95.

Bertelsen, H., Eriknauer, K., Bottcher, K., Christensen, J.S., Stougaard, P., Hansen, O.C., Jorgensen, F., 2006. Process for Manufacturing Tagatose. United States Patent 69912923.

Bize, C., Blanzat, M., Rico-Lattes, I., 2010. Self-assembled structures of catanionic associations: how to optimize vesicle formation? Journal of Surfactants and Detergents 13, 465–473.

Bosshart, A., Wagner, N., Lei, L., Panke, S., Bechtold, M., 2015. Highly efficient production of rare sugars D-psicose and L-tagatose by two engineered D-tagatose epimerases. Biotechnology and Bioengineering. http://dx.doi.org/10.1002/bit.25547.

Boudebbouze, S., Maguin, E., Rhimi, M., 2011. Bacterial L-arabinose isomerases: industrial application for D-tagatose production. Recent Patents on DNA and Gene Sequences 5, 194–201.

Buemann, B., Toubro, S., Raben, A., Blundell, J., Astrup, A., 2000. The acute effect of D-tagatose on food intake in human subjects. British Journal of Nutrition 84, 227–231.

Cammà, C., Fiorello, F., Tinè, F., Marchesini, G., Fabbri, A., Pagliaro, L., 1993. Lactitol in the treatment of hepatic encephalopathy. Digestive Diseases and Science 38, 916–922.

Cardelle-Cobas, A., Martinez-Villaluenga, C., Villamiel, M., Olano, A., Corzo, N., 2008a. Synthesis of oligosaccharides derived from lactulose and pectinex ultra SP-L. Journal of Agricultural and Food Chemistry 56, 3328–3333.

Cardelle-Cobas, A., Corzo, N., Villamiel, M., Olano, A., 2008b. Isomerization of lactose-derived oligosaccharides: a case study using sodium aluminate. Journal of Agricultural and Food Chemistry 56, 10954–10959.

Cardelle-Cobas, A., Corzo, N., Martínez-Villaluenga, C., Agustín, O., Villamiel, M., 2011a. Effect of reaction conditions on lactulose-derived trisaccharides obtained by transgalactosylation with β-galactosidase of *Kluyveromyces lactis*. European Food Research and Technology 233, 89–94.

Cardelle-Cobas, A., Corzo, N., Olano, A., Peláez, C., Requena, T., Ávila, M., 2011b. Galactooligosaccharides derived from lactose and lactulose: influence of structure on *Lactobacillus, Streptococcus* and *Bifidobacterium* growth. International Journal of Food Microbiology 149, 81–87.

Cardelle-Cobas, A., Olano, A., Corzo, N., Villamiel, M., Collins, M., Kolida, S., Rastall, R.A., 2012. In vitro fermentation of lactulose-derived oligosaccharides by mixed fecal microbiota. Journal of Agricultural and Food Chemistry 60, 2024–2032.

Cheetham, P.S.J., Wootton, A.N., 1993. Bioconversion of D-galactose into D-tagatose. Enzyme and Microbial Technology 15, 105–108.

Cheng, L., Mu, W., Zhang, T., Jiang, B., 2010a. An L-arabinose isomerase from *Acidothermus cellulolytics* ATCC 43068: cloning, expression, purification, and characterization. Applied Microbiology and Biotechnology 86, 1089–1097.

Cheng, L., Mu, W., Jiang, B., 2010b. Thermostable L-arabinose isomerase from *Bacillus stearothermophilus* IAM 11001 for D-tagatose production: gene cloning, purification and characterization. Journal of the Science of Food and Agriculture 90, 1327–1333.

Chen, Q., Zhang, W., Zhang, T., Jiang, B., Mu, W., 2015. Characterization of an epilactose-producing cellobiose 2-epimerase from *Thermoanaerobacterium saccharolyticum*. Journal of Molecular Catalysis B: Enzymatic 116, 39–44.

Cheon, J., Kim, S.B., Park, S.W., Han, J.K., Kim, P., 2009. Characterization of L-arabinose isomerase in *Bacillus subtilis*, a GRAS host, for the production of edible tagatose. Food Biotechnology 23, 8–16.

Chia, Y.N., Latusek, M.P., Holles, J.H., 2008. Catalytic wet oxidation of lactose. Industrial Engineering Chemical Research 47, 4049–4055.

Choi, H.J., Kim, C.S., Kim, P., Jung, H.C., Oh, D.K., 2004. Lactosucrose bioconversion from lactose and sucrose by whole cells of *Paenibacillus polymyxa* harboring levansucrase activity. Biotechnology Progress 20, 1876–1879.

Chouayekh, H., Bejar, W., Rhimi, M., Jelleli, K., Mseddi, M., Bejar, S., 2007. Characterization of an L-arabinose isomerase from the *Lactobacillus plantarum* NC8 strain showing pronounced stability at acidic pH. FEMS Microbiology Letters 277, 260–267.

Climent, M.J., Corma, A., Iborra, S., 2011. Converting carbohydrates to bulk chemicals and fine chemicals over heterogeneous catalysts. Green Chemistry 13, 520–540.

Corzo-Martínez, M., Copoví, P., Olano, A., Moreno, F.J., Montilla, A., 2012. Synthesis of prebiotic carbohydrates derived from cheese whey permeate by a combined process of isomerisation and transgalactosylation. Journal of the Science of Food and Agriculture 93, 1591–1597.

Dhariwal, A., Mavrov, V., Schroeder, I., 2006. Production of lactobionic acid with process integrated electrochemical enzyme regeneration and optimisation of process variables using response surface methods (RSM). Journal of Molecular Catalysis B: Enzymatic 42, 64–69.

Díez-Municio, M., Herrero, M., Olano, A., Moreno, F.J., 2014. Synthesis of novel bioactive lactose-derived oligosaccharides by microbial glycoside hydrolases. Microbial Biotechnology 7, 315–331.

Dobbs, C.M., Bell, L.N., 2010. Storage stability of tagatose in buffer solutions of various compositions. Food Research International 43, 382–386.

Drakoularakou, A., Hasselwander, O., Ouwehand, A., 2007. Lactitol, an emerging prebiotic: functional properties with a focus on digestive health. Food Science and Technology Bulletin: Functional Foods 3, 71–80.

Drummond, C.J., Wells, D., 1998. Nonionic lactose and lactitol based surfactants: comparison of some physico-chemical properties. Colloids and Surfaces A: Physicochemical and Engineering Aspects 141, 131–142.

Ekeberg, D., Morgenlie, S., Stenstrøm, Y., 2002. Base catalysed isomerisation of aldoses of the *arabino* and *lyxo* series in the presence of aluminate. Carbohydrate Research 337, 779–786.

de la Fuente, M.A., Juárez, M., de Rafael, D., Villamiel, M., Olano, A., 1999. Isomerization of lactose catalyzed by alkaline-substituted sepiolites. Food Chemistry 66, 301–306.

Fan, J., Shang, Y., Yuan, Y., Yang, J., 2010. Preparation and characterization of chitosan/galactosylated hyaluronic acid scaffolds for primary hepatocytes culture. Journal of Material Sciences – Materials in Medicine 21, 319–327.

Fan, C., Xu, W., Zhang, T., Zhou, L., Jiang, B., Mu, W., 2015. Engineering of *Alicyclobacillus hesperidum* L-arabinose isomerase for improved catalytic activity and reduced pH optimum using random and site-directed mutagenesis. Applied Biochemistry and Biotechnology. http://dx.doi.org/10.1007/s12010-015-1828-3.

Feng, W., Jing, X., Li, L., Ji, P., 2011. Protein adsorption on functionalized carbon nanotubes with a lactobionic amide amphiphile. Industrial and Engineering Chemistry Research 50, 11608–11613.

Finney, M., Smullen, J., Foster, H.A., Brokx, S., Storey, D.M., 2007. Effects of low doses of lactitol on faecal microflora, pH, short chain fatty acids and gastrointestinal symptomology. European Journal of Nutrition 46, 307–314.

Freimund, S., Huwig, A., Giffhorn, F., Köpper, S., 1996. Convenient chemo-enzymatic synthesis of D-tagatose. Journal of Carbohydrate Chemistry 15, 115–120.

Fujimaru, T., Park, J.H., Lim, J., 2012. Sensory characteristics and relative sweetness of tagatose and other sweeteners. Journal of Food Science 77, S323–S328.

Fujita, K., Ito, T., Kishino, E., 2009. Characteristics and applications of lactosucrose. Proceedings of the Research Society of Japan Sugar Refineries' Technologists 57, 13–21.

Gänzle, M.G., Haase, G., Jelen, P., 2008. Lactose: crystallization, hydrolysis and value-added derivatives. International Dairy Journal 18, 685–694.

Gee, J.M., Johnson, I.T., 2005. Dietary lactitol fermentation increases circulating peptide YY and glucagon-like peptide-1 in rats and humans. Nutrition 21, 1036–1043.

Givry, S., Duchiron, F., 2008a. Optimization of culture medium and growth conditions for production of L-arabinose isomerase and D-xylose isomerase by *Lactobacillus bifermentans*. Microbiology 77, 281–287.

Green, B.A., Yu, R.J., Van Scott, E.J., 2009. Clinical and cosmeceutical uses of hydroxyacids. Clinics in Dermatology 27, 495–501.

Grondal, C., Jeanty, M., Enders, D., 2010. Organocatalytic cascade reactions as a new tool in total synthesis. Nature Chemistry 2, 167–178.

Guerrero, C., Vera, C., Illanes, A., 2013. Optimisation of synthesis of oligosaccharides derived from lactulose (fructosyl-galacto-oligosaccharides) with β-galactosidases of different origin. Food Chemistry 138, 2225–2232.

Guerrero, C., Vera, C., Conejeros, R., Illanes, A., 2015. Transgalactosylation and hydrolytic activities of commercial preparations of β-galactosidase for the synthesis of prebiotic carbohydrates. Enzyme and Microbial Technology 70, 9–17.

Gutiérrez, L.F., Hamoudi, S., Belkacemi, K., 2012a. Lactobionic acid: a high value-added lactose derivative for food and pharmaceutical applications. International Dairy Journal 26, 103–111.

Gutiérrez, L.F., Hamoudi, S., Belkacemi, K., 2012b. Effective gold catalyst supported on mesoporous silica decorated by ceria for the synthesis of high value lactobionic acid. Applied Catalysis A: General 425–426, 213–223.

Han, W.C., Byun, S.H., Kim, M.H., Sohn, E.H., Lim, J.D., Um, B.H., Kim, C.H., Kang, S.A., Jang, K.H., 2009. Production of lactosucrose from sucrose and lactose by a levansucrase from *Zymomonas mobilis*. Journal of Microbiology and Biotechnology 19, 1153–1160.

Hatano, Y., Man, M.K., Uchida, Y., Crumrine, D., Scharschmidt, T.C., Kim, E.G., Mauro, T.M., Feingold, K.R., Elias, P.M., Holleran, W.M., 2009. Maintenance of an acidic stratum corneum prevents emergence of murine atopic dermatitis. Journal of Investigative Dermatology 129, 1824–1835.

Hernández-Hernández, O., Montañés, F., Clemente, A., Moreno, F.J., Sanz, M.L., 2011. Characterization of galactooligosaccharides derived from lactulose. Journal of Chromatography A 1218, 7691–7696.

Hong, Y.H., Lee, D.W., Lee, S.J., Choe, E.A., Kim, S.B., Lee, Y.H., Cheigh, C.I., Pyun, Y.R., 2007. Production of D-tagatose at high temperatures using immobilized *Escherichia coli* cells expressing L-arabinose isomerase from *Thermotoga neapolitana*. Biotechnology Letters 29, 569–574.

Hong, Y.H., Lee, D.W., Pyun, Y.R., Lee, S.H., 2011. Creation of metal-independent hyperthermophilic L-arabinose isomerase by homologous recombination. Journal of Agricultural and Food Chemistry 59, 12939–12947.

Hu, C., Li, L., Zheng, Y., Rui, L., Hu, C., 2011. Perspectives of biotechnological production of L-ribose and its purification. Applied Microbiology and Biotechnology 92, 449–455.

Hua, L., Nordkvist, M., Nielsen, P.M., Villadsen, J., 2007. Scale-up of enzymatic production of lactobionic acid using the rotary jet head system. Biotechnology and Bioengineering 97, 842–849.

Huwig, A., Emmel, S., Jäkel, G., Giffhorn, F., 1997. Enzymatic synthesis of L-tagatose from galactitol with galactitol dehydrogenase from *Rhodobacter sphaeroides* D. Carbohydrate Research 305, 337–339.

Ito, S., Taguchi, H., Hamada, S., Kawauchi, S., Ito, H., Senoura, T., Watanabe, J., Nishimukai, M., Ito, S., Matsui, H., 2008. Enzymatic properties of cellobiose 2-epimerase from *Ruminococcus albus* and the synthesis of rare oligosaccharides by the enzyme. Applied Microbiology and Biotechnology 79, 433–441.

Ito, S., 2009. Features and applications of microbial sugar epimerases. Applied Microbiology and Biotechnology 84, 1053–1060.

Ito, S., Hamada, S., Ito, H., Matsui, H., Ozawa, T., Taguchi, H., Ito, S., 2009. Site-directed mutagenesis of possible catalytic residues of cellobiose 2-epimerase from *Ruminococcus albus*. Biotechnology Letters 31, 1065–1071.

Itoh, H., Izumori, K., 1996. Enzymatic production of L-tagatose and L-fructose from L-sorbose and L-psicose, respectively. Journal of Fermentation and Bioengineering 81, 351–353.

Jørgensen, F., Hansen, O.C., Stougaard, P., 2004. Enzymatic conversion of D-galactose to D-tagatose: heterologous expression and characterisation of a thermostable L-arabinose isomerase from *Thermoanaerobacter mathranii*. Applied Microbiology and Biotechnology 64, 816–822.

Jung, E.S., Kim, H.J., Oh, D.K., 2005. Tagatose production by immobilized recombinant *Escherichia coli* cells containing *Geobacillus stearothermophilus* L-arabinose isomerase mutant in a packed-bed bioreactor. Biotechnology Progress 21, 1335–1340.

Karam, G., Compagnon, P., Hourmant, M., Despins, P., Duveau, D., Noury, D., Boudjema, K., 2005. A single solution for multiple organ procurement and preservation. Transplant International 18, 657–663.

Kenichi, O.E., Kimura, T., 2014. Equol Production Accelerating Composition. US Patent 8822432 B2.

Kim, P., Yoon, S.H., Seo, M.J., Oh, D.K., Choi, J.H., 2001. Improvement of tagatose conversion rate by genetic evolution of thermostable galactose isomerase. Biotechnology and Applied Biochemistry 34, 99–102.

Kim, B.C., Lee, Y.H., Lee, H.S., Lee, D.W., Choe, E.A., Pyun, Y.R., 2002. Cloning, expression and characterization of L-arabinose isomerase from *Thermotoga neapolitana*: bioconversion of D-galactose to D-tagatose using the enzyme. FEMS Microbiology Letters 212, 121–126.

Kim, H.J., Ryu, S.A., Kim, P., Oh, D.K., 2003a. A feasible enzymatic process for D-tagatose production by an immobilized thermostable L-arabinose isomerase in a packed-bed bioreactor. Biotechnology Progress 19, 400–404.

Kim, J.W., Kim, Y.W., Roh, H.J., Kim, H.Y., Cha, J.H., Park, K.H., Park, C.S., 2003b. Production of tagatose by a recombinant thermostable L-arabinose isomerase from *Thermus* sp. IM6501. Biotechnology Letters 25, 963–967.

Kim, P., 2004. Current studies on biological tagatose production using L-arabinose isomerase: a review and future perspective. Applied Microbiology and Biotechnology 65, 243–249.

Kim, H.J., Hyun, E.K., Kim, Y.S., Lee, Y.J., Oh, D.K., 2006a. Characterization of an *Agrobacterium tumefaciens* D-psicose 3-epimerase that converts D-fructose to D-psicose. Applied and Environmental Microbiology 72, 981–985.

Kim, H.J., Kim, J.H., Oh, H.J., Oh, D.K., 2006b. Characterization of a mutated *Geobacillus stearothermophilus* L-arabinose isomerase that increases the production rate of D-tagatose. Journal of Applied Microbiology 101, 213–221.

Kim, H.J., Uhm, T.G., Kim, S.B., Kim, P., 2010a. *Escherichia coli* arabinose isomerase and *Staphylococcus aureus* tagatose-6-phosphate isomerase: which is a better template for directed evolution of non-natural substrate isomerization? Journal of Microbiology and Biotechnology 20, 1018–1021.

Kim, J.H., Prabhu, P., Jeya, M., Tiwari, M.K., Moon, H.J., Singh, R.K., Lee, J.K., 2010b. Characterization of an L-arabinose isomerase from *Bacillus subtilis*. Applied Microbiology and Biotechnology 85, 1839–1847.

Kim, S.B., Park, S.W., Song, S.H., Lee, K.P., Oh, D.K., Lim, B.C., Kim, H.J., 2013a. Manufacturing Method of Tagatose Using Galactose Isomerization of High Yield. US Patent 8481720.

Kim, S.B., Kim, J.H., Lee, Y.M., Kim, J.H., Park, S.W., Lee, K.P., 2013b. Process for Manufacturing Tagatose Using Soy Oligosaccharide. US Patent 8377658.

Kim, B.J., Hong, S.H., Shin, K.C., Jo, Y.S., Oh, D.K., 2014b. Characterization of a F280N variant of L-arabinose isomerase from *Geobacillus thermodenitrificans* identified as a D-galactose isomerase. Applied Microbiology and Biotechnology 98, 9271–9281.

Kim, S.B., Lee, Y.M., Park, S.W., Kim, J.H., Song, S.H., Lee, K.P., Kim, H.W., Choi, H.J., 2014a. Arabinose Isomerase Expressed from Corynebacterium Genus and Tagatose Manufacturing Method by Using it. US Patent 8802393.

Kim, H.J., Oh, D.K., 2005. Purification and characterization of an L-arabinose isomerase from an isolated strain of *Geobacillus thermodenitrificans* producing D-tagatose. Journal of Biotechnology 120, 162–173.

Kimura, Y., Nagata, Y., Bryant, C.W., Buddington, R.K., 2002. Nondigestible oligosaccharides do not increase accumulation of lipid soluble environmental contaminants by mice. The Journal of Nutrition 132, 80–87.

Kiryu, T., Kiso, T., Nakano, H., Ooe, K., Kimura, T., Murakami, H., 2009. Involvement of *Acetobacter orientalis* in the production of lactobionic acid in Caucasian yogurt ("Caspian Sea yogurt") in Japan. Journal of Dairy Science 92, 25–34.

Klewicki, R., 2007. The stability of gal-polyols and oligosaccharides during pasteurization at a low pH. LWT – Food Science and Technology 40, 1259–1265.

Kokoh, K.B., Alonso-Vante, N., 2006. Electrocatalytic oxidation of lactose on gold nanoparticle modified carbon in carbonate buffer. Journal of Applied Electrochemistry 36, 147–151.

Kummel, K.F., Brokx, S., 2001. Lactitol as a functional prebiotic. Cereal Foods World 46, 424–429.

Kuusisto, J., Tokarev, A.V., Murzina, E.V., Roslund, M.U., Mikkola, J.P., Murzin, D.Y., Salmi, T., 2007. From renewable raw materials to high value-added fine chemicals-catalytic hydrogenation and oxidation of D-lactose. Catalysis Today 121, 92–99.

Krewinkel, M., Kaiser, J.M.M., Rentschler, E., Kuschel, B., Hinrichs, J., Fischer, L., 2015. Novel cellobiose 2-epimerases for the production of epilactose from milk ultrafiltrate containing lactose. Journal of Dairy Science 98, 3665–3678.

Kim, J.E., Kim, Y.S., Kang, L.W., Oh, D.K., 2012. Characterization of a recombinant cellobiose 2-epimerase from *Dictyoglomus turgidum* that epimerizes and isomerizes β-1,4- and α-1,4-gluco-oligosaccharides. Biotechnology Letters 34, 2061–2068.

Krewinkel, M., Gosch, M., Rentschler, E., Fischer, L., 2014. Epilactose production by 2 cellobiose 2-epimerases in natural milk. Journal of Dairy Science 97, 155–161.

Lærke, H.N., Jensen, B.B., 1999. D-Tagatose has low small intestinal digestibility but high large intestinal fermentability in pigs. The Journal of Nutrition 129, 1002–1009.

Lee, D.W., Jang, H.J., Choe, E.A., Kim, B.C., Lee, S.J., Kim, S.B., Hong, Y.H., Pyun, H.R., 2004. Characterization of a thermostable L-arabinose (D-galactose) isomerase from the hyperthermophilic eubacterium *Thermotoga maritima*. Applied and Environmental Microbiology 70, 1397–1404.

Lee, S.J., Lee, D.W., Choe, E.A., Hong, Y.H., Kim, S.B., Kim, B.C., Pyun, Y.R., 2005. Characterization of a thermoacidophilic L-arabinose isomerase from *Alicyclobacillus acidocaldarius*: role of Lys-269 in pH optimum. Applied and Environmental Microbiology 71, 7888–7896.

Lee, J.H., Lim, J.S., Song, Y.S., Kang, S.W., Park, C., Kim, S.W., 2007. Optimization of culture medium for lactosucrose (G-beta-D-galactosylsucrose) production by *Sterigmatomyces elviae* mutant using statistical analysis. Journal of Microbiology and Biotechnology 17, 1996–2004.

Levin, G.V., 2002. Tagatose, the new GRAS sweetener and health product. Journal of Medicinal Food 5, 23–36.

Levin, G.V., Lu, Y., 2007. D-Tagatose as an Anti-biofilm Agent. US Patent 7189351 B2.

Li, W., Xiang, X., Tang, S., Hu, B., Tian, L., Sun, Y., Ye, H., Zeng, X., 2009. Effective enzymatic synthesis of lactosucrose and its analogues by β-D-galactosidase from *Bacillus circulans*. Journal of Agricultural and Food Chemistry 57, 3927–3933.

Li, Z., Gao, Y., Nakanishi, H., Gao, K., Cai, L., 2013. Biosynthesis of rare hexoses using microorganisms and related enzymes. Beilstein Journal of Organic Chemistry 9, 2434–2445.

Li, W., Yu, S., Zhang, T., Jiang, B., Mu, W., 2015. Recent novel applications of levansucrases. Applied Microbiology and Biotechnology 99, 6959–6969.

Li, Y., Zhu, Y., Liu, A., Sun, Y., 2011. Identification and characterization of a novel L-arabinose isomerase from *Anoxybacillus flavithermus* useful in D-tagatose production. Extremophiles 15, 441–450.

Liang, M., Chen, M., Liu, X., Zhai, Y., Liu, X.W., Zhang, H., Xiao, M., Wang, P., 2012. Bioconversion of D-galactose to D-tagatose: continuous packed bed reaction with an immobilized thermostable L-arabinose isomerase and efficient purification by selective microbial degradation. Applied Microbiology and Biotechnology 93, 1469–1474.

Lifran, E.V., Hourigan, J.A., Sleigh, R.W., 2009. Lactose derivatives: turning waste into functional foods. Australian Journal of Dairy Technology 64, 89–93.

Lim, B.C., Kim, H.J., Oh, D.K., 2007. High production of D-tagatose by the addition of boric acid. Biotechnology Progress 23, 824–828.

Lim, B.C., Kim, H.J., Oh, D.K., 2008. Tagatose production with pH control in a stirred tank reactor containing immobilized L-arabinose isomerase from *Thermotoga neapolitana*. Applied Biochemistry and Biotechnology 149, 245–253.

Lin, W.J., Chen, T.D., Liu, C.W., 2009. Synthesis and characterization of lactobionic acid grafted pegylated chitosan and nanoparticle complex application. Polymer 50, 4166–4174.

Lindell, S.L., Compagnon, P., Mangino, M.J., Southard, J.H., 2005. UW solution for hypothermic machine perfusion of warm ischemic kidneys. Transplantation 79, 1358–1361.

Livesey, G., Brown, J.C., 1996. D-tagatose is a bulk sweetener with zero energy determined in rats. The Journal of Nutrition 126, 1601–1609.

López-Sanz, S., Montilla, A., Moreno, F.J., Villamiel, M., 2015. Stability of oligosaccharides derived from lactulose during the processing of milk and apple juice. Food Chemistry 183, 64–71.

Lu, Y., Levin, G.V., 2002. Removal and prevention of dental plaque with D-tagatose. International Journal of Cosmetic Science 24, 225–234.

Lu, Y., Levin, G.V., Donner, T.W., 2008. Tagatose, a new antidiabetic and obesity control drug. Diabetes, Obesity and Metabolism 10, 109–134.

Ludwig, R., Ozga, M., Zámocky, M., Peterbauer, C., Kulbe, K.D., Haltrich, D., 2004. Continuous enzymatic regeneration of electron acceptors used by flavoenzymes: cellobiose dehydrogenase-catalyzed production of lactobionic acid as an example. Biocatalysis and Biotransformation 22, 97–104.

Majumder, A., Sultan, A., Jersie-Christensen, R.R., Ejby, M., Gregers Schmidt, B., Lahtinen, S.J., Jacobsen, S., Svensson, B., 2011. Proteome reference map of *Lactobacillus acidophilus* NCFM and quantitative proteomics towards understanding the prebiotic action of lactitol. Proteomics 11, 3470–3481.

Manzo, R.M., Simonetta, A.C., Rubiolo, A.C., Mammarella, J., 2013. Screening and selection of wild strains for L-arabinose isomerase production. Brazilian Journal of Chemical Engineering 30, 711–720.

Marín-Manzano, M.C., Abecita, L., Hernández-Hernández, O., Sanz, M.L., Montilla, A., Olano, A., Rubio, L.A., Moreno, F.J., Clemente, A., 2013. Galacto-oligosaccharides derived from lactulose exert a selective stimulation on the growth of *Bifidobacterium animalis* in the large intestine of growing rats. Journal of Agricultural and Food Chemistry 61, 7560–7567.

Martínez-Castro, I., Olano, A.Q., 1980. Influence of thermal processing on carbohydrate composition of milk, formation of epilactose. Milchwissenschaft 35, 5–8.

Martínez-Villaluenga, C., Cardelle-Cobas, A., Corzo, N., Olano, A., Villamiel, M., Jimeno, M.L., 2008. Enzymatic synthesis and identification of two trisaccharides produced from lactulosa by transgalactosylation. Journal of Agricultural and Food Chemistry 56, 557–563.

Men, Y., Zhu, Y., Zhang, L., Kang, Z., Izumori, K., Sun, Y., Ma, Y., 2014. Enzymatic conversion of D-galactose to D-tagatose: cloning, overexpression and characterization of L-arabinose isomerase from *Pediococcus pentosaceus* PC-5. Microbiological Research 169, 171–178.

Mendoza, M.R., Olano, A., Villamiel, M., 2005. Chemical indicators of heat treatment in fortified and special milks. Journal of Agricultural and Food Chemistry 53, 2995–2999.

Mirescu, A., Prüße, U., 2007. A new environmental friendly method for the preparation of sugar acids via catalytic oxidation on gold catalysts. Applied Catalysis B: Environmental 70, 644–652.

Miyamoto, Y., Ooi, T., Kinoshita, S., 2000. Production of lactobionic acid from whey by *Pseudomonas* sp. LS13-1. Biotechnology Letters 22, 427–430.

Montañés, F., Fornari, T., Martín-Álvarez, P.J., Corzo, N., Olano, A., Ibáñez, E., 2007. Selective recovery of tagatose from mixtures with galactose by direct extraction with supercritical CO_2 and different cosolvents. Journal of Agricultural and Food Chemistry 54, 8340–8345.

Montañés, F., Corzo, N., Olano, A., Reglero, G., Ibáñez, E., Fornari, T., 2008. Selective fractionation of carbohydrate complex mixtures by supercritical extraction with CO_2 and different co-solvents. The Journal of Supercritical Fluids 45, 189–194.

Moore, M.C., 2006. Drug evaluation: tagatose in the treatment of type 2 diabetes and obesity. Current Opinion in Investigational Drugs 7, 924–935.

Mu, W., Chen, Q., Wang, X., Zhang, T., Jiang, B., 2013a. Current studies on physiological functions and biological production of lactosucrose. Applied Microbiology and Biotechnology 97, 7073–7080.

Mu, W., Li, Q., Fan, C., Zhou, C., Jiang, B., 2013b. Recent advances on physiological functions and biotechnological production of epilactose. Applied Microbiology and Biotechnology 97, 1821–1827.

Murakami, H., Seko, A., Azumi, M., Nakano, H., Kitahara, S., 2004. Microbial production of lactobionic acid by *Burkholderia cepacia*. Journal of Applied Glycoscience 51, 149–154.

Murakami, H., Seko, A., Azumi, M., Kiso, T., Kiryu, T., Kitahata, S., Shimada, Y., Nakano, H., 2006. Microbial conversion of lactose to lactobionic acid by resting cells of *Burkholderia cepacia* No. 24. Journal of Applied Glycoscience 53, 7–11.

Murzina, E.V., Tokarev, A.V., Kordás, K., Karhu, H., Mikkola, J.P., Murzin, D.Y., 2008. D-Lactose oxidation over gold catalysts. Catalysis Today 131, 385–392.

Manjasetty, B.A., Chance, M.R., 2006. Crystal structure of *Escherichia coli* L-arabinose isomerase (ECAI), the putative target of biological tagatose production. Journal of Molecular Biology 360, 297–309.

Nakakuki, T., 2002. Present status and future of functional oligosaccharide development in Japan. Pure and Applied Chemistry 74, 1245–1251.

Nath, A., Verasztó, B., Basak, S., Koris, A., Kovács, Z., Vatai, G., 2015. Synthesis of lactose-derived nutraceuticals from dairy waste whey – a review. Food and Bioprocess Technology. http://dx.doi.org/10.1007/s11947-015-1572-2 Published Online: 11 September.

Nishimukai, M., Watanabe, J., Taguchi, H., Senoura, T., Hamada, S., Matsui, H., Yamamoto, T., Wasaki, J., Hara, H., Ito, S., 2008. Effects of epilactose on calcium absorption and serum lipid metabolism in rats. Journal of Agricultural and Food Chemistry 56, 10340–10345.

Nordkvist, M., Nielsen, P.M., Villadsen, J., 2007. Oxidation of lactose to lactobionic acid by a *Microdochium nivale* carbohydrate oxidase: kinetics and operational stability. Biotechnology and Bioengineering 97, 694–707.

Oh, H.J., Kim, H.J., Oh, D.K., 2006. Increase in D-tagatose production rate by site-directed mutagenesis of L-arabinose isomerase from *Geobacillus thermodenitrificans*. Biotechnology 28, 145–149.

Oh, D.K., 2007. Tagatose: properties, applications, and biotechnological processes. Applied Microbiology and Biotechnology 76, 1–8.

Olano, A., Corzo, N., 2009. Lactulose as a food ingredient. Journal of the Science of Food and Agriculture 89, 1987–1990.

Ojima, T., Saburi, W., Sato, H., Yamamoto, T., Mori, H., Matsui, H., 2011. Biochemical characterization of a thermophilic cellobiose 2-epimerase from a thermohalophilic bacterium, *Rhodothermus marinus* JCM9785. Bioscience, Biotechnology, and Biochemistry 75, 2162–2168.

Ojima, T., Saburi, W., Yamamoto, T., Mori, H., Matsui, H., 2013. Identification and characterization of cellobiose 2-epimerases from various aerobes. Bioscience, Biotechnology, and Biochemistry 77, 189–193.

Padilla, B., Ruiz-Matute, A.I., Belloch, C., Cardelle-Cobas, A., Corzo, N., Manzanares, P., 2012. Evaluation of oligosaccharide synthesis from lactose and lactulose using β-galactosidases from *Kluyveromyces* isolated from artisanal cheeses. Journal of Agricultural and Food Chemistry 60, 5134–5141.

Park, C.S., Kim, J.E., Choi, J.G., Oh, D.K., 2011. Characterization of a recombinant cellobiose 2-epimerase from *Caldicellulosiruptor saccharolyticus* and its application in the production of mannose from glucose. Applied Microbiology and Biotechnology 92, 1187–1196.

Park, C.S., Kim, J.E., Lee, S.H., Kim, Y.S., Kang, L.W., Oh, D.K., 2013. Characterization of a recombinant mannobiose 2-epimerase from *Spirochaeta thermophile* that is suggested to be a cellobiose 2-epimerase. Biotechnology Letters 35, 1873–1880.

Park, N.H., Choi, H.J., Oh, D.K., 2005. Lactosucrose production by various microorganisms harboring levansucrase activity. Biotechnology Letters 27, 495–497.

Patel, S., Goyal, A., 2012. The current trends and future perspectives of prebiotics research: a review. 3 Biotech 2, 115–125.

Patra, F., Tomar, S.K., Arora, S., 2009. Technological and functional applications of low-calorie sweeteners from lactic acid bacteria. Journal of Food Science 74, R16–R23.

Peng, J., Wang, K., Tan, W., He, X., He, C., Wu, P., Liu, F., 2007. Identification of live liver cancer cells in a mixed cell system using galactose-conjugated fluorescent nanoparticles. Talanta 71, 833–840.

Pilgrim, A., Kawase, M., Ohashi, M., Fujita, K., Murakami, K., Hashimoto, K., 2001. Reaction kinetics and modeling of the enzyme-catalyzed production of lactosucrose using β-fructofuranosidase from *Arthrobacter* sp. K-1. Bioscience, Biotechnology and Biochemistry 65, 758–765.

Pilgrim, A., Kawase, M., Matsuda, F., Miura, K., 2006. Modeling of the simulated moving-bed reactor for the enzyme-catalyzed production of lactosucrose. Chemical Engineering Science 61, 353–362.

Playne, M.J., Crittenden, R.G., 2009. Galacto-oligosaccharides and other products derived from lactose. In: Fox, P.F., McSweeney, P.L.H. (Eds.), Advanced Dairy Chemistry, third ed. Springer, New York, pp. 121–201.

Poonperm, W., Takata, G., Izumori, K., 2008. Polyol conversion specificity of *Bacillus pallidus*. Bioscience, Biotechnology, and Biochemistry 72, 231–235.

Prabhu, P., Tiwari, M.K., Jeya, M., Gunasekaran, P., Kim, I.W., Lee, J.K., 2008. Cloning and characterization of a novel L-arabinose isomerase from *Bacillus licheniformis*. Applied Microbiology and Biotechnology 81, 283–290.

Rao, D., Gullapalli, P., Yoshihara, A., Jenkinson, S.F., Morimoto, K., Takata, G., Akimitsu, K., Tajima, S., Fleet, G.W.J., Izumori, K., 2008. Direct production of L-tagatose from L-psicose by *Enterobacter aerogenes* 230S. Journal of Bioscience and Bioengineering 106, 473–480.

Rastall, R.A., 2010. Functional oligosaccharides: application and manufacture. Annual Review of Food Science and Technology 1, 305–339.

Rehman, S., Adnan, M., Khalid, N., Shaheen, L., 2011. Calcium supplements: an additional source of lead contamination. Biological Trace Element Research 143, 178–187.

Rentschler, E., Schuh, K., Krewinkel, M., Baur, C., Claaßen, W., Meyer, S., Kuschel, B., Stressler, T., Fischer, L., 2015. Enzymatic production of lactulose and epilactose in milk. Journal of Dairy Science 98, 6767–6775.

Rhimi, M., Bajic, G., Ilhammami, R., Boudebbouze, S., Maguin, E., Haser, R., Aghajari, N., 2011. The acid-tolerant L-arabinose isomerase from the mesophilic *Shewanella* sp. ANA-3 is highly active at low temperatures. Microbial Cell Factories 10, 96.

Rhimi, M., Ilhammami, R., Bajic, G., Boudebbouze, S., Maguin, E., Haser, R., Aghajari, N., 2010. The acid tolerant L-arabinose isomerase from the food grade *Lactobacillus sakei* 23K is an attractive D-tagatose producer. Bioresource Technology 101, 9171–9177.

Ricca, E., Brucher, B., Schrittwieser, J.H., 2011. Multi-enzymatic cascade reactions: overview and perspectives. Advanced Synthesis and Catalysis 353, 2239–2262.

Roberfroid, M.B., 2008. Prebiotics: concepts, definition, criteria, methodologies and products. In: Gibson, G.R., Roberfroid, M. (Eds.), Handbook of Prebiotics. Taylor & Francis, Boca Raton, pp. 39–68.

Roberts, M.W., Wright, J.T., 2012. Nonnutritive, low caloric substitutes for food sugars: clinical implications for addressing the incidence of dental caries and overweight/obesity. International Journal of Dentistry. http://dx.doi.org/10.1155/2012/625701.

Rodriguez-Fernandez, M., Cardelle-Cobas, A., Villamiel, M., Banga, J.R., 2011. Detailed kinetic model describing new oligosaccharides synythesis using different β-galactosidases. Journal of Biotechnology 153, 116–124.

Roh, H.J., Yoon, S.H., Kim, P., 2000. Preparation of L-arabinose isomerase originated from *Escherichia coli* as a biocatalyst for D-tagatose production. Biotechnology Letters 22, 197–199.

Rollini, M., Manzoni, M., 2005. Bioconversion of D-galactitol to tagatose and dehydrogenase activity induction in *Gluconobacter oxydans*. Process Biochemistry 40, 437–444.

Ruiz Matute, A.I., Cardelle-Cobas, A., García-Bermejo, A.B., Montilla, A., Olano, A., Corzo, N., 2013. Synthesis, characterization and functional properties of galactosylated derivatives of chitosan through amide formation. Food Hydrocolloids 33, 245–255.

Ryu, S.A., Kim, C.S., Kim, H.J., Baek, D.H., Oh, D.K., 2003. Continuous D-tagatose production by immobilized thermostable L-arabinose isomerase in a packed-bed bioreactor. Biotechnology Progress 19, 1643–1647.

Saarela, M., Hallamaa, K., Mattila-Sandholm, T., Mättö, J., 2003. The effect of lactose derivatives lactulose, lactitol and lactobionic acid on the functional and technological properties of potentially probiotic *Lactobacillus* strains. International Dairy Journal 13, 291–302.

Saari, P., Häkkä, K., Jumppanen, J., Heikkilä, H., Hurme, M., 2010. Study on industrial scale chromatographic separation methods of galactose from biomass hydrolysates. Chemical Engineering Technology 33, 137–144.

Saburi, W., Yamamoto, T., Taguchi, H., Hamada, S., Matsui, H., 2010. Practical preparation of epilactose produced with cellobiose 2-epimerase from *Ruminococcus albus* NE1. Bioscience, Biotechnology, and Biochemistry 74, 1736–1737.

Salonen, N., Nyyssölä, A., Salonen, K., Turunen, O., 2012. *Bifidobacterium longum* L-arabinose isomerase-overexpression in *Lactococcus lactis*, purification, and characterization. Applied Biochemistry and Biotechnology 168, 392–405.

Salonen, N., Salonen, K., Leisola, M., Nyyssölä, A., 2013. D-Tagatose production in the presence of borate by resting *Lactococcus lactis* cells harboring *Bifidobacterium longum* L-arabinose isomerase. Bioprocess and Biosystem Engineering 36, 489–497.

Sato, H., Saburi, W., Ojima, T., Taguchi, H., Mori, H., Matsui, H., 2012. Immobilization of a thermostable cellobiose 2-epimerase from *Rhodothermus marinus* JCM9785 and continuous production of epilactose. Bioscience, Biotechnology, and Biochemistry 76, 1584–1587.

Schaafsma, G., 2008. Lactose and lactose derivatives as bioactive ingredients in human nutrition. International Dairy Journal 18, 458–465.

Seki, N., Saito, H., 2012. Lactose as a source for lactulose and other functional lactose derivatives. International Dairy Journal 22, 110–115.

Senoura, T., Taguchi, H., Ito, S., Hamada, S., Matsui, H., Fukiya, S., Yokota, A., Watanabe, J., Wasaki, J., Ito, S., 2009. Identification of the cellobiose 2-epimerase gene in the genome of *Bacteroides fragilis* NCTC 9343. Bioscience, Biotechnology, and Biochemistry 73, 400–406.

Silvério, S.C., Macedo, E.A., Teixeira, J.A., Rodrigues, L.R., 2015. Perspectives on the biotechnological production and potential applications of lactosucrose: a review. Journal of Functional Foods 19, 74–90.

Skytte, U.P., 2006. Tagatose. In: Mitchell, H. (Ed.), Sweeteners and Sugar Alternatives in Food Technology. Blackwell Publishing, Oxford, pp. 262–294.

Staudigl, P., Haltrich, D., Peterbauer, C.K., 2014. L-Arabinose isomerase and D-xylose isomerase from *Lactobacillus reuteri*: characterization, coexpression in the food grade host *Lactobacillus plantarum*, and application in the conversion of D-galactose and D-glucose. Journal of Agricultural and Food Chemistry 62, 1617–1624.

Sui, L., Nomura, R., Dong, Y., Yamaguchi, F., Izumori, K., Tokuda, M., 2007. Cryoprotective effects of D-allose on mammalian cells. Cryobiology 55, 87–92.

Suzuki, T., Nishimukai, M., Shinoki, A., Taguchi, H., Fukiya, S., Yokota, A., Saburi, W., Yamamoto, T., Hara, H., Matsui, H., 2010. Ingestion of epilactose, a non-digestible disaccharide, improves postgastrectomy osteopenia and anemia in rats through the promotion of intestinal calcium and iron absorption. Journal of Agricultural and Food Chemistry 58, 10787–10792.

Szilagyi, A., 2010. Functional disaccharides: lactulose, lactitol and lactose. In: Cho, S.S., Finocchiaro, E.T. (Eds.), Handbook of Prebiotics and Probiotics Ingredients. Health Benefits and Food Applications. CRC Press, Boca Raton, pp. 95–122.

Taguchi, H., Senoura, T., Hamada, S., Matsui, H., Kobayashi, Y., Watanabe, J., Wasaki, J., Ito, S., 2008. Cloning and sequencing of the gene for cellobiose 2-epimerase from a ruminal strain of *Eubacterium cellulosolvens*. FEMS Microbiology Letters 287, 34–40.

Takata, G., Poonperm, W., Rao, D., Souda, A., Nishizaki, T., Morimoto, K., Izumori, K., 2007. Cloning, expression, and transcription analysis of L-arabinose isomerase gene from *Mycobacterium smegmatis* SMDU. Bioscience, Biotechnology, and Biochemistry 71, 2876–2885.

Taniguchi, Y., Mizote, A., Kohno, K., Iwaki, K., Oku, K., Chaen, H., Fukuda, S., 2007. Effects of dietary lactosucrose (4^G-β-D-galactosylsucrose) on the IgE response in mice. Bioscience, Biotechnology, and Biochemistry 71, 2766–2773.

Taylor, T., 2006. Evaluation of the Bulk Sweetener D tagatose and the High Intensity Sweetener Splenda as Sugar Replacers in Cookies. Auburn University, USA (Ph.D. thesis).

Taylor, T.P., Fasina, O., Bell, L.N., 2008. Physical properties and consumer liking of cookies prepared by replacing sucrose with tagatose. Journal of Food Science 73, S145–S151.

Teramoto, F., Rokutan, K., Sugano, Y., Oku, K., Kishino, E., Fujita, K., Hara, K., Kishi, K., Fukunaga, M., Morita, T., 2006. Long-term administration of 4G-beta-D-galactosylsucrose (lactosucrose) enhances intestinal calcium absorption in young women: a randomized, placebo-controlled 96-wk study. Journal of Nutritional Science and Vitaminology 52, 337–346.

Tokarev, A.V., Murzina, E.V., Mikkola, J.P., Kuusisto, P., Kustov, L.M., Murzin, D.Y., 2007. Application of *in situ* catalyst potential measurements for estimation of reaction performance: lactose oxidation over Au and Pd catalysts. Chemical Engineering Journal 134, 153–161.

Torres, P.R., Manzo, R.M., Rubiolo, A.C., Batista-Viera, F.D., Mammarella, E.J., 2014. Purification of an L-arabinose isomerase from *Enterococcus faecium* DBFIQ E36 employing a biospecific affinity strategy. Journal of Molecular Catalysis B: Enzymatic 102, 99–105.

Van Hecke, W., Bhagwat, A., Ludwig, R., Dewulf, J., Haltrich, D., Van Langenhove, H., 2009. Kinetic modeling of a bi-enzymatic system for efficient conversion of lactose to lactobionic acid. Biotechnology and Bioengineering 102, 1475–1482.

Van Hecke, W., Haltrich, D., Frahm, B., Brod, H., Dewulf, J., Van Langenhove, H., 2011. A biocatalytic cascade reaction sensitive to the gas–liquid interface: modeling and upscaling in a dynamic membrane aeration reactor. Journal of Molecular Catalysis B Enzymatic 68, 154–161.

van Velthuijsen, J.A., 1979. Food additives derived from lactose: lactitol and lactitol palmitate. Journal of Agricultural and Food Chemistry 27, 680–686.

Vastenavond, C.M., Bertelsen, H., Hansen, S.J., Laursen, R.S., Saunders, J., Eriknauer, K., 2012. Tagatose (D-tagatose). In: O'Brien Nabors, L. (Ed.), Alternative Sweeteners, fourth ed. CRC Pres, Boca Raton, pp. 198–221.

Villamiel, M., Corzo, N., Martínez-Castro, I., Olano, A., 1996. Chemical changes during microwave treatment of milk. Food Chemistry 56, 385–388.

Wagner, N., 2014. Integration of Biocatalysis and Simulated Moving Bed Chromatography for the High-yield Production of Rare Sugars. ETH Zurich, Switzerland (Doctor of Science thesis).

Wanarska, M., Kur, J., 2012. A method for the production of D-tagatose using a recombinant *Pichia pastoris* strain secreting β-D-galactosidase from *Arthrobacter chlorophenolicus* and a recombinant L-arabinose isomerase from *Arthrobacter* sp. 22c. Microbial Cell Factories 11, 113.

Watanabe, J., Nishimukai, M., Taguchi, H., Senoura, T., Hamada, S., Matsui, H., Yamamoto, T., Wasaki, J., Hara, H., Ito, S., 2008. Prebiotic properties of epilactose. Journal of Dairy Science 91, 4518–4526.

Xu, Z., Li, S., Fu, F., Li, G., Feng, X., Xu, H., Ouyang, P., 2012. Production of D-tagatose, a functional sweetener, utilizing alginate immobilized *Lactobacillus fermentum* CGMCC2921 cells. Applied Biochemistry and Biotechnology 166, 961–973.

Xu, Z., Li, S., Feng, X., Zhan, Y., Xu, H., 2014a. Function of aspartic acid residues in optimum pH control of L-arabinose isomerase from *Lactobacillus fermentum*. Applied Microbiology and Biotechnology 98, 3987–3996.

Xu, Z., Li, S., Feng, X., Liang, J., Xu, H., 2014b. L-Arabinose isomerase and its use for biotechnological production of rare sugars. Applied Microbiology and Biotechnology 98, 8869–8878.

Xu, Z., Qing, Y.J., Li, S., Feng, X.H., Xu, H., Ouyang, P.K., 2011. A novel L- arabinose isomerase from *Lactobacillus fermentum* CGMCC2921 for D-tagatose production: gene cloning, purification and characterization. Journal of Molecular Catalysis B: Enzymatic 70, 1–7.

Yamanaka, K., 1975. L-Arabinose isomerase from *Lactobacillus gayonii*. Methods in Enzymology 41, 458–461.

Yanahira, S., Suguri, T., Yakabe, T., Ikeuchi, Y., Hanagata, G., Deya, E., 1992. Formation of oligosaccharides from lactitol by *Aspergillus oryzae* β-D-galactosidase. Carbohydrate Research 20, 151–159.

Yang, S.T., Silva, E.M., 1995. Novel products and new technologies for use of a familiar carbohydrate, milk lactose. Journal of Dairy Science 78, 2541–2562.

Yoshida, M., Yamada, M., Nishitani, T., Takada, G., Izumori, K., Kamitori, S., 2007. Crystal structures of D-tagatose 3-epimerase from *Pseudomonas cichorii* and its complexes with D-tagatose and D-fructose. Journal of Molecular Biology 374, 443–453.

Yoshimura, T., Ishihara, K., Esumi, K., 2005. Sugar-based gemini surfactants with peptide bonds-synthesis, adsorption, micellization, and biodegradability. Langmuir 21, 10409–10415.

Zamocky, M., Ludwig, R., Peterbauer, C., Hallberg, B.M., Divne, C., Nicholls, P., Haltrich, D., 2006. Cellobiose dehydrogenase – a flavocytochrome from wood-degrading, phytopathogenic and saprotrophic fungi. Current Protein and Peptide Science 7, 255–280.

Zhan, Y., Xu, Z., Li, S., Liu, X., Xu, L., Feng, X., Xu, H., 2014. Coexpression of β-D-galactosidase and L-arabinose isomerase in the production of D-tagatose: a functional sweetener. Journal of Agricultural and Food Chemistry 62, 2412–2417.

Zhang, H., Jiang, B., Pan, B., 2007. Purification and characterization of L-arabinose isomerase from *Lactobacillus plantarum* producing D-tagatose. World Journal of Microbiology and Biotechnology 23, 641–646.

Zhang, Y.W., Prabhu, P., Lee, J.K., Kim, I.W., 2009. Enhanced stability of *Bacillus licheniformis* L-arabinose isomerase by immobilization with alginate. Preparative Biochemistry and Biotechnology 40, 65–75.

Zhang, Y.W., Prabhu, P., Lee, J.K., 2010. Alginate immobilization of recombinant *Escherichia coli* whole cells harboring L-arabinose isomerase for L-ribulose production. Bioprocess and Biosystem Engineering 33, 741–748.

Zhang, Y.W., Jeya, M., Lee, J.K., 2011. Enhanced activity and stability of L-arabinose isomerase by immobilization on aminopropyl glass. Applied Microbiology and Biotechnology 89, 1435–1442.

Chapter 7

Technical and Economic Analysis of Industrial Production of Lactose-Derived Prebiotics With Focus on Galacto-Oligosaccharides

F. Scott, C. Vera and R. Conejeros

Galacto-oligosaccharides (GOS) are currently produced at industrial scale to fulfill the increasing demand of this probiotic, which is expected to continue increasing during the following decade. It is produced by enzymatic synthesis from lactose using β-galactosidase as catalyst, and the raw product is further purified and concentrated. As in any chemical or biochemical process, GOS production can be analyzed as a series of stages or unit operations interconnected by mass and energy flows. This will be the subject of Section 7.1, where process stages and alternatives for GOS production are analyzed.

Section 7.2 presents a case study of GOS production from lactose in a stand-alone (or grass-roots) plant. The case study includes several GOS purification alternatives and shows the effects of its selection over GOS purity and GOS minimum acceptable selling price.

Finally, Section 7.3 discusses the possibility of integration between GOS production and an existing dairy industry operation.

7.1 TECHNOLOGIES FOR INDUSTRIAL PRODUCTION OF GALACTO-OLIGOSACCHARIDES

Annual growth of GOS production has been estimated in 10%, being the total world production in 2009 of 21,000 tons (Affertsholt-Allen, 2007; Playne and Critterden, 2009). Being GOS the lactose-derived product with the largest market share in the functional food market, this will be the focus of this chapter. In this regard, a more recent study by Grand View Research, Inc. indicates that global GOS market was 94,100 tons in 2013 and it is expected to reach 175,700 tons by 2020, considering

Lactose-Derived Prebiotics. http://dx.doi.org/10.1016/B978-0-12-802724-0.00007-X

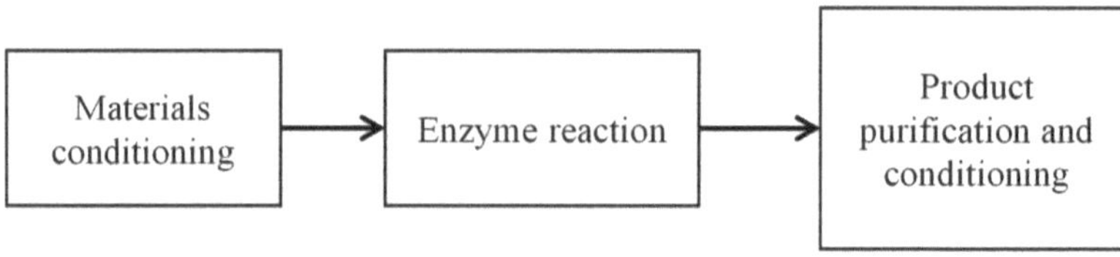

FIGURE 7.1 Block diagram for the GOS production process.

an annual growth rate of 9.3% from 2014 to 2020. The increase in GOS production is supported by a recent report by The Dutch Friesland Campina Domo and the Australian Warrnambool Cheese & Butter Factory announcing the opening in Australia of a Great Ocean Ingredients plant with a production capacity of 15,000 metric tons of Vivinal GOS, representing a large scale plant (FrieslandCampina Domo and Warrnambool Cheese & Butter Factory Open New Plant in Australia, 2009, Grand View Research Market Research and Consulting, 2015). This plant sets an upper limit on plant capacity for the analysis in Section 7.2.

GOS production process typically considers a number of stages which include lactose conditioning, enzyme reaction, thermal inactivation and product purification and conditioning. The operations considered for each stage vary in the existing commercial processes, but they can be broadly represented in the block diagram in Fig. 7.1.

7.1.1 Lactose Conditioning

Raw materials conditioning will depend on whether food grade lactose or whey is used. Current market prices place lactose cost at roughly US$ 500/ton, which makes the former option more convenient, thus avoiding lactose purification from whey. Thus, material conditioning considers lactose dissolution, and pH and temperature adjustment to the reaction conditions. Values for these operating conditions depend on the origin of the enzyme selected for the reaction of GOS synthesis.

Solubility of lactose in water at ambient temperature is low (close to 20 g of α-lactose hydrate per 100 g of water at 20°C). Typical reaction conditions for GOS synthesis involve at least 30% w/v lactose at 40–60°C (roughly 37 g of lactose per 100 g of water). To facilitate lactose dissolution and avoid precipitation, lactose is dissolved in water at temperatures ranging between 80 and 90°C. At this high temperature, nearly 150 g of α-lactose hydrate can be dissolved in 100 g of water; however, once the solution is cooled to 40°C, only 30 g are soluble in 100 g of water. The ratio between actual concentration and saturation concentration (in this case 150/30 or five) is called supersaturation factor. For supersaturation factors between one and 1.6, the solution is metastable and crystallization does not occur unless induced by seeding with crystals. At supersaturation factors above 2.1, spontaneous crystallization occurs rapidly (Walstra et al., 2005). Since crystals are not accessible for reaction, this is an undesirable effect.

Depending on the holding time at this temperature, the mixture is sanitized. Once dissolved, the solution is cooled to the reaction temperature and pH is adjusted to the value required by the specific enzyme used in the process. pH conditioning is accomplished using GRAS additives, such as citrate and citric acid and sodium carbonate.

7.1.2 Enzyme Selection

Synthesis of GOS from lactose is catalyzed by the enzyme β-galactosidase in a kinetically controlled reaction. Therefore, there is a competition between hydrolysis rates of lactose and oligosaccharides and the formation rates of oligosaccharides by transgalactosylation. Thus, higher water concentrations promote lactose hydrolysis rates to glucose and galactose and higher lactose concentrations promote higher rates of oligosaccharides formation (Vera et al., 2011). This fact has two important consequences on the design of the reaction stage: maximum GOS yield is attained at an intermediate lactose conversion in a batch reactor and high initial lactose concentrations promote higher GOS yields. As for the former, maximum GOS yield is attained at different lactose conversions depending on the source of the enzyme used. In batch mode and using β-galactosidase from *Aspergillus oryzae*, 28% yield is attained at a 58% lactose conversion using an initial concentration of lactose monohydrate of 40% (w/v) at 47°C (Vera et al., 2011). When enzyme from *Bacillus circulans* is used at 60°C and 10% (w/v) initial lactose concentration, 54% GOS yield was achieved (Gosling et al., 2009); a high GOS yield of 75% was attained at a 90% lactose conversion at 50% w/v initial lactose and 80°C using a β-galactosidase from the thermophilic archaeon *Thermus caldophilus*. For a throughout list of lactose conversion versus GOS yield see Torres et al. (2010).

Selection of an adequate β-galactosidase must be done considering many factors besides GOS yield. The source of the enzyme affects the distribution of the produced oligosaccharides (Nakayama and Amachi, 2002): while the β-galactosidase from *A. oryzae* produces mostly tri-saccharides (GOS-3), lower levels of tetra-saccharides (GOS-4) and negligible levels of di- (GOS-2) and penta-saccharides (GOS-5) and higher, the enzyme from *Kluyveromyces marxianus* produces important levels of GOS-2 and the one from *B. circulans* produces a more evenly distributed product profile ranging from GOS-2 to GOS-5 (Panesar et al., 2006). Since the prebiotic potential of GOS is mainly associated with GOS-3 and GOS-4, the enzymes of choice for industrial productions are those favoring short GOS chains above GOS-2. Yet other relevant factors in enzyme selection are its GRAS status and commercial availability (Table 7.1).

7.1.3 Galacto-Oligosaccharides Production

Synthesis of GOS at industrial level is carried out in batch mode according to GRAS notices from several producers (van Vliet, 2007; Spherix Consulting, Inc., 2010); hence this will be the focus of the description of this stage.

TABLE 7.1 Commercially Available β-Galactosidases as Reviewed in the Last Quarter of 2015

Enzyme Source	Commercial Name	Supplier
Yeasts		
Kluyveromyces lactis	Maxilact	DSM food Specialties, Delft, the Netherlands
	β-Galactosidase	Sigma–Aldrich, UK
	Lactozyme	Novozymes A/S, Bagsvaerd, Denmark
	HA-Lactase	Chr. Hansen A/S, Denmark
Kluyveromyces sp.	Enzeco Lactase NL	Enzyme Development Corporation, New York, USA
Fungi		
Aspergillus niger	Maxilact A4	DSM food Specialties, Delft, the Netherlands
Aspergillus oryzae	Enzeco fungal lactase	Enzyme Development Corporation, New York, USA
	Biolactase	Biocon (US) Inc., Lexington, USA
	Lactase 849MDP	Biocatalyst
	Lactase 14DS	Amano enzyme Inc., Japan
Bacteria		
Escherichia coli	β-Galactosidase	Sigma–Aldrich, UK
Bacillus circulans	Biolacta FN5	Amano enzyme Inc., Japan

Enzymes, Except β-galactosidase from *Bacillus circulans* and *Escherichia coli*, have GRAS status according to the US Code of Federal Regulations Title 21, part 173, 184 and Opinion Letters From FDA. GOS produced using *β*-Galactosidase from *B. circulans* are considered GRAS according to GRAS Notice No. GRN 000236.

Batch reactor volumes are on the order of cubic meters to tens of cubic meters; for these sizes, reactors can be shop-fabricated and transported to the plant. Since GOS are intended for human consumption, sterility must be observed. Hence, reactors need to be designed as steam-sterilizable tanks with a clean-in-place (CIP) system. To withstand pressure during sterilization, ASME Boiler and Pressure Vessel Code can be used to design the tank. Moreover, the inner surface of the tank needs to be stainless austenite steel (generally 316) with surface treatment, such as electro-polishing, to avoid material adherence.

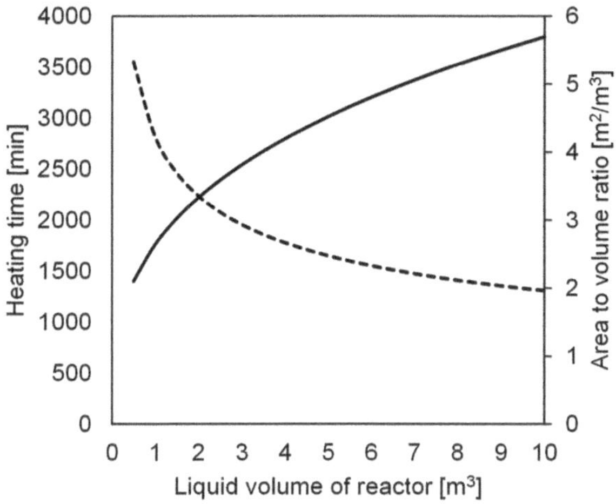

FIGURE 7.2 Calculated heating time (*continuous line*) as a function of reactor liquid volume for a 40% lactose solution heated with 2 bar saturated steam in the jacket side. An overall heat transfer coefficient of $500\,Wm^{-2}K^{-1}$ was considered. *Dashed line* represents the ratio of jacket surface to reactor volume.

Welding must produce joints free of pits and smooth to minimize accumulation of material. Guidelines for the design of bioreactors, including selection of construction materials and welding techniques, can be found in the ASME′ Bioprocessing Equipment Standard.

Agitation must be provided to promote heat transfer and ensure a homogeneous reactor content during reaction and to guarantee lactose dissolution during the loading phase. Among the factors determining impeller selection, viscosity plays a key role. A 40% w/w lactose solution shows a modest viscosity of 7 cP at 20°C and 2.2 cP at 60°C; due to these low viscosities common impellers used in bioreactors, such as propellers or turbines, are preferred (see Figure 10.57 in Towler and Sinnot, 2007).

The amount of heat transferred in a given amount of time (the heat duty) from the reactor content depends on the operations performed on it. As the heat of reaction of GOS production is negligible, the most demanding operations are heating for lactose dissolution, and cooling after the reaction has been stopped by thermal or chemical inactivation. Heat transfer can be provided by a jacket, an internal coil, or by an external heat exchanger. This decision depends on the calculated area required for the estimated heat duty and on the reactor size. For jacketed reactors, heat transfer area decreases as the volume of the reactor increases. This implies that for a given heat exchange operation, i.e., heating in Fig. 7.2, the time required to complete it increases with reactor volume. If time is an important constraint, extra heat exchange area can be provided by an internal coil, although fouling and cleaning problems are to be expected, or by an external heat exchanger, where the liquid is withdrawn from the reactor by a pump, sent to the exchanger and returned to the reactor. Heating and cooling times can be adjusted to any given value by selecting the appropriate area for the heat exchanger and the fluid circulation velocity in the recirculation loop. The ability to control fluid velocity also allows for fouling minimization.

7.1.4 Product Purification

Reaction products include GOS, monosaccharides, disaccharides, ions, and enzyme (protein). Purification techniques are intended to increase GOS purity, this is, the mass ratio of GOS to total solids. A number of strategies are reported for GOS purification, which include adsorption in activated carbon, diafiltration, nanofiltration, ion exchange fractionation, molecular exclusion chromatography, moving bed chromatography, supercritical extraction and selective fermentation (see Section 4.4). Existing commercial processes use only a number of these.

- Color, protein, and monosaccharide removal by adsorption: Color and protein removal can be performed by adsorption using activated carbon. For color removal of GOS syrups typical dosage range from 14 to 33 g of activated carbon per kg of carbohydrates (Matsumoto et al., 1995; Hernández et al., 2009), similar to the dosages used in the sugar refining industry. Proteins can be removed by adsorption in Celite or Perlite as in the case of Purimune GOS (GTC Nutrition) and GOS-1000-P (King Prebiotics) production, and also by ultrafiltration as in the case of Bimuno GOS (Clasado). Activated carbon, due to its hydrophobic nature, can be used to remove mono and disaccharides from GOS reaction mixtures. In this system, GOS are preferentially adsorbed to the active carbon while mono and disaccharides are eluted. This is caused by the higher hydrophobicity of GOS compared to di- and monosaccharides. Hernández et al. (2009) investigated the purification of Vivinal-GOS on activated carbon. The GOS preparation was dissolved in ethanolic solutions ranging from 1% to 15% v/v ethanol; sugars and GOS were eluted using 50% v/v ethanol. Results showed that as the ethanol concentration in the solvent increased the purity of the GOS product also increased. However, GOS yield decreased with increasing ethanol concentration in the solvent. In fact, although monosaccharides were completely removed when Vivinal-GOS was dissolved in 10% v/v ethanol solution and a product with 92% purity was attained, only 52% of GOS-3 and 89% of the GOS-4 were recovered. Since 30% of GOS was lost using this purification process, its economic merit is questionable.
- Demineralization: Ion and salts removal from the GOS mixture can be performed using weak acid and basic resins. Ion exchange using resin columns is commonly applied in the sugar industry for juice softening, thus reducing scaling in the evaporators and anti-scaling agents consumption, or for the production of demineralized liquid sugar (Kearney, 2002; Alexandratos, 2009;

Zaganiaris, 2011). Since saccharides are weak electrolytes they show little interaction in its natural form with ion-exchange resins in aqueous medium (Sanz and Martínez-Castro, 2007); thus, ion exchange is an effective process for ion removal without affecting GOS yield. Moreover, ion exchange does not affect the carbohydrate profile of the GOS product. Ion exchange resins can be purchased as packaged units from several companies such as Purolite (www.purolite.com), Novasep (www.novasep.com), DOW (www.dow.com, exchange resins), and Amalgamated Research LLC (www.arifractal.com, engineering, design and installation). Ion exchange is used in the production of Yakult's Oligomate 55N/55NP, Purimune and GOS-1000-P.

- Chromatographic fractionation: Size exclusion chromatography is commonly applied for semi- or preparative fractionation of GOS at laboratory scale. This technique was described in Section 4.4, hence, this section will be devoted to the industrial applications of chromatographic systems based in the simulated moving bed (SMB) principle. In traditional batch elution chromatography, the solution containing the solutes to be separated flows through a column packed with a porous material that retains the sugars differentially according to its molecular size or affinity for the packing material. Afterward, the retained solute is eluted by desorbing it from the packed bed. Hence, the column is not productive during the elution phase since no sugar solution can be fed during this period. On the other hand, in a true moving bed system, a conceptual device that cannot be economically constructed, but useful to illustrate the principle behind SMB, the solid and liquid phases flow countercurrently in the column. In such system, the solution to be separated is fed at the middle of the column while the eluent is fed at its bottom. Considering that the solution is composed of GOS and glucose and that glucose has a higher affinity for the solid phase. Now, the solid phase continuously moves down in the column while the eluent flow (water in this case) moves upwards. Provided that the GOS and glucose solution is fed at the middle of the column, the more adsorbed solute (glucose) will move downwards with the solid while the less adsorbed solute (GOS) will move upwards with the eluent. Before the bottom end of the column a continuous flow of solution enriched in glucose (the extract) can be withdrawn carrying the more adsorbed solute. From this extraction point down to the bottom of the column the solid phase is washed with eluent removing glucose (that moves upwards to the extraction point). Once the washed solid reaches the bottom of the column, it is recycled to the top. Meanwhile, the less adsorbed solute (GOS) has moved upwards with the eluent; even the small amount that adsorbs to the solid has been washed and moves with the eluent stream, and the less adsorbed solute can be recovered in a stream at an extraction point near to the top of the column (the raffinate). From this extraction point up to the top of the column the amount of GOS that flows with the eluent is adsorbed in the solid flowing downwards. Thus, the amount of liquid recovered at the top of the column (eluent flow + feed flow - extract flow -raffinate flow) is recycled to the bottom of the column. Thereby, a true moving bed is a continuous chromatographic system that uses less eluent since a fraction of it is recycled, and requires less solid since washing and feeding are performed simultaneously. However, it is very difficult to circulate solids in a column. Thereby, SMB is implemented using several fixed bed columns connected in series forming a closed ring, with an appropriate shift of the injection and collection points. The inlet and outlet positions are rotated at regular time intervals in the direction of the liquid flow, mimicking an apparent flow of the solid phase in the opposite direction. SMB chromatography has been applied to large scale separation tasks especially in the petrochemical (Broughton, 1984) and sugar industries. In the latter, it is used for fructose separation from glucose in the production of high-fructose syrup. Recently SMB chromatography was applied for GOS purification starting with regular Vivinal-GOS syrup. Results indicate that GOS and sugars can be almost completely fractionated (99% GOS content in the purified stream) at the operating conditions used in the study with Vertex Eurokat, a sulfonated cross-linked styrene-divinylbenzene resin (Wiśniewski et al., 2013). However, those conditions, selected to maximize GOS purity and recovery, affected the productivity of the SMB chromatography. A throughout description and design guidelines for SMB chromatography can be found elsewhere (Nicoud, 2000).

- GOS purification by selective fermentation: GOS preparation (raw GOS) can be purified by selective fermentation using yeasts, such as *Saccharomyces cerevisiae*, *Kluyveromyces lactis* and *K. marxianus*. Unlike *Kluyveromyces* strains, *S. cerevisiae* is unable to ferment lactose, but does ferment galactose and glucose. Two difficulties hinder the utilization of fermentation as a cost-effective purification process. First, since it is not desirable to include nutrients for yeast growth (salts, minerals and yeast extract) to avoid product contamination and subsequent purification stages, cells need to be grown in a propagation fermenter, concentrated by centrifugation and used as a whole-cell biocatalyst under nongrowing conditions (bioconversion). The mass ratio of yeast to sugars determines the time required for the bioconversion, as cells are not growing during the purification stage due to the absence of nutrients others than the carbon source. Thus, depending on the value of this operational parameter, the amount of sugars and nutrients required for yeast propagation can represent an important production cost. Second, GOS are produced at high initial lactose concentrations, typically over 40% w/w lactose; such high sugar concentrations may be inhibitory for the yeast metabolism, thus for an efficient fermentation the raw GOS entering the fermenter needs to be diluted or the amount of yeast per gram of total sugars needs to be increased when conventional batch fermentation is used (Guerrero et al., 2014). GOS preparations have been purified to 50% GOS using *S. cerevisiae* at 50 mg of dry yeast per gram of total sugars (Hernández et al., 2009). Using cells of *K. lactis* immobilized in calcium alginate, GOS purity of 97% could be achieved in 18 h using 2.5 g of cells per gram of sugar, which is an extremely high yeast to sugars ratio (Li et al., 2008). This high concentration of cells can only be economically feasible if the yeast can be reused in several cycles; however, results showed that the purity rapidly decreased after two cycles of cell reuse. Guerrero et al. (2014) analyzed the purification of GOS at different initial total sugar concentration at the beginning of the fermentation considering the effect of fermentation time and cells to total sugars mass ratio. Results show that in order to achieve a GOS purity above 90%, total carbohydrates concentration needs to be below 20% at 0.5 g of cells per gram of total sugars. Considering that cells recycle proved to be infeasible, a lower cells to carbohydrates mass ratio needs to be used in order to reduce the expenditures in nutrients for yeast propagation. Thus, at a 0.1, g/g yeast to carbohydrates ratio, the total initial concentration of carbohydrates at the beginning of the fermentation needs to be reduced to 10%. This implies that the raw GOS leaving the synthesis reactor needs to be significantly diluted, affecting the energy consumption at the evaporation stage and increasing water consumption. Yakult Oligomate GOS and King Prebiotics GOS-1000-P consider the use of *K. lactis* in their processes. Thus, a purification process using fermentation is considered in Section 7.2 as one of the alternative processes.

7.1.5 Commercial Products Purity and Characteristics

Yakult GOS products are: Oligomate 55, a syrup with 55% dry matter of oligosaccharides, Oligomate 55P in powder form with the same oligosaccharide content, and TOS-100, a purified powder containing 99% oligosaccharides. Nissin Sugar Manufacturing Co. Ltd. produces Cup-Oligo H-70, a syrup with 52% w/v of oligosaccharides and a powder Cup-Oligo P with 70% of oligosaccharides. Friesland Foods Domo products are TOS-syrup, with 75% w/v content of GOS, and Vivinal GOS, a syrup with 75% w/v of solids and 59% of GOS (Playne and Critterden, 2009).

As shown above, commercial products contain about 50% w/v of GOS in syrup form and about 60–70% in powder form. Products with a content of GOS higher than 90% (w/w) are commercialized by some manufacturers, even though some of these high-purity products have been apparently discontinued. Besides the differences in purity amongst the commercially offered products, there are differences also in the β-type linkages of the oligosaccharide chain due to the different enzymes used in their production (Torres et al., 2010).

Table 7.2 presents a summary of the main processing stages used in commercial GOS production.

7.2 A CASE STUDY ON GALACTO-OLIGOSACCHARIDES PRODUCTION AS A STAND-ALONE PROCESS

7.2.1 Galacto-Oligosaccharides Production Processes Under Analysis

Five processes for GOS production from edible lactose are analyzed in this section. Processes differ both in the technology used for product purification, in product

TABLE 7.2 Characteristics of Commercial GOS Production Processes

Parameter	Bimuno GOS (Clasado)	Vivinal (Friesland Foods Domo)	Purimune (GTC Nutrition)	Oligomate 55N/55NP (Yakult)	GOS-1000-P King Prebiotics
Purity (dw% GOS)	46–60	57	90–92	>=55	>=99
Lactose source	Food grade lactose	Food grade lactose	Food grade lactose	Food grade lactose	Food grade lactose
β-galactosidase source	*Bifidobacterium bifidum*	*Bacillus circulans*	*B. circulans* LOB377	*Sporobolomyces singularis*	*B. circulans*
Reaction stopping method	Protein removal	pH (citric acid addition)	Thermal inactivation (80°C)	Thermal inactivation	pH (citric acid addition)
Purification	Ultrafiltration (10 kDa membrane cut)	Protein adsorption and color removal using activated carbon	• Protein adsorption (celite) • color Removal (active carbon) • ions And salts removal (ion exchange) • chromatography (Sugars removal)	• Lactose Hydrolysis (*Kluyveromyces lactis* β-gal.) • decoloration using powdered activated carbon, ultrafiltration to remove proteins and active carbon, demineralization with ion exchange resins, microfiltration to remove microorganism	• Fermentation with Kluyveromyces lactis, • cell Removal using disk separator • decoloration (activated carbon) • protein Removal (perlite) • ion Exchange
Concentration	Evaporation	Evaporation	Evaporation and spray drying	Evaporation	

purity and in the source of the β-galactosidase used for GOS synthesis. Processes whose tag begins with AO use β-galactosidase from *A. oryzae* as enzyme source for GOS production. Process AO-Ch includes filtration using activated carbon and celite for color and protein removal, ion exchange and SMB chromatography to achieve high GOS purity. Process AO-SF (SF stands for selective fermentation) uses the same enzyme and operations for color, protein and ions removal as AO-Ch, but SMB chromatography is replaced by a fermentation stage using *K. marxianus* to selectively remove glucose, galactose and lactose achieving high GOS purity. Process alternatives whose tag starts with BC use lactose from *B. circulans* at the GOS production stage. Process BC-LP does not include any enrichment purification stage, this is, ion chromatography and filtration are applied to remove salts, protein and color, but lactose or monosaccharides are not removed. Details regarding yields and process considerations are shown in Table 7.3.

Enzyme activities and prices are shown in Table 7.4 along with prices for other chemicals and separation aids.

Each process alternative begins with edible lactose dissolution at 90°C and 40% w/w. Heat is provided by an external plate heat exchanger that also allows cooling the mixture to reaction temperature (40°C for GOS synthesis using *A. oryzae* β-galactosidase and 45°C using *B. circulans* β-galactosidase). pH is adjusted using citric acid and Na_2HPO_4. Once the reaction time indicated in Table 7.3 is achieved, enzymes are thermally inactivated by heating to 90°C. GOS synthesis is conducted in batch reactors with a maximum volume of $10\,m^3$

TABLE 7.3 Characteristics of the GOS Production Processes Under Evaluation

	AO-Ch	AO-SF	BC-LP	BC-Ch	BC-SF
Enzyme Source	***Aspergillus oryzae***		***Bacillus circulans***		
GOS synthesis[a]	**Conversions at:**				
	4h Reaction time		48h Reaction time		
$Lac \rightarrow Glu + Gal + H_2O$:	9.46%		5.98%		
$2Lac \rightarrow GOS3 + Glu$:	24.84%		31.32%		
$3Lac \rightarrow GOS4 + 2Glu$:	14.06%		26.40%		
Fermentation with *Kluyveromyces marxianus*	No	Yes	No	No	Yes
Active carbon and celite filtration	Yes	Yes	Yes	Yes	Yes
Ion-exchange	Yes	Yes	Yes	Yes	Yes
SMB chromatography	Yes	No	No	Yes	No

[a]*Yields were taken from Sanz-Valero (2009). Conversions refer to lactose as the limiting reactant.*

TABLE 7.4 Prices for Lactose, Enzyme and Chemicals Required for GOS Production Along With Electricity and Natural Gas Prices

Material	Price	Units
Lactose	1	US$/kg
Deproteinized whey powder		
Natural gas	10	US$/MMBtu
Electricity	80	US$/MWh
Aspergillus oryzae β-gal (106,742 IU/$g_{protein}$)[a]	350	US$/kg
Bacillus circulans β-gal (5500 IU/$g_{protein}$)[a]	800	US$/kg
Citric acid[b]	3.2	US$/kg
Na_2HPO_4[b]	1.85	US$/kg
Celite[c]	300	US$/MT
Activated carbon[d]	5	US$/kg

[a]*Enzyme prices and specific activity from Sanz-Valero (2009).*
[b]*ICIS, Indicative Chemical Prices A-Z. Available in: http://www.icis.com/chemicals/channel-info-chemicals-a-z/.*
[c]*Nevada Bureau of Mines and Geology, 2009. The Nevada Mineral Industry 2008, Mackay School of Earth Sciences and Technology.*
[d]*Woods (2007).*

scheduled to achieve the plant capacity. Further processing depends on the purification strategy.

Each process includes sequential celite and activated carbon filtration using precoats of the aforementioned filtration aids at dosages of 0.1 kg of celite per m^3 of reaction solution and activated carbon at 1 kg per 50 kg of initial lactose. Activated carbon is regenerated for 10 cycles using flue gas from boiler (Tipnis and Harriott, 1986), and after the specified number of cycles of use is disposed as solid waste. After filtration, salts are removed using cationic and anionic ion exchange columns operated in sequence.

BC-Ch and AO-Ch production processes (see Fig. 7.3) include an SMB chromatography stage to remove monosaccharides and lactose. The process was designed according to the method published by Wiśniewski et al. (2013) using the affinity constants reported by Sanz-Valero (2009). The design incorporates the amount of water required to elute GOS and disaccharides and the effect of the number of stages (which determines the columns length) on GOS purity and yield. SMB chromatography produces a stream containing small amounts of GOS along with monosaccharides and lactose (the extract), and a stream containing most of the GOS (the raffinate). Both streams are concentrated in triple-effect evaporators up to 75% total solid content. Extract, containing the

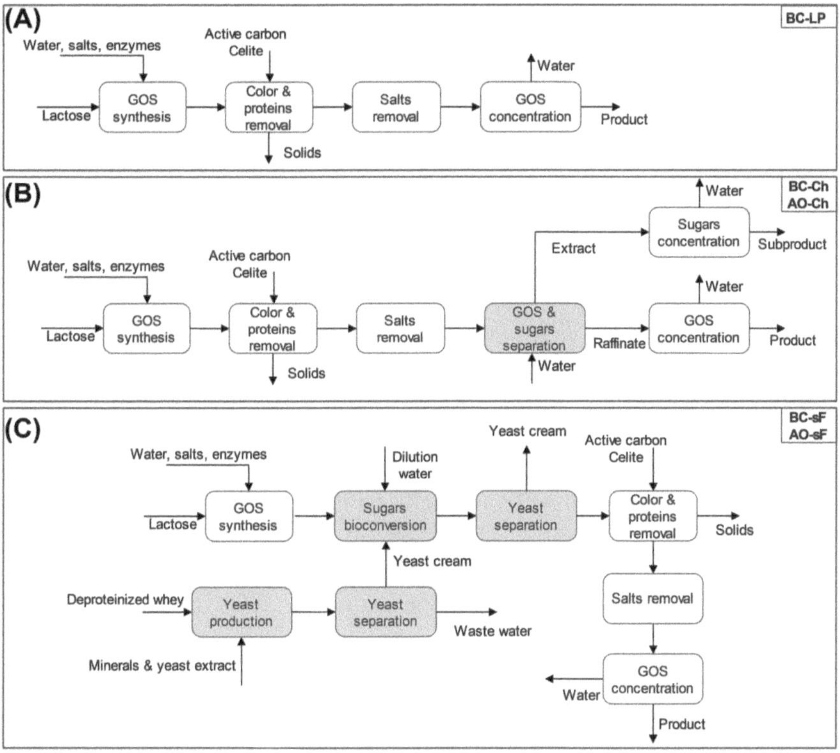

FIGURE 7.3 Block diagrams for alternative GOS production processes. (A) shows the unit operations required for producing a low GOS content product, considering only salts removal using ion exchange and concentration. (B) depicts a process for high purity GOS production including activated carbon and celite filtration, ion exchange, SMB chromatography and evaporation. Finally, (C) presents a simplified process flow diagram for product purification using selective fermentation.

non-reacted lactose and monosaccharides is sold as animal feed at a price equivalent to 20% of the food grade lactose price.

For AO-SF, SMB chromatography is replaced by a fermentation using *K. marxianus* immediately after β-galactosidase is inactivated by a heat shock. Once the reactor is cooled to the fermentation temperature (30°C), yeast cells are inoculated and the fermentation proceeds until the desired purity is achieved. Yields for lactose, glucose and galactose fermentations were taken from the experimental study of Guerrero et al. (2014). Operating conditions for the fermentation stage were selected as 10% total carbohydrate content, which requires diluting the stream of crude GOS leaving the reactor (Fig. 7.3C), and adding 1 g of *K. marxianus* cells (dry weight) per 10 g of carbohydrates. Depending on the desired purity, the fermentation time corresponds to 4, 10, or 24 h yielding GOS purities of 0.48, 0.75, and 0.97, respectively. Unfortunately, yeast cells cannot be reused in multiple fermentation cycles, hence, each fermentation requires the generation of fresh yeast. This process requires a carbon source, assumed to be provided by deproteinized whey powder, a nitrogen source, ammonium sulfate, other salts (KH_2PO_4 and $MgSO_4$) and yeast extract. A biomass yield on lactose of 0.38 was assumed for biomass production in fed-batch mode (Lukondeh et al., 2005), and biomass yields on KH_2PO_4 and $MgSO_4$ were calculated based on a biomass molecular composition represented by $CH_{1.78}O_{0.66}N_{0.158}P_{0.009}S_{0.0035}K_{0.0015}$ (Löser et al., 2015). An extra 20% over the values calculated using the stoichiometric yields of biomass in nitrogen, sulfur, and potassium were considered for the fermentation. Yeast extract was assumed to be provided by *in-house* production from the yeast separated after the purification stage.

7.2.2 Process Simulation

Each process alternative was simulated using the chemical process simulator Aspen HYSYS. Thermodynamic properties of water, glucose, citric acid, and other salts are provided in Aspen Thermodynamic databases. Properties of lactose and GOS were calculated in HYSYS using the PCSAFT thermodynamic model with parameters taken from Held et al. (2013).

7.2.3 Economic Assessment

The project is assumed to break ground in 2015, and starts its operation one year after. The total project duration is 21 years. 60% of the total capital investment is financed by a 10 years bank loan (at an 8% annual interest rate) and the remaining is financed by equity at 20% discount rate.

Once process simulations produce mass and energy balances for each process, equipments are designed and delivered cost for each one is estimated using cost correlations (Couper et al., 2005; Woods, 2007). Then factors are applied to determine direct, indirect, and overhead costs, allowing the estimation of the total capital investment (TCI). TCI, along with operating expenses, are used in a discounted cash flow rate of return (DCFROR) analysis to calculate a plant gate price of the product, not pure GOS, but the product at a given GOS purity. This price represents the minimum admissible value of the product (MPSP, minimum product selling price) in order to obtain a net present value of zero for a 20% internal rate of return after taxes (25% income tax). MPSP is used to assess the effect of important process variables and process configurations over manufacturing and investment costs. Considering that the real product is pure GOS, and not the sugars along with them, a minimum GOS selling price (MGSP) is also calculated. This value represents the minimum acceptable GOS selling price at the plant-gate considering that the product is priced only based on its GOS content.

7.2.4 Results

The processes described in the preceding sections were simulated in Aspen HYSYS to calculate mass and energy balances for the most relevant operations involved in the production of GOS at different levels of purity. Figure 7.4

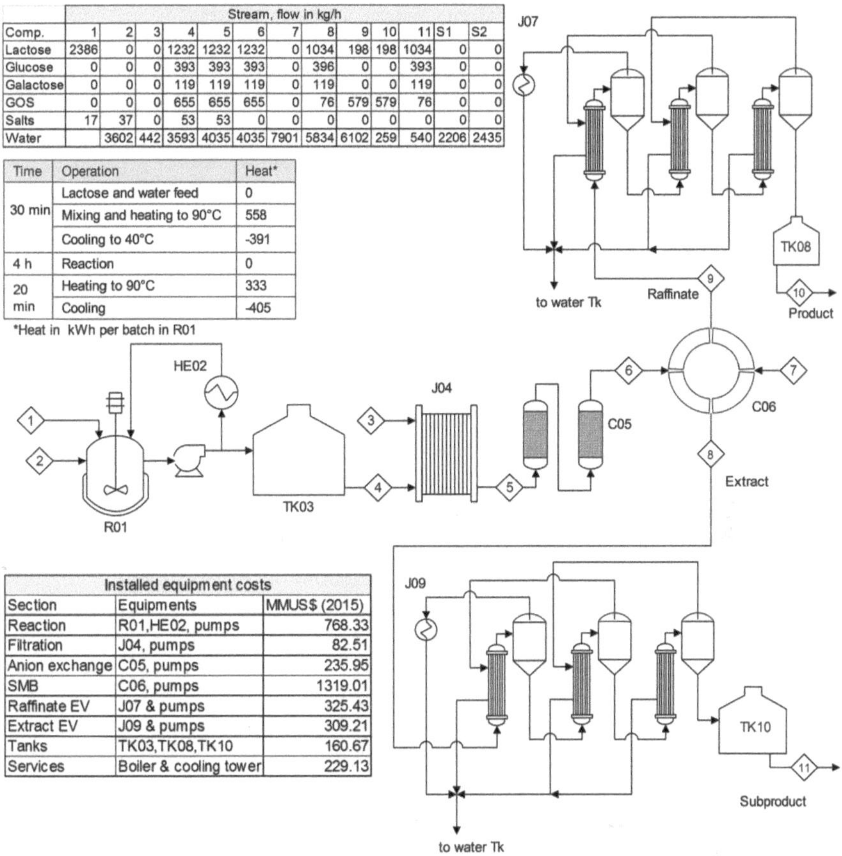

	Stream, flow in kg/h												
Comp.	1	2	3	4	5	6	7	8	9	10	11	S1	S2
Lactose	2386	0	0	1232	1232	1232	0	1034	198	198	1034	0	0
Glucose	0	0	0	393	393	393	0	396	0	0	393	0	0
Galactose	0	0	0	119	119	119	0	119	0	0	119	0	0
GOS	0	0	0	655	655	655	0	76	579	579	76	0	0
Salts	17	37	0	53	53	0	0	0	0	0	0	0	0
Water		3602	442	3593	4035	4035	7901	5834	6102	259	540	2206	2435

Time	Operation	Heat*
30 min	Lactose and water feed	0
	Mixing and heating to 90°C	558
	Cooling to 40°C	-391
4 h	Reaction	0
20 min	Heating to 90°C	333
	Cooling	-405

*Heat in kWh per batch in R01

Installed equipment costs		
Section	Equipments	MMUS$ (2015)
Reaction	R01,HE02, pumps	768.33
Filtration	J04, pumps	82.51
Anion exchange	C05, pumps	235.95
SMB	C06, pumps	1319.01
Raffinate EV	J07 & pumps	325.43
Extract EV	J09 & pumps	309.21
Tanks	TK03,TK08,TK10	160.67
Services	Boiler & cooling tower	229.13

FIGURE 7.4 Simplified process flow diagram of GOS production using *Aspergillus oryzae* β-galactosidase and purification using SMB chromatography. The upper left table shows components mass balances for the most relevant streams. Mid-left table shows times and heat requirements for the batch reaction of GOS synthesis. Low-left table presents the installed equipment costs in 2015 1000 US$, lumped by section. Mass balances and installed equipment prices are shown for a plant processing capacity of 20,000 metric tons of edible lactose per year.

shows the results of mass balances for a GOS production process using *A. oryzae* β-galactosidase and product purification using SMB chromatography at a plant capacity of 20,000 metric tons of edible lactose per year. The plant produces GOS with a purity of 75% at a global yield of 24% (pure dry GOS over lactose), thus achieving a GOS production of 4819 metric tons per year as pure GOS, corresponding to 8625 metric tons per year of product with 75% purity and 75% solids. Moreover, the plant produces a concentrated sugars stream with 75% of solids that is sold as animal feed. The cost of installed equipment rises to 3.43 million US$ of 2015, while total direct plant costs and fixed capital investment reaches 8.85 and 12.39 million US$ respectively. Considering a working capital of 2.19 million US$, the total capital investment for this plant rises to 14.58 million US$.

7.2.5 Minimum Product Selling Price Versus Purity

Product purity after the reaction of GOS synthesis depends on the conversion values for the different reactions involved. As shown in Table 7.2, GOS production using *B. circulans* β-galactosidase allows achieving a higher conversion of lactose to GOS than with the enzyme from *A. oryzae*. This fact influences the investment and operating costs of the purification processes, thus influencing MPSP as shown in Fig. 7.5.

When MPSP is considered and if no carbohydrate fractionation is performed (process BC-LP), MPSP of a 40% w/w GOS product reaches 1.9 US$ per kg. For the case of the highly purified product with 90% GOS, MPSP is 5 US$ per kg for the BC-Ch process, and 6 US$/kg for the AO-Ch process. Since the β-galactosidase from *A. oryzae* is less expensive per unit of activity than the one from *B. circulans*, processes using the former enzyme show less MPSP unless high purities are required. On the other hand, if it is assumed that the product will be priced by its GOS content, the MGSP for different processes shows a very different trend (compare Fig. 7.5 A and B). In processes using SMB chromatography, an optimal MGSP exists with respect to purity. This optimum value is caused by the tradeoff between two factors: as the number of stages in the SMB chromatographic system increases (and thus the purity of the product), capital and operating costs rise but the GOS recovery yield also increases as more GOS are obtained in the raffinate.

7.2.6 Galacto-Oligosaccharide Cost Versus Plant Capacity

Plant capacity influences MPSP by making less investment to be required per unit of GOS produced per year as production capacity increases. The importance

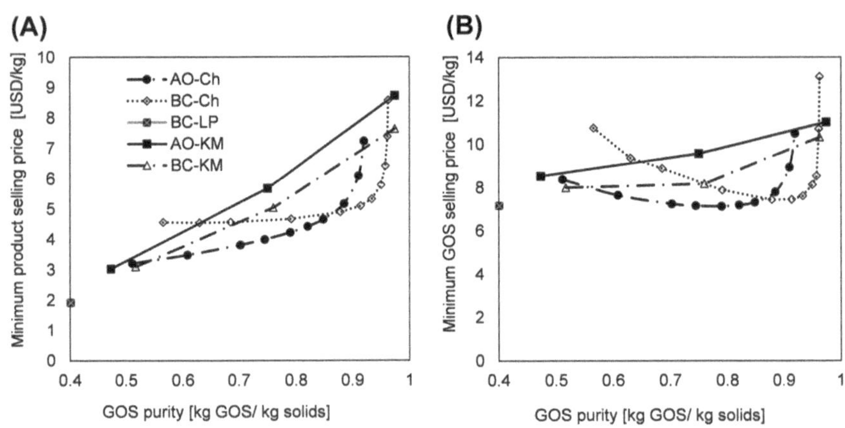

FIGURE 7.5 (A) Minimum product selling price (MPSP) versus product purity for different process alternatives for GOS production. BC-LP process is only represented as a point at 40% GOS, since no purification steps are applied. The product corresponds to a syrup with 75% solids. (B) MPSP of pure and dry GOS.

of the economy of scale depends on how much of the MPSP is determined by the total capital investment; in this case, the effect is not important over plant capacities exceeding 5000 tons of pure GOS per year (See Fig. 7.6).

7.2.7 MPSP Distribution

Regardless of the enzyme selected or the purification technology used, lactose is the main contributor to MPSP, accounting for nearly 40% of it for processes using β-galactosidase from *B. circulans* and nearly 60% when β-galactosidase from *A. oryzae* is used. This difference is explained by the higher GOS yield of the synthesis reaction when β-galactosidase from *B. circulans* is used, thus requiring less lactose to produce the same amount of GOS.

The second factor determining MPSP is the total capital investment financing (investment) followed by fixed costs (including salaries and overheads) and enzyme cost, especially for those processes where *B. circulans* β-galactosidase is used. MPSP distribution for each process alternative for GOS production is shown in Fig. 7.7. As shown, the costs attributed to the separation aids greatly increase for processes were purification by fermentation is used. This is explained by the significant amounts of deproteinized whey permeate and salts required for yeast production.

Lactose price controls product price, especially for those processes producing purified GOS preparations, where the separated sugars must be sold at a fraction of the lactose price, or simply disposed. On the other hand, for nonpurified GOS, monosaccharides and lactose are sold along with GOS. This allows

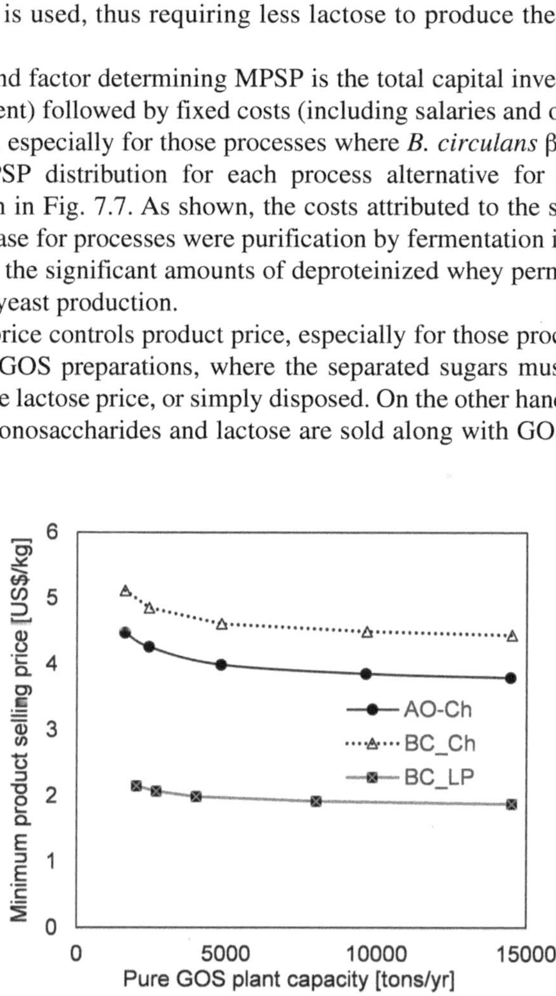

FIGURE 7.6 Minimum product selling price (MPSP) versus plant capacity for 75% GOS purity (processes AO-Ch and BC-Ch) and 40% GOS purity (process BC-LP). Product contains 75% solids.

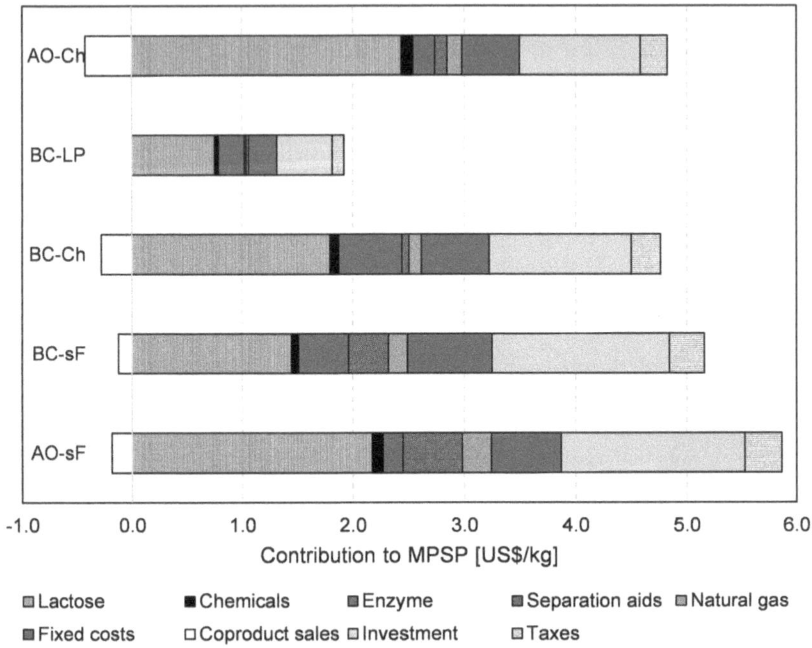

FIGURE 7.7 Minimum product selling price (MPSP) distribution for alternative GOS production processes considering a product purity of 75% and 75% solids content (GOS syrup).

achieving a lower MPSP, which might explain why high purity products are no longer available in the market.

7.3 A CASE STUDY ON GALACTO-OLIGOSACCHARIDES PRODUCTION WITHIN AN EXISTING DAIRY INDUSTRY

This section deals with the possibilities for the valorization of by-products from a cheese factory. Cheese factories produce sweet whey as a by-product after draining of the curd. It typically contains 92% water, 4.6% lactose, 0.8% proteins, 0.8% fat, 0.8% salts, lactic acid, and cheese traces. Although in the past whey was considered as a waste stream and an environmental nuisance, nowadays whey is valued as a source of high quality proteins and lactose. Depending on plant whey production, this by-product can be used as animal feed, for the production of whey powder or as starting material for its fractionation into lactose and proteins (see Chapter 1: Lactose: Production and Upgrading). Although determining which products are more profitable requires a detailed analysis in each cheese factory, general guidelines can be found in the literature considering whey production capacity, local regulations and market prices. Peters (2005) analyzed the possibilities of whey utilization for small and large cheese factories from an economic point of view. He found out that for plants processing

100,000 kg of milk per day or less, drying of whey is not economical unless whey powder is sold at a price of 0.70 US$/kg or above. On the other hand, a large plant processing 2.7 million kg of milk per day can profitably dry whey even at a whey powder price of 0.40 US$/kg. This economy of scale is influenced by the costs of the equipment required for water removal and its scaling factor. For example, a typical scaling factor for an evaporator is 0.54 (Table 6.4 in Peters et al., 2003); thus, if the capacity of the production facility is increased four times, the equipment cost only doubles (approximately). Thus in this section it will be assumed that cheese plants with a capacity below 5000 kg/h of whey does not produce whey powder or other more refined products. Hence, an analysis will be performed to determine plant capacities (in terms of whey production) and market conditions that allow an existing plant, already producing whole whey powder, to retrofit its installations to produce edible lactose and whey powder concentrate with an 80% protein content (WPC80). The project is assumed to be internally financed with an internal return rate of 15% in a 20 year horizon. While WPC80 is expected to be sold at market price, the lactose produced is intended to be used for GOS production. Clearly, if the minimum price of the lactose produced by this hypothetical plant upgrade is superior to the lactose market price, then the retrofitting project should not be undertaken, and if GOS are to be produced, lactose must be purchased in the market. It is important to highlight that since whey is already being used to produce whey powder, it has a price. This price will be estimated by calculating the operating costs and assuming that the investment once made to produce it has already been depreciated.

Composition of solids in edible lactose is taken as 94.05% lactose, 0.095% proteins, 0.855% salts and other compounds, being the remaining 5% water. For WPC80, a composition of 4.88% lactose, 81.86% proteins, 8.28% fat, 0.98% salts, and 4% water is considered (Navarro da Silva et al., 2015). In the original plant (prior to retrofitting), it is assumed that after fat separation using centrifuges, whey is concentrated using a triple effect evaporator consuming 800 kJ per kg of evaporated water, that removes enough water to achieve a 30% solid content and then spray dried, consuming 4500 kJ per kg of water evaporated in the spray drying tower (Walstra et al., 2005). Heat is assumed to be provided by natural gas at a price of 10 US$ per MMBTU. Hence, the operating costs per kg of processed whey rises to 0.86 US$ cents. This amount is discounted from the sales of whey powder in order to calculate the value of whey permeate. For example, if a price of 0.7 US$ per kg of whey powder is assumed, then the price of whey is 4.3 US$ cents/kg.

After retrofitting, the plant produces lactose with a yield of 79% based on the lactose content of whey and WPC with a yield of 99% based on the protein content of whey (Navarro da Silva et al., 2015). Operating cost includes heat for water evaporation and WPC spray drying (as in the whey powder production process), electricity for ultrafiltration, reverse osmosis and centrifugation. After fat removal from whey using the existing centrifuges, whey is fed to a

TABLE 7.5 Equipment Used in the Retrofitting of a Plant Producing Whey Powder to Upgrade its Production to Lactose and Whey Protein Concentrate

Equipment	Scaling Variable	Cost Correlation	Cost at Base Plant Capacity [MMUS$][a]
Ultrafiltration	Inlet flow	0.8	0.84
Reverse osmosis	Inlet flow	0.85	2.32
Crystallizer	Volume	0.68	1.92
Centrifuge	Crystal mass flow	1	0.3

[a]Base plant capacity is 20,000 kg/h of whey. Equipment costs are calculated from cost correlations in Woods (2007), except for the centrifuge cost which was taken from Couper et al. (2005).

diafiltration system with ultrafiltration (UF) membranes. The retentate contains 22% solids and the system is operated with a concentration factor equal to 57 at 4 bar of transmembrane pressure. The UF retentate, rich in proteins, is spray dried producing WPC80. UF permeate is fed to a reverse osmosis system operated at 30 bar achieving 18% solid content. This stream is further concentrated up to 50% solids in the existing triple effect evaporator and fed to a crystallizer operated at 15°C with 18 h residence time. Lactose crystals are separated and washed in two centrifugal decanters.

To calculate the minimum price at which the produced lactose must be sold in order to meet the demanded profitability of the retrofitting project, the sizes of the equipments shown in Table 7.5 were calculated using a simplified simulation based on yields and product compositions.

Once installed equipment costs are estimated, the total capital investment for the retrofitting project is estimated adding 30% for instrumentation, 68% for piping, 11% for electrical systems, 33% for engineering and supervision and 50% for equipment not included in the analysis (tanks and pumps). Operating costs are estimated considering an electricity price of 80 US$/MWh and gas price of 10 US$/MMBTU.

Results are shown in Fig. 7.8, where the minimum selling lactose price is presented as a function of plant capacity and whey powder price. It was assumed that the price per kg of the protein concentrate (WPC80) is five times the price per kg of the whey powder. The bands in gray correspond to the 25th and 75th percentiles of the lactose price since January 2014 to November 2015. Thus, for plant capacities and whey permeate prices that produce minimum lactose prices outside the bands, the retrofitting project is not profitable. Thus, if GOS are to be produced, at the current prices of whey powder and lactose (November

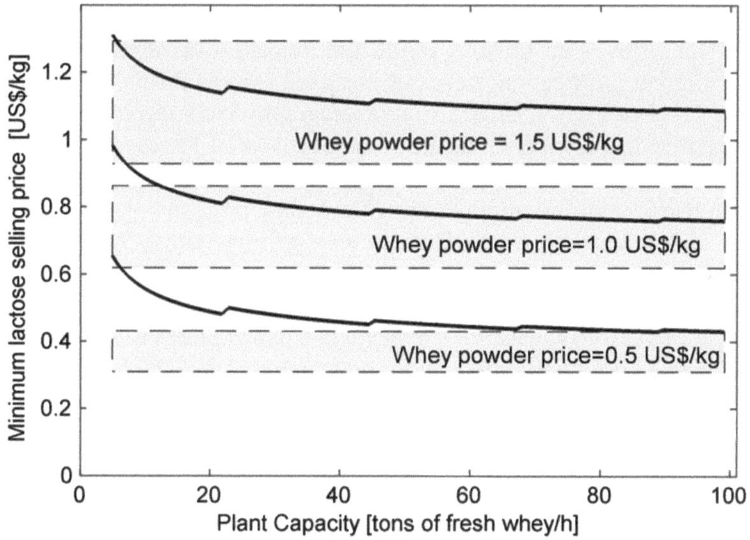

FIGURE 7.8 Minimum lactose selling price as a function of whey purification plant capacity and whey powder price (competing product).

2015), it is more convenient to purchase the lactose in the market and keep on producing whey powder. If whey powder price rises (and consequently WPC80 prices too), the retrofitting project is increasingly more profitable, thus lowering the minimum lactose selling price below the lactose market prices (gray bands in Fig. 7.8). A final consideration in the analysis is that the bands representing the 25th and 75th percentiles of lactose market price are always below the price of the whey powder. If this situation is reversed, it can be more convenient to produce lactose (and WPC80) than to buy it in the market to produce GOS.

As shown in Fig. 7.8, plant capacity plays an important role in determining the minimum lactose selling price, especially for plants processing less than 50,000 kg per hour of fresh whey.

7.4 CONCLUSION

In this chapter, the technologies and processes for GOS production have been analyzed and economically assessed. The production processes are characterized by the source of the enzyme used in the GOS synthesis stage and by the purification process performed to enrich the GOS content of the product. At indicative prices of 4 US$ per kg of GOS products with low purities, say 50%, the processes analyzed show minimum selling prices of 3 US$ per kg, while this value reaches 2 US$ per kg for a product with 40% GOS content obtained without GOS enrichment. Thus, at the indicative market price, every process analyzed can generate profits beyond the minimum values required to achieve an internal return rate of 20%. As the purity of the product increases so do the prices. Albeit this is an obvious result, the trajectory of prices versus purity for the different purification processes analyzed is quite different. While GOS price for processes based on SMB chromatography increases exponentially as product purity increases (as a result of the exponential increase in the number of stages in SMB chromatography), processes based on selective fermentation show a linear increase in minimum product selling price (as a result of a linear increase in fermentation time). Thus, at high purities (above 95%), fermentation (or bioconversion) is a more economical purification process compared to SMB chromatography. A challenge in selective fermentation is reducing the mass ratio of yeast to sugars, thus reducing the expenditures in sugar and other nutrients required for cell propagation. Reducing this cost might allow a selective fermentation process that can outcompete SMB chromatography.

As for the possibilities of integration between existing cheese production plants and GOS production facilities, the analysis performed in Section 7.3 showed that for small cheese plants (below 5.000 kg/h of fresh whey) the investment in whey upgrading to even simple products, such as whey powder, cannot be economically justified. Large plants which were assumed to be already producing whey powder can economically justify an upgrading project to produce lactose and whey protein concentrate depending on plant capacity, protein concentrate price and whey price. At the current whey powder price (0.5 US$/kg), the retrofitting project produces lactose that must be sold at a minimum price higher than the current lactose price band (percentiles 25th and 75th of historical 2014–2015 prices); thus, the retrofitting project is not economical and lactose for GOS production should be bought in the market, unless amounts above 80,000 kg/h of fresh whey are available. At higher whey powder prices, and thus higher whey protein concentrate prices, the minimum lactose selling price reaches values that are competitive with the lactose market price, especially for plants with capacities above 50,000. In summary, the decision to finance a retrofitting project, for upgrading an existing whey powder producing plant toward lactose and whey protein concentrate production, depends on market prices and plant capacity, among other factors such as the internal return rate demanded for the project. Each cheese producer needs to perform a technical and economic analysis considering at least the variables presented in this chapter.

REFERENCES

Affertsholt-Allen, T., 2007. Market developments and industry challenges for lactose and lactose derivatives. In: IDF Symposium "Lactose and its Derivatives." Moscow. Available from http://lactose.ru/present/1Tage_Affertsholt-Allen.pdf.

Alexandratos, S.D., 2009. Ion-exchange resins: a retrospective from industrial and engineering chemistry research. Industrial and Engineering Chemistry Research 48, 388–398.

Broughton, D., 1984. Production-scale adsorptive separations of liquid mixtures by simulated moving-bed technology. Separation Science and Technology 19, 723–736.

Couper, J.R., Penney, W.R., Fair, J.R., 2005. Chemical Process Equipment: Selection and Design 2nd Edition. Gulf Professional Publishing, Burlington.

FrieslandCampina Domo and Warrnambool Cheese & Butter Factory Open New Plant in Australia, 2009. https://www.frieslandcampina.com/en/news/2009-05-27-frieslandcampina-domo-and-warrnambool-cheese-butter-factory-open-new-plant-in-australia/.

Gosling, A., Alftrén, J., Stevens, G.W., Barber, A.R., Kentish, S.E., Gras, S.L., 2009. Facile pre-treatment of *Bacillus circulans* β-galactosidase increases the yield of galactosyl oligosaccharides in milk and lactose reaction systems. Journal of Agricultural and Food Chemistry 57, 11570–11574.

Grand View Research Market Research & Consulting, January 2015. Galacto-oligosaccharides (GOS) market analysis by application (food & beverage, prebiotic dietary supplements) and segment forecasts to 2020. http://www.grandviewresearch.com/industry-analysis/galacto-oligosaccharides-gos-market. isbn:978-1-68038-300-3.

Guerrero, C., Vera, C., Novoa, C., Dumont, J., Acevedo, F., Illanes, A., 2014. Purification of highly concentrated galacto-oligosaccharide preparations by selective fermentation with yeasts. International Dairy Journal 39, 78–88.

Held, C., Sadowski, G., Carneiro, A., Rodríguez, O., Macedo, E.A., 2013. Modeling thermodynamic properties of aqueous single-solute and multi-solute sugar solutions with PC-SAFT. AIChE Journal 59, 4794–4805.

Hernández, O., Ruiz-Matute, A.I., Olano, A., Moreno, F.J., Sanz, M.L., 2009. Comparison of fractionation techniques to obtain prebiotic galactooligosaccharides. International Dairy Journal 19, 531–536.

Kearney, M., 2002. Chromatographic applications in the cane sugar industry. International Sugar Journal 104, 194–203.

Li, Z., Xiao, M., Lu, L., Li, Y., 2008. Production of non-monosaccharide and high-purity galactooligosaccharides by immobilized enzyme catalysis and fermentation with immobilized yeast cells. Process Biochemistry 43, 896–899.

Löser, C., Urit, T., Gruner, E., Bley, T., 2015. Efficient growth of *Kluyveromyces marxianus* biomass used as a biocatalyst in the sustainable production of ethyl acetate. Energy, Sustainability and Society 5, 2. http://dx.doi.org/10.1186/s13705-014-0028-2.

Lukondeh, T., Ashbolt, N.J., Rogers, P.L., 2005. Fed-batch fermentation for production of *Kluyveromyces marxianus* FII 510700 cultivated on a lactose-based medium. Journal of Industrial Microbiology and Biotechnology 32, 284–288.

Matsumoto, K., Kobayashi, Y., Kan, T., 1995. Method of Preparing Galactooligosaccharides. European Patent 0363214B1.

Nakayama, T., Amachi, T., 2002. β-Galactosidase enzymology. In: Flickinger, M.C., Drew, S.W. (Eds.), Encyclopedia of Bioprocess Technology. John Wiley and Sons Inc., New York.

Navarro da Silva, A., Perez, R., Rodrigues Minim, V.P., Martins, D.D.S., Minim, L.A., 2015. Integrated production of whey protein concentrate and lactose derivatives: what is the best combination? Food Research International 73, 62–74.

Nicoud, R.M., 2000. Simulated moving-bed chromatography for biomolecules. Separation Science and Technology 2, 475–509.

Panesar, P.S., Panesar, R., Singh, R.S., Kennedy, J.F., Kumar, H., 2006. Microbial production, immobilization and applications of β-D-galactosidase. Journal of Chemical Technology and Biotechnology 81, 530–543.

Peters, M., Timmerhaus, K., West, R., 2003. Plant Design and Economics for Chemical Engineers, fifth ed. McGraw-Hill Education, New York.

Peters, R.H.A., 2005. Economic aspects of cheese making as influenced by whey processing options. International Dairy Journal 15, 537–545.

Playne, M.J., Crittenden, R.G., 2009. Galacto-oligosaccharides and other products derived from lactose. In: Fox, P.F., McSweeney, P.L.H. (Eds.), Advanced Dairy Chemistry. Lactose, Water, Salts and Minor Constituents, vol. 3. Springer, New York, pp. 121–202.

Sanz, M.L., Martínez-Castro, I., 2007. Recent developments in sample preparation for chromatographic analysis of carbohydrates. Journal of Chromatography A 1153, 74–89.

Sanz-Valero, J.I., 2009. Production of Galacto-Oligosaccharides From Lactose by Immobilized β-Galactosidase and Posterior Chromatographic Separation. Ph.D. Thesis Ohio State University.

Spherix Consulting, Inc, 2010. Generally Recognized as Safe (GRAS) Determination for the Use of Galacto-Oligosaccharides (GOS) in Foods and Infant Formulas. Yakult Pharmaceutical Industry Co., Ltd, p. 156.

Tipnis, P.R., Harriott, P., 1986. Thermal regeneration of activated carbons. Chemical Engineering Communications 46, 11–28.

Torres, D.P.M., Gonçalves, M.D.P.F., Teixeira, J.A., Rodrigues, L.R., 2010. Galacto-oligosaccharides: production, properties, applications, and significance as prebiotics. Comprehensive Reviews in Food Science and Food Safety 9, 438–454.

Towler, G., Sinnott, R.K., 2007. Chemical engineering design: principles, practice and economics of plant and process design. Butterworth-Heinemann, Oxford.

van Vliet, R., 2007. GRAS exemption claim for galacto-oligosaccharides. GRAS Notification Program. p. 159.

Vera, C., Guerrero, C., Illanes, A., Conejeros, R., 2011. A pseudo steady-state model for galacto-oligosaccharides synthesis with β-galactosidase from *Aspergillus oryzae*. Biotechnology and Bioengineering 108, 2270–2279.

Walstra, P., Wouters, J.T.M., Geurts, T.J., 2005. Dairy Science and Technology, second ed. CRC Press, Boca Raton.

Wiśniewski, Ł., Antošová, M., Polakovič, M., 2013. Simulated moving bed chromatography separation of galacto-oligosaccharides. Acta Chimica Slovaca 6, 206–210.

Woods, D.R., 2007. Rules of Thumb in Engineering Practice. Wiley-VCH Verlag, Weinheim.

Zaganiaris, E.J., 2011. Ion exchange resins and synthetic adsorbents in food processing. BoD-Books on Demand France, Paris.

Chapter 8

Future Trends and Concluding Remarks

A. Illanes

8.1 WHEY AND LACTOSE AS RAW MATERIALS

Lactose is a plentiful material contained in milk whey and which is often in surplus with respect to demand, creating a problem of proper management. Only medium-to-large size cheese factories can economically dry whey, which is an absolute requirement for its delivery because of its high water content and perishability. Whey price is remarkably variable so companies producing it face uncertainty about its profitable marketing, while the disposal of surplus whey to treatment plants is costly because of its high oxygen demand and low dissolved solids content.

Whey proteins are a valuable asset and membrane technology has been developed to the point of becoming a sound technology to recover them. This operation leads to the production of whey permeate, which is a diluted solution of lactose solution and salts whose proper management does not differ significantly from whole whey handling. Whey permeate and also whole whey are the raw materials for industrial lactose production for food and pharmaceutical use, being commodities sold at modest prices, usually below US$1000/ton dry product.

A platform for lactose upgrading seems an appealing strategy to whey and whey permeate producers since valuable products can be obtained from them with a much higher added value. Of course this will imply fractionation and eventual chemical transformation and purification, and among the many options for lactose upgrading, lactose-derived nondigestible oligosaccharides (NDOs) are increasingly attractive.

Such a platform should be considered strategic in the sense of bringing together economic and environmental considerations to optimally balance them. Production of lactose-derived compounds is presumably sound from the perspective of life cycle and carbon footprint analyses since they generate from an underutilized, sometimes a surplus, raw material not compromising the use of extra arable land.

Lactose-Derived Prebiotics. http://dx.doi.org/10.1016/B978-0-12-802724-0.00008-1

8.2 FUNCTIONAL FOODS, NONDIGESTIBLE OLIGOSACCHARIDES, AND PREBIOTICS

Functional food consumption and health eating habits are now sustained trends in response to the constraints imposed by modern lifestyle where fast foods and extensively processed foods have progressively replaced homemade cooking, leading to perceivable health problems like obesity, irritable bowel syndrome, heart conditions, and diabetes. Such problems affect persons of all ages and socioeconomic conditions so that health feeding is no longer an issue concerning only developed nations, having also a profound impact on developing countries where adverse effects of improper feeding may be stronger because of economic constraints. In this scenario, probiotics and prebiotics have been considered as major components in health-promoting foods. In consonance with the development of a global market for functional foods, probiotics and prebiotics have spread from the East to the West and from the North to the South.

NDOs are considered as dietary fiber (the name is rather misleading since they are for the most part nonfibrous materials) and, as such, several health-promoting effects are associated, mostly related to good intestinal functioning. There are several lactose-derived NDOs, notably galacto-oligosaccharide (GOS) and lactulose, that are considered as prebiotics based on sound scientific evidence.

Prebiotics are compounds that promote a healthy bacterial population in the lower intestine where they are metabolized after passing intact by the upper gastrointestinal tract. Requirements for a compound to be considered prebiotic are stringent so that only a few NDOs comply with all of them. At this time, only fructans (inulin and fructo-oligosaccharides) and galactans (GOS and lactulose) are properly considered as prebiotics, but this situation may change as new evidence arises and more knowledge is gathered on the complex functioning of the bowel system; indeed, xylo-oligosaccharides, isomalto-oligosaccharides, soybean oligosaccharides, gluco-oligosaccharides, and resistant starch are prebiotic candidates that are already being used, mostly in Asian countries, as health-promoting food components.

Lactose-derived prebiotics are gaining increased importance within the functional food market and exhibit distinct features among prebiotics, both in terms of health-promoting and functional effects. GOS and lactulose are properly considered as prebiotics, but lactosucrose, lactitol, lactobionic acid, and tagatose are candidate prebiotics. However, most of them have potential beyond that condition as functional food ingredients and also in nonfood applications, some of which are increasingly important.

8.3 LACTOSE-DERIVED PREBIOTICS: GALACTO-OLIGOSACCHARIDES

GOS is an outstanding lactose-derived functional food ingredient with a wide spectrum of applications, mostly related to its prebiotic condition, even though it can also confer desirable functional properties to the food matrices containing it. The GOS market has expanded considerably alongside the expansion of the functional food market; GOS production was 15,000 tons in 1995 (mostly in Japan) and its actual market is close to 100,000 tons, being produced in Asia, Europe, and America, with a forecast figure of 175,000 tons for 2020. As prebiotic, GOS exhibits a distinctive feature by mimicking the effect of human milk oligosaccharides. Being rather irreplaceable in this function, GOS plays a key role in supplementing formulas for newborns and follow-on formulas. This is quite probably the most important niche for GOS utilization whose market still has ample room for development. Stability under conditions of food processing and prolonged storage is another distinctive feature of GOS, which represents an advantage with respect to prebiotic competitors.

Chemical synthesis of GOS is poorly specific so that a very heterogeneous mixture of molecules with different sizes and linkages is produced, so it is not a technological option. The synthesis of GOS is exclusively performed by biocatalysis due to its high selectivity and also because it is conducted under mild reaction conditions in aqueous medium leading to a simple process well matching the principles of green chemistry.

β-galactosidases in whole cell or isolated enzyme format are used as catalysts for GOS synthesis, the latter being mostly used at industrial level. This is quite fortunate since β-galactosidases are commodity enzymes with a long tradition of safe use in the dairy industry for producing low-lactose milk and dairy products by catalyzing the hydrolysis of lactose into its monosaccharide components. In the case of GOS synthesis the objective is the opposite: the formation of β-type linkages instead of its hydrolytic cleavage. This requires arresting the hydrolytic potential of the enzyme by reducing the water activity in the reaction medium. The most used strategy to that purpose is the use on nonaqueous organic solvents as reaction medium, but this is inapplicable in this case because of the low solubility of lactose and poor β-galactosidase stability in such media, with the additional drawback of increased costs and environmental burden. Hydrolytic activity can be arrested though working at (very) high lactose concentration in a kinetically controlled reaction of transgalactosylation in which lactose acts both as donor and acceptor of the galactosyl moiety; water will compete as acceptor, leading to hydrolysis, but this is not favorable at such working conditions.

GOS synthesis has been conducted mostly with well-established commercial β-galactosidase preparations coming from microbial strains with generally recognized as safe (GRAS), or equivalent, status, but also thermophilic and probiotic microorganisms have been used. Non-GRAS organisms cannot be used in food, but the option is the cloning of such genes into suitable GRAS hosts if the use of recombinant organisms is not banned. Several efforts have been devoted to developing enzyme catalysts more compliant with the conditions required for GOS synthesis (very high lactose concentration, high temperature, reduced hydrolytic activity), using molecular tools like site-directed mutagenesis and directed evolution; its technological impact is still to be appraised.

Enzyme immobilization has been a powerful tool for designing robust catalysts performing well and allowing prolonged use under harsh reaction conditions. Immobilized enzymes have contributed significantly to industrial biocatalysis, as exemplified by the massive production of high-fructose syrup with immobilized glucose (xylose) isomerase, acrylamide with nitrile hydratase–containing immobilized cells, and semisynthetic β-lactam antibiotics with immobilized penicillin acylase, plus a large number of smaller-scale processes of organic synthesis leading to bioactive molecules or precursors catalyzed mostly by immobilized lipases in nonaqueous media. Synthesis of GOS has been profusely studied with immobilized β-galactosidases aiming to increase enzyme efficiency of use and the consequent increase in GOS specific productivity. However, its practical impact seems to be low and it is doubtful that immobilized enzymes are actually being used for industrial production of GOS. This is most probably so because β-galactosidases are commodity enzymes, some of them being quite robust even in soluble format, so that the impact of enzyme cost on production cost may be of lesser significance than other items of the process. In fact, raw material and downstream processing are the most sensitive cost items in GOS synthesis (see Chapter 7: Technical and Economic Analysis of Industrial Production of Lactose-Derived Prebiotics). Immobilization has, however, some important features for GOS synthesis since catalyst separation and management are much simpler.

Purification is a key issue in GOS production as the consequence of the modest product yields (usually not exceeding 40%) and lactose conversion (usually not exceeding 60%) attainable. Level of purity is dictated by the final use, but commercial products usually contain 50–60% GOS on dry basis, the rest being unreacted lactose, glucose, and galactose. This purity may be acceptable in some food applications but not in special products, like baby formulas and products for elderly people, and also for persons suffering diabetes or lactose intolerance. Purification of GOS implies the removal of lactose and monosaccharides so that classical operations of sugar refining have been used.

Simulated moving bed (SMB) chromatography is considered the state-of-the-art technology for oligosaccharide fractionation and it is most probably used at industrial scale for GOS purification, despite being a complex and costly operation; however, when high levels of purity are required, the number of fixed bed columns may be exceedingly high (see Chapter 7: Technical and Economic Analysis of Industrial Production of Lactose-Derived Prebiotics) and cost may turn prohibitive. Excessive dilution and low throughput are also potential disadvantages of SMB so that other strategies of GOS purification have been developed.

Membrane fractionation by nanofiltration and selective fermentation (bioconversion) with yeast strains are the most relevant. Membrane separation is increasingly being used at industrial scale and much technology has been developed regarding equipment design and membrane fabrication. However, purification of GOS by nanofiltration is a complex task because the difference in molecular size between the product (in the range from 500 to 1000 Da) and the contaminant sugars to be removed (342 and 180 Da) is rather small, and because of the reduction of membrane selectivity that occurs at high solids concentration, making it necessary to dilute the raw GOS stream at least twice. Prehydrolysis of residual lactose in raw GOS to its monosaccharide components helps in improving GOS purification by establishing a broader molecular size gap between the product and the contaminant sugars. Compromises between GOS recovery and sugars removal, and between selectivity and throughput, are inevitable and set up the limits for nanofiltration as an operation for GOS purification.

Selective fermentation (bioconversion) of unwanted sugars with yeasts is in principle a very attractive option for the purification of raw GOS. If using *Saccharomyces* strains, only monosaccharides will be removed so a prehydrolysis of lactose will be required for further purification. In the case of using *Kluyveromyces* strains, monosaccharides and lactose will be removed as well. Both options have pros and cons: when using *Saccharomyces*, an additional step will be required for the prehydrolysis of lactose (usually with *Kluyveromyces lactis* β-galactosidase) but the reward will be that high rates of unwanted sugars removal can be attained and plain baker's yeast and even spent brewer's yeast can be used. When using *Kluyveromyces*, prehydrolysis can be avoided but rates of sugars removal will be slower and, due to their lower osmotolerance, raw GOS containing from 40% to 50% w/w solutes will have to be diluted at least twice; nonetheless, GOS of high purity can be obtained, all glucose and most galactose and residual lactose being removed. The main drawback of GOS purification by selective fermentation (bioconversion) is the rather high biomass/carbohydrate ratio required for effective removal; if cells are allowed to grow, extra nutrients will be required and a significant amount of fermentation products (mostly ethanol and glycerol, besides cell biomass) will have to be removed from the purified GOS, making downstream processing more complex and costly. As mentioned in "Chapter 7, Technical and Economic Analysis of Industrial Production of Lactose-Derived Prebiotics," this strategy may be a good alternative to chromatographic purification in the case of producing GOS of high purity.

GOS production cost is strongly dependent on the cost of the raw material (whole whey, whey permeate, or edible lactose) and downstream operations that are pressed by the low GOS yields obtained by lactose transgalactosylation. Efforts to increase yield by improving catalyst performance have been for the most part unsuccessful, but protein-engineering strategies are promising for the construction of mutant enzymes with strongly improved synthesis to hydrolysis activity ratios. Feasibility of using such mutant enzymes will face both technological and regulatory challenges.

8.4 LACTOSE-DERIVED PREBIOTICS: LACTULOSE

Lactulose is a lactose-derived disaccharide whose primary use refers to the treatment of chronic constipation and hepatic encephalopathy. However, its prebiotic condition has opened a wide spectrum of applications as a functional ingredient in health foods, now representing more than 70% of the total lactulose market, which has been estimated close to 50,000 tons/year. Lactulose is produced exclusively by chemical isomerization of lactose with alkaline catalysts, despite being a rather complex and poorly specific process requiring pure lactose as substrate and generating unwanted side products like epilactose, galactose, tagatose, isosaccharic acids, and colored compounds. Downstream processing is critical and represents a major portion of the operating cost since several steps of downstream processing are required to attain the required purity, which is high in the case of pharmaceutical use. Purification is done by using different chromatographic operations, with the drawback of low throughputs and considerable consumption of water; supercritical carbon dioxide extraction and differential precipitation with organic solvents and borates have also been evaluated although their application at productive level is doubtful.

The synthesis of lactulose by enzyme catalysis appears as an interesting technological option to overcome the limitations inherent to chemical synthesis; however, a competitive enzymatic process for lactulose production has not been developed yet. Lactulose has been synthesized by the kinetically controlled reaction of transgalactosylation of fructose with lactose catalyzed by β-galactosidase. Inevitably a GOS-lactulose mixture is produced since lactose acts both as galactose donor and acceptor, so this is not an adequate strategy for producing lactulose devoid of GOS, since an impractical excess of fructose will be required to such purpose. However, this strategy is appealing if a GOS-lactulose prebiotic mixture is the target; GOS–lactulose ratio can be controlled at will, according to the lactose–fructose ratio used as substrate. The obvious strategy of a one-substrate enzymatic conversion of lactose into its isomer lactulose was tried unsuccessfully until quite recently when it was demonstrated that a thermophilic cellobiose 2-epimerase was able to catalyze the direct isomerization of lactose into lactulose at acceptable yield producing epilactose as side product. Since then, valuable information has accumulated showing that this may be the way to go to displace the ongoing chemical process by a greener biocatalytic process; safety aspects and commercial availability of the enzyme are aspects to be dealt with. Purification of enzymatically produced lactulose has been studied using the same strategies outlined herein for GOS purification.

Lactulose is recognized mostly as a mild laxative within the pharmaceutical sector. However, in recent years its use as a functional ingredient in health foods has gained in importance, now representing, at least in quantitative terms, its most important use. Enzymatic synthesis of lactulose, despite its potential advantages over chemical synthesis, is not yet considered a mature technology. Production by fructose transgalactosylation with lactose is an option for producing lactulose-GOS prebiotic mixes but not for producing pure lactose. Production of lactulose by direct isomerization of lactose with cellobiose 2-epimerase should gain industrial status as technology matures and compliance with green chemistry principles becomes a substantial issue. A cost-basis analysis on such enzymatic process may shed light as to the opportunity of biocatalysis as a sound technology for lactulose production.

Index

'*Note*: Page numbers followed by "f" indicate figures and "t" indicate tables.'

F

G

M

N

O

X

Y

Z

FSC
www.fsc.org
MIX
Papier | Fördert gute Waldnutzung
FSC® C083411

Kolibri 360 GmbH
Ferdinand-Jühlke-Straße 7
99095 Erfurt, Deutschland
produktsicherheit@kolibri360.de